ADVANCES IN

GEOPHYSICS

VOLUME 19

Contributors to This Volume

Enrique A. Caponi
Gottfried Hänel
Joanne Simpson

Advances in
GEOPHYSICS

Edited by
H. E. LANDSBERG

Institute for Fluid Dynamics and Applied Mathematics
University of Maryland, College Park, Maryland

J. VAN MIEGHEM

Royal Belgian Meteorological Institute
Uccle, Belgium

Editorial Advisory Committee

BERNARD HAURWITZ   R. STONELEY
ROGER REVELLE   URHO A. UOTILA

VOLUME 19

1976

Academic Press • **New York**   **San Francisco**   **London**
A Subsidiary of Harcourt Brace Jovanovich, Publishers

ACADEMIC PRESS, INC.
111 Fifth Avenue, New York, New York 10003

United Kingdom Edition published by
ACADEMIC PRESS, INC. (LONDON) LTD.
24/28 Oval Road, London NW1

LIBRARY OF CONGRESS CATALOG CARD NUMBER: 52–12266

ISBN 0–12–018819–8

PRINTED IN THE UNITED STATES OF AMERICA

CONTENTS

LIST OF CONTRIBUTORS

Numbers in parentheses indicate the pages on which the authors' contributions begin.

Enrique A. Caponi,* *Institute for Fluid Dynamics and Applied Mathematics, University of Maryland, College Park, Maryland* (189)

Gottfried Hänel, *Institut für Meteorologie, Johannes Gutenberg-Universität, Mainz, Germany* (73)

Joanne Simpson, *Center for Advanced Studies and Department of Environmental Sciences, University of Virginia, Charlottesville, Virginia* (1)

* *Present address*: Laboratorio de Hidráulica Aplicada/INCYTH, Casilla de Correo 21, 1802 Aeropuerto Ezeiza, Prov. de Buenos Aires, Argentina.

PRECIPITATION AUGMENTATION FROM CUMULUS CLOUDS AND SYSTEMS: SCIENTIFIC AND TECHNOLOGICAL FOUNDATIONS, 1975

JOANNE SIMPSON

*Center for Advanced Studies
and
Department of Environmental Sciences
University of Virginia, Charlottesville, Virginia*

1. INTRODUCTION

Cumulus clouds and convective systems produce about three-fourths of the rain in the tropics and subtropics and a significant fraction of the precipitation in the extratropical United States, especially in the crucial growing seasons. Moreover, in many areas, the rain or snow-producing cyclonic storms of winter are embedded with convective cloud bands (Elliott *et al.*, 1971; Gagin and Neumann, 1974), which are responsible for the heaviest precipitation (Austin and Houze, 1972).

Convective processes also drive damaging atmospheric storms, such as the hurricane, tornado, hailstorm, and squall line. Furthermore, these clouds play a key part in the machinery running the large-scale planetary wind systems (Malkus, 1962) and affect the income of solar and the outgo of terrestrial radiation. Hence, their predictable modification, if feasible, offers enormous benefits to man's livelihood and life itself. There are also many potentially useful impacts upon the environment, ranging from drought mitigation, increased food production, and pollution scavenging, to more effective power generation and utilization, since water resources are essential for many types of mineral extraction and power conversions. With the impending food and energy shortages the assessment of cumulus modification becomes an urgent national and international concern.

Attempts at cumulus modification have been underway for more than 25 years (Simpson, 1970). A few dozen randomized experiments and many more operational applications, some of the latter alleged to exceed the scientific foundations, have proliferated on a worldwide basis. Most of the efforts have been directed toward precipitation augmentation, although major projects are in progress with the goals of suppressing hail and lightning. One research program aimed at reducing hurricane destruction by seeding its cloud systems has been in the full-scale field phase intermittently for 13 years.

Out of all this activity, only about a half dozen weather modification experiments involving cumulus have attained clear sucess. Success is defined here as a demonstrated effect, statistically significant, of the treatment in the sense intended by the experimenters or postulated by their modification hypotheses, or both, which is generally accepted as such by a majority of the scientific community.[1]

Ironically, it is much more expensive, difficult, and time-consuming to test a weather modification hypothesis by a proper scientific experiment than it is to apply it operationally. This discrepancy has led the field into many of the problems in which it finds itself today. Some of these problems are inadequate support at governmental levels (particularly federal) and undiminishing or only slowly diminishing controversy regarding both feasibility and application.

A major problem has been that for many years not even expert atmospheric scientists recognized the extreme difficulties involved in establishing a weather modification hypothesis. Even today, only a minority are aware of their full enormity. For a key example, natural rainfall within seedable situations commonly varies by factors of 10–1000 while the largest seeding effects claimed or established by man have never exceeded a factor of three; often a demonstrated increase of 10 % would be of priceless value.

The tremendous natural variability or "noise level" is intimately related to the basic difficulty in evaluating a treatment effect, namely, *it is necessary to estimate what the cloud system would have done* had the treatment not been applied. The prediction of unmodified convective precipitation still remains more of an art than a science, particularly in undisturbed situa-

---

[1] As expressed, for example in the recent evaluation by a committee of The National Academy of Sciences (1973). This review, however, was criticized as too optimistic by Mason (1974) and as too conservative by many practicing modifiers.

tions. Clearly, man cannot await solution of the forecast problem to develop and evaluate modification experiments. How statistical techniques together with model simulations have been combined to produce approximate "control" estimates will be a major subject of this article. A necessary basis for these developments is improved understanding of cloud processes.

Only recently has the physical basis of cumulus modification been publicly addressed (Braham and Squires, 1974) in a summary of the status of cloud physics. This summary was distilled from lengthy reports composed by a majority of the U.S. scientific community working on cloud problems.

Significant advances in the scientific basis of cumulus modification have indeed been made since World War II, and particularly in the past decade (see Simpson *et al.*, 1974, Panel 1 Report), but glaring gaps in the crucial necessary knowledge still remain. It is now possible to outline how many of these gaps can be filled, but they are presently impeding progress toward sound application.

The purpose of this article is to provide an assessment of the current scientific and technological foundations of cumulus modification, pointing out both the pertinent advances and the relevant gaps.

2. Assessment of the Scientific Basis for the Modification of Cumulus Clouds and Systems

2.1 Aerosols, Microphysics, and Precipitation Processes

Documentation of the nature and role of aerosols that are important in cloud processes has advanced from a nearly uncharted frontier in 1945 to a sizable body of knowledge with only a few important gaps remaining in 1975. The differences between Aitken nuclei, cloud condensation nuclei (CCN), and ice nuclei (IN) have been recognized. Quantitative relationships between CCN spectra and droplet concentrations during the development stage of clouds have been established (Twomey and Warner, 1967; Fitzgerald, 1972) and, with these, the demonstration of important microphysical differences between clouds of maritime versus continental origin has been clearly made.

CCN are hygroscopic particles in the radius range 10^{-5} to 10^{-2} mm. At low levels over the ocean, about 20 % of these nuclei are sea salt. In continental air, the same number of salt particles constitute only 0.5–2 % of the much larger total number of CCN. The remainder are apparently ammonium sulfate and other primarily natural impurities, although smoke and man-made pollution can add greatly to their number locally. These may hasten or delay the initiation of rain, depending on the size distribution and physical-chemical nature of the polluting aerosol. Considerable progress in documenting the complex effects upon cloud processes of urban aerosols is beginning to be published, in particular by the Metropolitan Meterological Experiment (METROMEX) participants (Huff and Changnon, 1972; Huff, 1973; Fitzgerald and Spyers-Duran, 1973).

Even without man, however, condensation nuclei are usually five to ten times more concentrated over landmasses than over the sea. Consequently, maritime clouds usually have to share their water mass among only 20–100 drops per cm^3, while continental clouds commonly have 100–500 drops per cm^3, making it much harder to grow rain-sized particles and thereby providing a self-enhancing feature for droughts.

Concerning ice nuclei, there are many dust-type substances (Mason, 1971) the effectiveness of which as a function of temperature and other variables has been documented in the laboratory. These substances have also been found imbedded in real snowflakes and in frozen cloud particles. These nuclei appear to possess diameters in the range from about 0.1 to 8 μm with the majority composed of clay silicates. Recently, evidence has begun appearing that smaller particles, in the Aitken size range, are capable of playing a role in the real ice processes in clouds (Sax and Goldsmith, 1972). A systematic, worldwide aerosol climatology is needed but presently does not exist.

The condensation process is well described today. Progress has been made in evaluating the condensation coefficient and its theoretical importance in determining the initial drop size distribution in clouds. There has also been development and some testing of the basic condensation equations establishing droplet growth rates. Numerical integration of these equations assuming given updraft and initial CCN spectra show that within normal cumulus lifetimes (5 to 30 min is typical) drops greater than about 40 μm diameter are unlikely to be formed by the condensation process alone.

The microphysics of warm (temperatures above 0°C), nonprecipitating small cumuli has been extensively measured by aircraft. Particularly noteworthy are the series of careful joint measurements of dynamical and physical variables published by Warner (1969a, b; 1970a; 1973a, b). An unsolved puzzle brought out by these measurements is the common bimodal spectrum of the small cloud droplets. This puzzle is an important one relative to precipitation formation in that the bimodality probably indicates the start of the drop spectrum broadening process, which is believed to be a precursor to the growth of precipitation.

Resolving this and related questions has been impeded by the difficulty of accurate in-cloud measurements.

Even in warm clouds, temperatures in the presence of moderate water contents cannot be measured to better than a few tenths of a degree centigrade, and liquid water measurements are just now becoming adequate to test theories. An obstacle in advancing on virtually all fronts of cloud physics is our present inability to measure supersaturation.

Cumulus precipitation formation not involving the ice phase has been extensively documented. Coalescence models have been developed and partially tested. Early models used the concept of continuous collection, with a single large "collector" drop falling through a uniform distribution of water (the small droplets). Recently, sophisticated stochastic collection models have been evolved (Telford, 1955; Twomey, 1964, 1966; Berry, 1967; Kovetz and Olund, 1969) in which drops of all sizes are allowed to collect each other. Important features of spectrum broadening in the time frame of a life cycle have been thereby obtained. Theoretical work on collection kernels has eliminated the sharp cutoff in collection at a specific drop size. Coalescence efficiencies have been calculated theoretically (e.g., Hocking and Jonas, 1970) and checked by laboratory experiments (Whelpdale and List, 1971; Levin *et al.*, 1973), as have the terminal velocities of liquid hydrometeors. Extensive studies have been made to determine whether and how coalescence efficiency might be enhanced at small drop sizes, for example, by turbulent shear in cloud (Tennekes and Woods, 1973) or by electrical effects. In unglaciated clouds, the extent and magnitude of electrical fields are not well known. Current laboratory evidence, however, indicates that at normal field strengths, coalescence cannot proceed effectively until collector drops have attained about 40 μ diameter (Mason, 1971; Neiburger *et al.*, 1974).

An important implication for the potential modification of coalescence rain is the indication from collection models of important microphysical differences between clouds with slow (<5 m sec^{-1}) versus rapid (>10 m sec^{-1}) updrafts. A difference may be the possibility of the so-called "Langmuir chain reaction." With the rapid updrafts, as raindrops grow large, they may become unstable and start breaking up. The drop size and conditions at which this breakup occurs are not well known, but by the time the precipitation drops have reached 6 mm diameter, the breakup probability is high. Each fragment can in turn grow to breakup size and repeat the process. If the cumulus updraft is strong enough to sustain the near-breaking drops (nearly 10 m sec^{-1} is the terminal velocity for a 6 mm raindrop) a rapid acceleration of rain growth would be possible (Langmuir, 1948). Both further modeling and further measurement work are required to ascertain the frequency and conditions for the natural occurrence of chain reactions and to assess man's chances of instigating them.

Resolutions of the conditions for "chain reaction" droplet growth is of utmost relevance to the potential of cumulus rain enhancement by methods intended to augment coalescence artificially. Most wholly warm (above freezing) cumuli process so little water that doubling their precipitation would produce only a few acre-feet per cloud (25–50 at most). Apparently, the best subjects for application of coalescence enhancement are: (1) clouds that re-form continually owing to orography or convergence lines otherwise maintained and (2) strong updraft cumuli where the "chain reaction" is a possibility.

Many field programs (Battan, 1963; Braham, 1958, 1964; Ludlam, 1951) have demonstrated the importance of coalescence in cumulus precipitation formation even in clouds where the ice phase is present, particularly in summer tropical air masses. The most important recent advances in our knowledge of the ice phase in cumuli have also revealed the most serious unresolved dilemma, namely, the large concentrations of natural ice at warm supercooled temperatures ($-5°C$ to $-15°C$) in some but not all situations. These large ice concentrations are commonly associated with huge discrepancies (often a factor of 10^3 to 10^4) between the concentrations of ice particles and measurements of the concentrations of ice nuclei active at such warm temperatures. Two alternative, not mutually exclusive, avenues of explanation are possible. First, it is possible that current methods of counting ice nuclei are either inaccurate or do not encompass all the actual in-cloud mechanisms of nucleation. Second, numerous ice multiplication processes could be occurring.

Evidence has been accumulating that the crystal-nucleus discrepancy is greatest where large drizzle drops have been formed by coalescence. Recent important laboratory experiments by Hallet and Mossop (1974) have produced rapid enough ice multiplication to account for the discrepancy, but under very specialized, complex conditions. The liquid water in the experimental cloud was about 1 gm m^{-3}. The greatest rate of ice splintering occurred when "graupel" particles with surface temperatures of about $-4.5°C$ fell through a supercooled laboratory cloud. This temperature is close to that at which ice crystals growing by diffusion show a maximum growth rate along their major axis and grow as needles or hollow columns.

During the accretion of 1 mg of rime (corresponding to a graupel particle of diameter about 2 mm), several hundred daughter particles may be shed. If each secondary ice particle grows and rimes in its turn, these authors calculate that there should be no difficulty in providing, within a reasonable time, the observed "multiplication" factor of about 10^4.

If these results are relevant to real clouds, the apparent absence of significant multiplication in the temperature range $-3°$ to $-8°C$ in some documented cases remains to be explained. Of particular interest is the modification situation in Israel (Gagin and Neumann, 1974), which exhibited close correspondence between concentrations of ice particles and ice nuclei in highly continental cumuli. For these tall vigorous clouds, it is suggested that riming may take place at temperatures colder than $-10°C$, so that the required conditions for rapid ice multiplication are not met. The main point for our purposes is that splinter production may hinge on the coincidence of several concomitant cloud conditions, namely, a specific range in ambient temperature, water content, and fall velocity of the particles.

This discovery reemphasizes the necessity for detailed understanding of cloud physics and measurements as essential background prior to modification attempts. The "ice multiplication" dilemma is a critical gap in the necessary scientific foundations of cumulus modification, particularly pertinent to those approaches that involve introducing artificial ice nuclei into supercooled clouds. The presence of natural ice concentrations of 10–100 particles per liter as frequently measured (Koenig, 1963; Mossop *et al.*, 1968; Mee and Takeuchi, 1968) might spell useless or negative expectations from some cloud-seeding hypotheses if applied to clouds with these concentrations of natural ice. Resolution of the problem involves the ability to measure water-ice distributions together with the cloud motion fields (a very difficult problem) over space and time variations of the draft structure. The improved aircraft instrumentation now available, with optical ice particle counters and inertial platform-gust probe systems, together with ground-based dual Doppler radars, offer concrete but expensive hope for progress. Nevertheless, the outstanding fact remains that the basic physics of glaciation processes in real clouds are not well understood and must be studied, together with the dynamic processes that regulate them, by model, laboratory experiments, and field measurements, if the mainstream of cumulus modification is to advance.

Despite the major dilemma just cited, some aspects of the cumulus ice phase have made progress. There have been laboratory studies of ice crystal growth rates and habits as functions of temperature. Formvar replications in real clouds frequently appear to agree with these results, particularly in less wet, less turbulent nontropical clouds. Junk ice and wet graupel appear common in supercooled tropical cumuli, although good replications are always difficult and are virtually lacking at the high levels and colder temperatures. Laboratory studies have been made of the terminal velocities of ice crystals and hailstones (List and Schemenauer, 1971). In the field, Doppler radars (Fig. 1) are at last providing data on the actual motions of frozen particles in real clouds.

Considerable progress has been made on documentation of the physical structure of hailstones and their embryos by crystallography, bore methods, stereomicroscopy, and other means (see List *et al.*, 1970, 1972). Isotopic analysis has been used to determine hailstone paths and origins, while infrared thermometers have revealed persistent lowered surface temperatures

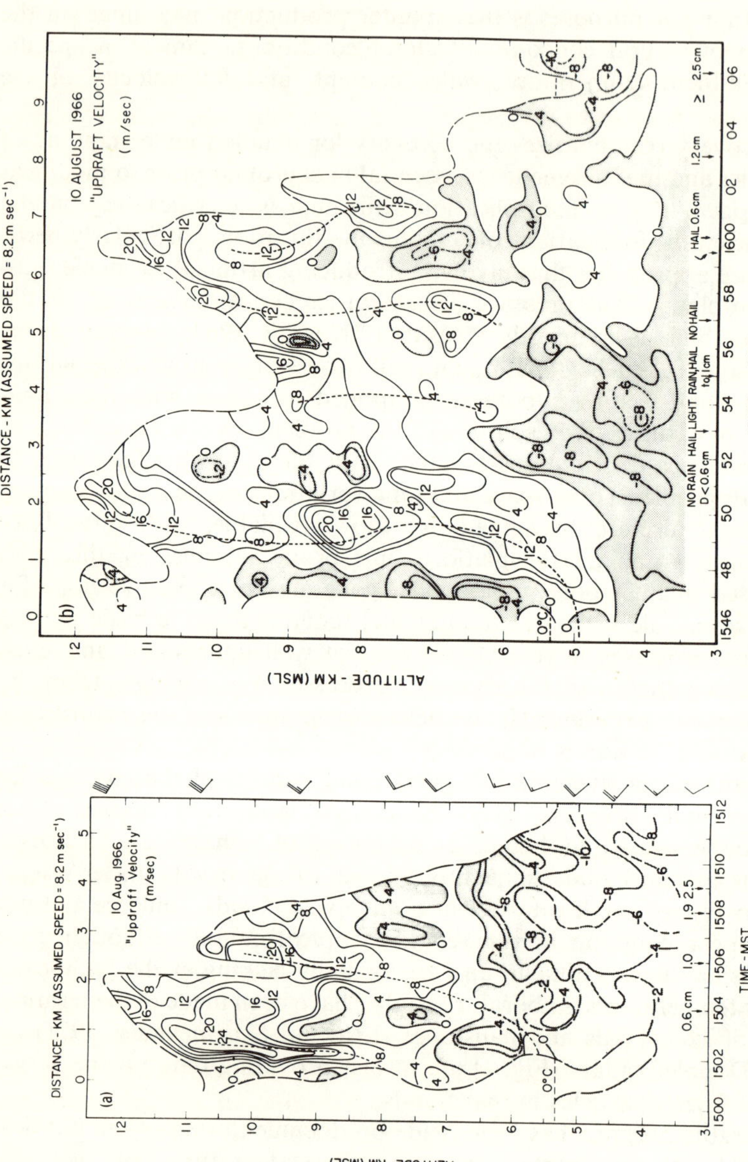

FIG. 1. Draft velocities in two Arizona hailstorms, as estimated by single doppler radar, with correction for particle fallspeeds. The distance scale (top) is constructed from the time scale using the mean wind in the cloud layer. Hailstone diameters actually observed are indicated by the arrows along the bottom. (After Battan, 1974.) (a) First hailstorm. Arrows to right are from nearby rawinsonde. (b) Second hailstorm. Note succession of updraft cores, suggesting buoyant "bubbles" or "thermals" entering at the upshear side.

in hail swathes. Size and fall speed distributions of hailstones reaching the ground are better documented by denser networks of hail pads, and theories of hail growth within clouds are slowly being evolved (Charlton and List, 1972a, b).

In the relevant areas of cloud electrification, the scientific foundations are incomplete, leaving critical gaps in the knowledge upon which to base precipitation modification efforts. The dominant electrifying mechanism, if indeed one dominates, has not yet been identified or agreed upon. In building up the observed field strengths, the relative roles of convective motions versus microphysical ice processes have not been established, nor has it been specified whether large (precipitation) or small (cloud) hydrometeors are the primary carriers of charge.

The firm, although rare, documentation of "warm" lightning (cf. Moore and Vonnegut, 1973) not only proves that ice processes cannot be invoked as the complete explanations for all thunderstorm events, but suggests an important precipitation-electrification relationship in all-liquid clouds. Accumulating "rain gush" measurements indicate that electric field growth and discharge may often precede rather than follow precipitation enhancement (Moore *et al.*, 1964; Battan and Theiss, 1970; Levin and Ziv, 1974). These important observations are finally beginning to be supported and extended by numerical simulations. Stochastic treatments of coalescence rain development now incorporate charge transfer and the rate of electrification in the growth of warm rain (Scott and Levin, 1974) with predicted dramatic positive or negative effects depending on conditions which can now be specified. Also, Sartor (1973) presents persuasive calculations and laboratory evidence suggesting that coalescence coefficients may be enhanced far above one when the strong electric fields prevail which are characteristic of thunderstorm conditions.

Of course, cloud electrification is well known to be most intense when ice hydrometeors are present. Over the years, many ice-phase electrification models have been proposed and are reviewed by Mason (1971). None incorporate and few consider cloud dynamics. Recently, the polarization-charging mechanism has received considerable theoretical, experimental, and simulation support (Ziv and Levin, 1974). This charging process is a result of collisions and separations of particles of different fall velocities in the presence of an ambient electric field. Clouds containing large ice particles and small drops are found most favorable for the growth of the electric field to values that produce lightning.

In nature, the most strongly electrified clouds are generally the most convectively active and the most deeply penetrative, so that the presence of the ice phase and intense convection commonly coincide, making difficult the assessment of their relative roles in electrification.

Real progress toward combining mixed phase microphysics, electrification, and some aspects of dynamics is just now on the horizon, with the development of the cumulus model simulations to be described next.

2.2. Cumulus Dynamics and Models

A major advance in cumulus science enabling sharper and sounder modification approaches has been the growing ability to simulate cloud processes on the computer. In the area of cumulus dynamics, the models provide the most revolutionary development, but have also encountered very difficult conceptual and practical obstacles to further progress. An excellent and exhaustive review on the basic science has just been prepared by Cotton (1975a) so that only those points most relevant to modification will be highlighted here. A mathematical summary of three-, two-, and one-dimensional models is contained in the Appendix, with a brief recapitulation of the pertinent assumptions, advantages, and shortcomings of each type.

In nature, most meteorologists believe that the motion field of a cumulus exerts strong, perhaps dominant, control upon its microphysics. The updraft strength and structure regulate the condensation rate, the latent heat release, the supersaturation, the embryonic hydrometeor forms, and how they are able to grow and interact. Downdrafts at the edge, beneath, and within cloud further regulate hydrometeor development, both by means of the environment (thermal, humidity, etc.) they provide, and by their effect on particle trajectories. All the ways and mechanisms by which cloud microphysics feeds back influences upon convective dynamics are not well documented or modeled at present. However, the feedback mechanisms fall into two categories: dynamic and thermodynamic. In the former category, a major impact is the buoyancy reduction exerted by the particles owing to their weight. When particles are falling at terminal speed, this effect is readily modeled; it can eliminate or even reverse the sign of the buoyancy with moderate contents of water substance. Other dynamic "drag" effects under nonequilibrium particle descents are usually ignored, although the range of validity of the assumption is not known.

Modeling work (Murray and Koenig, 1972) suggests that thermodynamic feedback of microphysics to the dynamics can be even more important than the "water loading" just described. In their two-dimensional simulation, evaporation of hydrometeors at cloud edges produces cooling and thereby induces downdrafts. These downdrafts play a critical role in the life cycle of the model cloud, which is not simulated realistically when cloud-edge evaporation is omitted. Qualitative comparison with observations confirm the probable importance of evaporative feedback, especially where the cloud environment is relatively dry. In tropical regions, and particularly in highly humid tropical disturbances, this thermodynamic feedback is likely to be minimized.

The most important role that cloud hydrometeors play in dynamics is, of course, the condensation/deposition heating released with the initial change from water vapor into liquid or solid. Cumulus buoyancy, i.e., acceleration field, is created by temperature differences which are in turn altered and redistributed by the motions. Therefore, the dynamics of a cumulus is basically so highly nonlinear that the problem was virtually intractable prior to the advent of computers.

Conceptually, the most significant advance in cumulus dynamics was the documentation of entrainment (Stommel, 1947). The demonstration that cumuli were diluting their buoyancy by dragging in outside air into their circulation, while at the same time shedding their moist lifeblood to their surroundings, explained why most trade cumuli topped out at levels where wet adiabatic "parcel" buoyancy was a maximum. The discovery revolutionized tropical meteorology, in that cumulus impacts upon large-scale circulations became identifiable (Malkus, 1962).

The documentation of entrainment also became a vital basis of the existing foundation for precipitation modification. The development of entrainment concepts initiated the "entity" approach to cumulus modeling and later to the idea of "dynamic" seeding (Simpson, 1970).

Using the newly available aircraft measurements of temperature and humidity made inside and directly outside of trade cumuli, Stommel devised a mathematical and graphical method of estimating the fractional rate of entrainment in the vertical (Malkus, 1954), which can be expressed mathematically as $(1/M)\,(dM/dz)$, where M is the mass present in the cloud element or entity, and z is the vertical coordinate. In small cumulus clouds, calculations from observations suggest that in about 1 km ascent, a cumulus takes in about as much air from its surroundings as the amount it already contained. Of course, observations have demonstrated that a cloud's mass does not commonly increase continually by the amount of air entrained. "Detrainment" or shedding of cloud air may occur. Lateral shedding related to vertical shear in the horizontal wind was first documented (Malkus, 1949, 1952) by aircraft measurements which found moist air and cloud-scale turbulence on the downshear sides of tropical cumuli. Later, time-lapse photography suggested vertical shedding, or "erosion" of cloudy material into a trailing wake. This concept was the cornerstone of the famous "bubble" idealization of cumuli at Imperial College (Ludlam and Scorer, 1953; Scorer and Ludlam, 1953; Malkus and Scorer, 1955). Regardless of whether cloudy air is shed or not, entrainment has been shown by both observations and models to be a most important brake on cumulus growth, particularly in warm environments.

It has been demonstrated, by measurements and calculations jointly, that the main destructive effect of entrainment in tropical air is buoyancy reduction by drying, while momentum braking exerts significant but secondary retardation. In the years immediately following the discovery of entrainment by Stommel, some meteorologists (e.g., Austin and Fleisher, 1948) proposed the concept of subdividing entrainment in terms of postulated mechanism, namely, "dynamic" and "turbulent." Dynamic entrainment is just that intake of mass into the updraft required to satisfy continuity. If the radius remains approximately constant, this influx only occurs through the lower part of a growing cloud, where the updraft speed increases upward. Model clouds based on this concept of entrainment are generally forced to detrain air above the updraft maximum, possibly a severe limitation. "Turbulent" entrainment was postulated to occur due to shear at cloud or draft edges and could proceed regardless of the sign of the draft or its derivative.

The recognition that entrainment or dilution by means of outside air entering through the top of growing towers was not emphasized until the postulation of the "bubble theory" by Ludlam and Scorer. Measurements on single "thermals" produced in the laboratory (Woodward, 1959) suggested about the same amount of environment fluid entering at the tops as through the side and bottom of the rising buoyant element. Unfortunately, clear documentation of entrainment processes in real clouds has proceeded so slowly that it is still not possible to evaluate the merits of these concepts.

The functional dependence of entrainment must be specified in order to incorporate the process into analytical or numerical models, or both. A vital early revelation from data by both the Thunderstorm Project (Byers and Braham, 1949) and tropical studies (Riehl and Malkus, 1958; Malkus, 1960) was that for huge towers, the proportional dilution rate had to be much less than that found by Stommel in small trade cumuli topping at 2–3 km. These observations, together with the famous series of Imperial College Laboratory experiments (cf. Scorer and Ronne, 1956), led to the postulate that to first order the entrainment rate was inversely proportional to the width of the rising element.

The $1/R$ postulate, still controversial, forms a cornerstone of the one-dimensional cumulus models. These models have provided a major impetus and foundation for a large class of modification experiments on a world-wide basis. They form, at the same time, a focus for considerable debate in the scientific community regarding their physical soundness and ability to predict measurable cumulus parameters in their observed relationships (cf. Warner, 1970b, 1972; Simpson, 1971, 1972a).

The other cornerstone of the one-dimensional models was provided jointly by aircraft and quantitative time-lapse photography of cumuli, mainly by the Woods Hole and Imperial College groups (see Malkus, 1962, and bibliography; Simpson *et al.*, 1965). Cumuli were likened to buoyant plumes, jets, or bubbles with vortical internal circulations. An equation for their rate of rise was formulated and integrated in height steps, postulating a rate of entrainment inversely dependent on dimension, thus permitting computerization of Stommel's method to calculate dilution, buoyancy, and condensate.

Early one-dimensional models were Lagrangian; that is, the coordinates were fixed to the rising tower (Simpson *et al.*, 1965), with extremely crude assumptions regarding the fallout of the condensate resulting after the ascent and entrainment calculation. Later models used Kessler's (1969) formulations of precipitation growth by autoconversion and collection (Simpson and Wiggert, 1969, 1971), while still later advances incorporated stochastic collection and the details of droplet and ice crystal growth (Cotton, 1972a, b).

Dynamically, steady state and time-dependent models have been added to the Lagrangian tower concept in an attempt to look at the entire vertical extent of a cloud, and in the latter some attempts have been made to take into account impacts upon the cloud of changes it induces in its environment (Cotton, 1975a), a factor omitted in previous versions.

In their early stages, the one-dimensional models generated a modification hypothesis that has led to a decade of work on "dynamic" cumulus seeding. Right away their predictions suggested that cumulus buoyancy could be increased by rapid induced freezing of supercooled water and that, under specified conditions of the environment, this buoyancy increase could lead to enhanced tower growth in the vertical. "Seedability" was defined as the model-predicted height difference between a heavily seeded (with artificial ice nuclei) tower and the same tower if left unseeded.

In 1965, the model and corresponding hypothesis regarding vertical growth were successfully tested in a randomized airborne pyrotechnic seeding experiment over the Caribbean (Simpson *et al.*, 1967) by the Experimental Meteorology Laboratory of the National Oceanic and Atmospheric Administration (EML, NOAA). These experiments firmly established the foundations and execution of dynamic seeding with regard to the increased growth of selected tropical clouds. They also demonstrated the usefulness of one-dimensional models and began to delineate their range of applicability.

The EML-NOAA dynamic seeding program was moved to south Florida in 1968, where it has continued intensively for 7 years (Simpson *et al.*, 1971, 1973; Simpson and Woodley, 1971) investigating the relationships between dynamic seeding and rainfall changes, with randomized experiments on isolated clouds, floating targets, and later in a large target area of 1.3×10^4 km^2.

The greatest conclusively demonstrated seeding factor on moving target rainfall was shown in the EML single cloud experiments in Florida, with 26 pairs of cases. The seeding factor was shown to be close to three, averaging about 340×10^3 m^3 per cloud (Simpson *et al.*, 1971). The extension of dynamic seeding in Florida to multiple cumuli and to the large fixed target area is still incomplete; however, the effort has already led to an increase in the available scientific and technological bases of rain modification that will be discussed herein.

In experimental programs, one-dimensional models are used in three main ways. First, they are employed in real time to screen out days of poor seedability from those of adequate seedability in order to launch the expensive experimental procedures. Second, the model is used to provide covariates and relationships between variables in a more powerful hypothesis testing during evaluation. Third, these models have been applied on a worldwide basis in developing "seeding climatologies" leading to estimates concerning the potential of water resource augmentation. Most of these "climatologies" must be applied with caution for the following reasons: Commonly, they consist only of one-dimensional model calculations upon all available soundings using a hierarchy of assumed cloud radii. Only rarely is an attempt made to ascertain the actual cloud populations and radii really occurring. Additionally, hierarchies of assumptions are made to relate the calculated seedability to potential increase in rain. This step is taken either with model calculations of rainfall, which in the current state of model realism is risky, or an empirical relation is used that has been constructed at one locality from limited data.

A considerably improved study of this type for Florida, with clearly explained limitations, has just been published by Holle (1974); the methodology is transferable to other regions.

Holle investigated the months of April and July in two south Florida areas of about 4400 km^2 each. First, he established an empirical relation between rain volume increase and model-predicted seedability for the single clouds measured in the EML experiment (Fig. 2). This regression gives the seed-control rainfall difference (in rain volume) as a function of seedability, which unfortunately requires a photogrammetric measurement[2] of tower radius. Then he used a special radar-population study in which the number of cumulus tops in the seedable height range had been tabulated

---

[2] Other authors have determined the tower radius by working backward from measured cloud top height, making the assumption that the radius-height relationships, or entrainment rates, used in the model are correct and have been adequately tested.

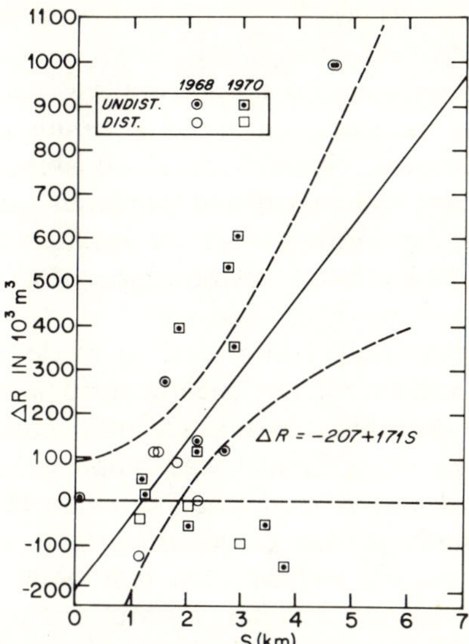

FIG. 2. Rainfall change (in 10^3 m^3) related empirically to seedability (km) based on results of single cloud seeding over south Florida during 1968 and 1970. Note that rainfall is indicated to decrease for seedabilities less than about 1 km. The 95% confidence band for the regression line is shown by the dashed lines. Different relations may apply to other areas, seasons of the year, or weather conditions. (After Holle, 1974.)

by hour. The soundings nearest in time were used to calculate seedability. The empirical regression was employed to obtain rain increase per cloud if seeded, and the number of seedable clouds was called upon to estimate total potential rain increase, which for south Florida is calculated as up to 10–12 % in some months.

An important limitation of this study so far is that the seedability-rainfall increase regression from single cloud experimentation has been used, while rain budget studies (Woodley *et al.*, 1971) show that merged complexes are by far the greatest rain producers for south Florida. Adequate regressions relating predictors to area-wide seeded rainfall increases may not be available in the next decade and, if they are evolved, they are likely to be multivariate and dependent on synoptic and meso-scale circulations and perhaps also on microphysical and aerosol conditions in a still unknown manner. Meanwhile "seedability" climatologies (cf. Weinstein, 1972) have been widely compiled using one-dimensional model calculations on available soundings. These calculations require an assumption about cloud radii. Therefore, the results provide information only on what the seeded and unseeded vertical growth of these arbitrarily chosen radii would be, if clouds of these dimensions were present and if the soundings used were representative of conditions near those clouds. Even if the one-dimensional models provide good "seedability" information, two links are missing that must be supplied to render the information useful to water management. First, the frequency of seedable clouds must be documented by radar or other means. Second, a relationship between seedability (which is only a height difference) and potential rain increase must be evolved, as was done by Holle (1974) for isolated Florida clouds. For cloud systems and areal rainfall, relationships of this sort are presently nonexistent.

Even as these models were being used to guide and evaluate experiments, the scientific validity of the one-dimensional approach was being challenged by some members of the scientific community. Two main criticisms were brought out. First, it was claimed that the $1/R$ postulate regarding entrainment was invalid. Second, evidence was cited that the models could not jointly predict cloud top heights and the liquid water contents within the towers. Recently the first point has been settled. A careful two-aircraft observational study by McCarthy (1974) has substantiated the $1/R$ law to first order when isolated actively rising towers are selected and their properties of water content and temperature measured very carefully, together with a painstaking measurement of radius.[3]

The second point remains more perplexing. Since the early 1960s data have been sought to test the one-dimensional models in a well-documented situation of unmodified cumuli, with hydrometeors mainly in the liquid phase. Such data are demanding to obtain, since other types of information are required in addition to those obtainable by aircraft penetrations. Tower dimensions and rise rates are key variables. To date, one of the few existing data sets is that obtained by Saunders (1965) near Barbados, with careful radar measurements and theodolite determinations of rise rates and radii. In most of the dozen cases available, the EML one-dimensional model (Simpson and Wiggert, 1969) gave tops and radar echoes within measurement accuracy, but tended to overestimate rise rates by 25–35 %.

Among the few other data collections available are sets of observations on smaller, drier clouds off the coast of Australia reported by Warner (1969a, b; 1970a; 1973a, b) and later by Cotton (1975b).

Figures 3–5 illustrate the difficulty for a case study made on a small oceanic cumulus near Australia (Cotton, 1975b). Figure 6 compares the case study–observed ratios of measured cloud water content to adiabatic. Three of the curves are from averages of many penetrations. The circled

---

[3] Thus invalidating earlier conclusions by Sloss (1967) which can now be shown to have been derived from measurements not carefully enough obtained or analyzed for the demanding problem involved.

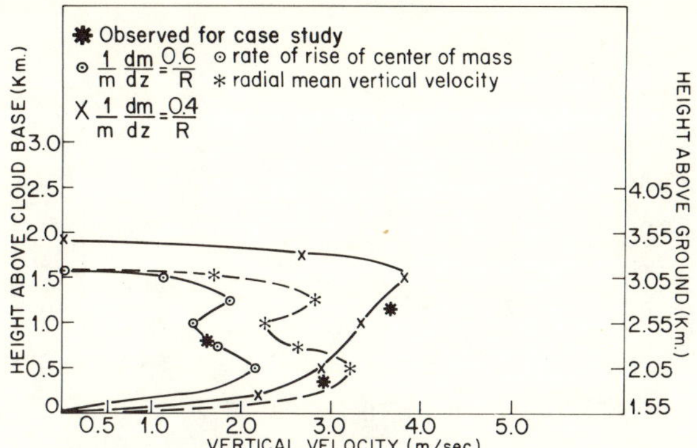

FIG. 3. Tower rise rates as a function of height. Stars are aircraft observations by Warner. Curves are one-dimensional Lagrangian model calculations with entrainment coefficients as indicated. (After Cotton, 1975b.)

and x-ed curve are one-dimensional model calculations, using very high entrainment coefficients as noted and still giving "too wet" clouds. It appears that with the Australian data, no one-dimensional model, even the more sophisticated time-dependent variety, is able to estimate the measured top heights and water contents simultaneously, even approximately accurately. The puzzle remains today as to whether the difficulty is mainly with the models or whether some of the difference may reside in the clouds or in the observational procedures.

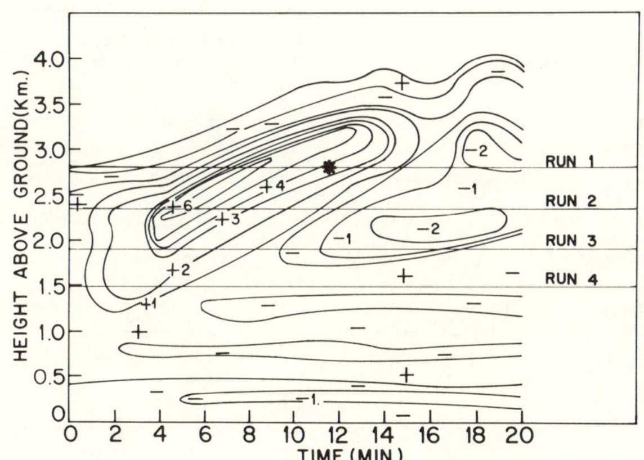

FIG. 4. Comparison of time-dependent one-dimensional model and aircraft-observed values of vertical velocities in m sec^{-1}. Star placed as described for Fig. 5. "Run" denotes aircraft traverse. (After Cotton, 1975b.)

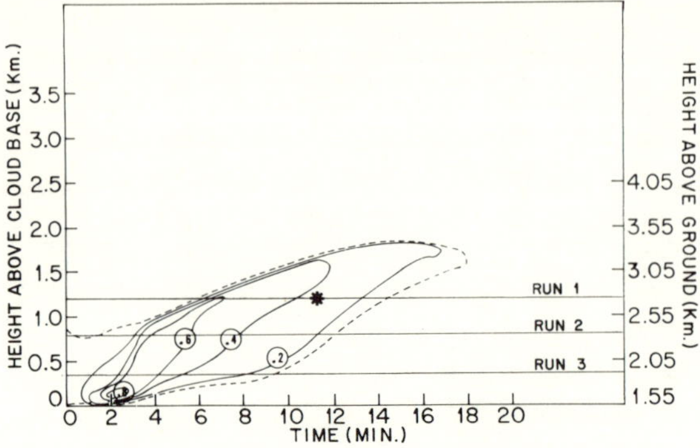

FIG. 5. Comparison of time-dependent one-dimensional model and aircraft observed values of Q/Q_a, namely, ratio of actual to adiabatic liquid water content. The star is placed at that time five minutes after the tower reaches the run level, estimated by the aircraft scientist (J. Warner) as time required for aircraft to reach tower after sighting. Star represents average water in updraft portion. "Run" denotes aircraft traverse. (After Cotton, 1975b.)

Cotton (1975a) concludes that the predictions of models based on the $1/R$ entrainment hypothesis neither represent the average data of a cloud over its lifetime nor of the average of a large body of clouds. At best, the predictions represent the actively growing stage of the cloud. Even so, he (Cotton, 1975b) could find no consistent pattern between the velocity and Q/Q_a predicted by the models and observed sample means (Australian

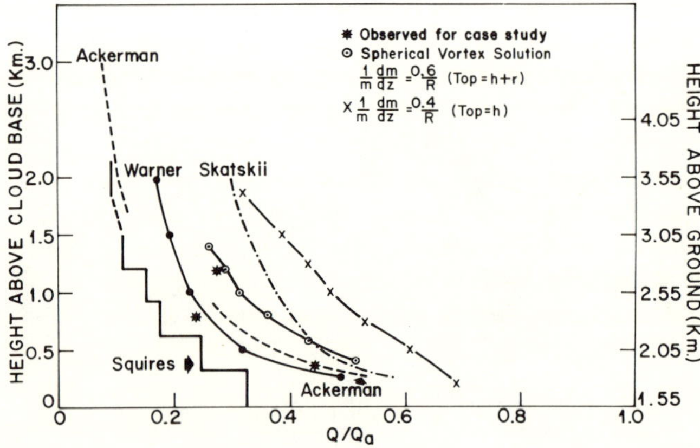

FIG. 6. Mean cloud ratio of Q/Q_a as observed by several workers, as modeled with two large entrainment coefficients (x and ⊙) and as observed in case study by Warner of one individual cumulus off Australia. (After Cotton, 1975b.)

small clouds). Other authors, modeling a sequence of parcels (Danielsen *et al.*, 1972) found good agreement with model predictions and most measured parameters except vertical velocities. Overprediction of vertical velocities and rise rates appears characteristic of these models generally. This fault has been attributed to the neglect of nonhydrostatic pressure forces that produce the toroidal motions in the cloud, thereby reducing its vertical motions, or to nonlinear interaction among cloud parcels.

It is noteworthy that the best agreement of one-dimensional model predictions with observations occurs in those cases where isolated, rounded, actively rising towers are selected for the test cases. Two examples are the measurements on real clouds by McCarthy (1974) and on a rocket-generated "cloud" by Lopez and Vickers (1973), a single thermal investigated under conditions of weak shear.

In any case, virtually all meteorologists concur that while one-dimensional models can continue a useful role in guiding and evaluating modification experiments, physically they serve, at best, only as a rough beginning to the proper simulation of cumulus processes. Their limitations must be kept clearly in mind at all times. At most, the $1/R$ entrainment relationship is a first-order approximation, useful for some types of computations, but perhaps disastrously inadequate for others. In particular, the assumption of "top hat" profiles or uniform property across the cloud cannot be adequate for hydrometeor growth. Hail models, for example, based on one-dimensional dynamics must be regarded at this time as "tuned models" (see Appendix), possibly extremely useful in the experimental situation for which they have been "tuned," rather than generally applicable, realistic physical simulations of cloud processes.

Most important, the actual process and mechanisms of entrainment are still far from clarification. Mixing processes remain a crucial gap in the scientific basis for weather modification, to which urgent and intensive research must be devoted. The key question regarding entrainment stands out, of necessity fraught with impact on modification, and that is, does entrainment depend mainly on cumulus-scale processes that are not too difficult to document, or does it depend on the intensity, scale, and distribution of smaller eddies within the clouds or at their boundaries? Or could it depend further on forced or free convective fluxes through cloud base? And here stands the frontier today.

From this point onward, the numerical simulation background of cumulus dynamics is still evolving, and it can perhaps be evolved more expeditiously with the motivation, data, and technology of modification to thrust it forward.

With the more sophisticated "field of motion" cumulus models, fifteen years of effort have been made which may now be starting to relate usefully to modification. Outstanding work on two-dimensional cumulus models has been conducted in the tropics (Murray, 1971; Murray and Koenig, 1972; Wiggert, 1972) and in a situation near a heated mountain barrier (Liu and Orville, 1969). The latter showed a "model shower" life cycle in beautiful qualitative agreement with the history of real thunderstorm cells.

Ice processes and simulated seeding are now being introduced into this model series. In the case of warm clouds, the shallow convection models of Arnason *et al.* (1968; Arnason and Greenfield, 1972) combine two-dimensional dynamics with the details of condensation, with a goal of simulating hygroscopic seeding.

Careful examination, however, shows important differences between two- and three-dimensional simulations under otherwise identical conditions. Two- and three-dimensional model cloud growth is highly sensitive to conditions in the layer below cloud base, to the initial shape, size, and intensity of the perturbation, to the outer boundaries, the grid spacings, the differencing scheme, and presence or absence of implicit damping therein.

Even more challenging, however, is the question of how to simulate the crucial mixing processes in field of motion models. Linear viscosities, once prevalent, have now been largely abandoned as inadequate and various nonlinear mixing schemes are being attempted in a series of fully three-dimensional cumulus models that are just reaching the early production stage. These models are of such complexity that their presentation and testing represent a difficult challenge. The danger of errors compensating to give realistic-looking results is ever present. Internal tests of sensitivity, consistency, and conservation, step-by-step on the computer are expensive. External tests against data are even more so, and unattractively laborious. Data bases must be evolved, where remote and direct sensing are used conjointly and modifiers and experimenters focus on the same clouds, which could readily be those subject clouds in a modification experiment.

A fruitful approach, somewhat out of fashion in the current computer age, is the admittedly simplified semianalytic model which is tested against conditions in a field or laboratory, specifically selected to highlight the model's key features. Excellent examples are the analytic work directed at understanding the mixing processes in dry thermals and plumes by Morton (1968) and, based thereon, the two-shell simulation of a cumulus by Lopez (1973). Lopez releases a succession of parcels from the immediate sub-cloud layer. He then parameterizes the postulated tendency of thermals and clouds to form toroidal circulations by ejecting mass from a central convective region to an outer shell when an upward decrease in vertical velocity is present. Entrainment and detrainment rates are formulated as a function of the turbulent intensity of the cloud parcel and of the environment respectively, based on the work of Morton. Unfortunately, this intriguing model has not yet received observational test.

Nevertheless, this type of approach may prove a valuable supplement to numerical experimentation in filling one of the most vital gaps in cumulus knowledge, namely what determines the spectrum of horizontal sizes of the active towers on a given occasion. This unsolved problem is of vital relevance to the modification of cumulus precipitation, since first, horizontal size and vertical penetration are well known to be related (Saunders, 1961). Second, natural hydrometeor structure and growth are related to the horizontal extent of the growing region and third, dynamic seedability is a function of tower dimension, as we have seen.

Two promising avenues of progress on this difficult problem have been recently reported. The first by Betts (1973) uses one-dimensional models, observations, and thermodynamic analysis to derive a preferred ratio between the convective layer depth and permitted tower radii. The second is fascinating two-dimensional simulation by Hill (1974). His results suggest that over land, early in the day cloud radii are governed by boundary layer convection below cloud, but later their sizes are controlled by cloud interactions of several sorts. These will be the subject of the next section. His predictions compared well with cloud spectrum measurements by Plank (1969) over Florida. Neither model yet treats the effect of convergence upon cloud spectrum, which most meteorologists believe leads to larger-width towers (cf. Malkus and Riehl, 1964).

In summary, advances in knowledge and simulation of individual cumulus dynamics and dynamical-physical interactions have led to significant progress in experimental concepts and approaches. Modeling has not only led to new modification hypotheses and means of testing them, but the model predictions have made it patently clear that different outcomes can follow the same treatment, depending on the initial conditions of the cloud-environment system. Formalization and general acceptance of this idea, long well known by expert practicing modifiers, has done much to bring cloud modification into the realm of respectable science.

Nevertheless, cautions are in order. The complexity of cumulus processes is so great that despite its admirable progress, modeling is still in its early infancy. All models are, of necessity, so tenuously poised upon hierarchies of oversimplifications and assumptions, that while their use in guiding and evaluating experiments is indispensable, the following additions or reservations are mandatory: (1) no model or models can substitute for or replace randomization in the foreseeable future; (2) tests of every link in the modeling chain should be made against actual data wherever possible; and (3) the results of the experiments should continually be fed back to test and improve the models, providing an improved scientific basis for the next series of experiments.

Particularly wide gaps in knowledge concern ice multiplication, the entrainment process, the role of subconvective-scale turbulence therein, and the specification of the radius spectrum of cumulus towers and its relation to convergence on the meso- and synoptic scales.

2.3. Cumulus Interactions, Groups, and Patterns

Cumulus clouds rarely occur in isolation. Nearly always, cumuli form and live their lives in groups, lines, patterns, and clusters (Fig. 7). The synoptic-scale disturbance is historically the best known organizer of convective systems. Figure 7 shows at least four types of these, from the inverted V's and hurricanes of the tropical oceans to the fronts and wave cyclones of mid-latitudes. An outstanding synoptic feature is four tropical storms. In the eastern Pacific, we see tropical storm Nanette (west) and tropical storm Monica (east). In the Atlantic, we have the remains of moderate hurricane Doria and extreme hurricane Beulah about to strike the Gulf Coast of the United States. In addition to the hurricanes, there is a tropical wave in the eastern Atlantic. Frontal systems are seen in both northern and southern hemispheres, with several bands of clouds parallel to the front itself in the cold air. Even larger planetary and "monsoonal" cumulus organization is seen by the satellite picture, particularly the cumulonimbus outlining the west coast of South America, and the extension and bulge of the Intertropical Convergence Zone (ITCZ) "hot towers" over Africa (the western portion of the ITCZ has been carried away by the hurricanes).

Going downward in size, the meso-scale is an important, fast-developing frontier of cloud organization, although sea breeze, mountain, and heated island patterns have been documented observationally for two decades (see Simpson, 1973, also illustrations and references).

Even in the absence of disturbances, however, some of the most important effects of cumuli, such as heavy rains and severe weather, occur when two or more cumulonimbus clouds join together to become a merged complex or "merger." For example, in Florida, a single isolated cumulonimbus thunderstorm may produce in its lifetime $100–3000 \times 10^3$ m^3 of rainfall, while a merger of two or more such clouds often rains $5000–50,000 \times 10^3$ m^3 (Woodley *et al.*, 1971) so that 50-100 mergers can account for a wet month's precipitation in south Florida.

Most large cumuli are, in fact, built up from the merger or interaction of adjacent and/or successive smaller ones. Yet the documentation and modeling of cumulus interactions, groups, and systems are now in roughly the same primitive state that individual cumulus modeling was 30 years ago at

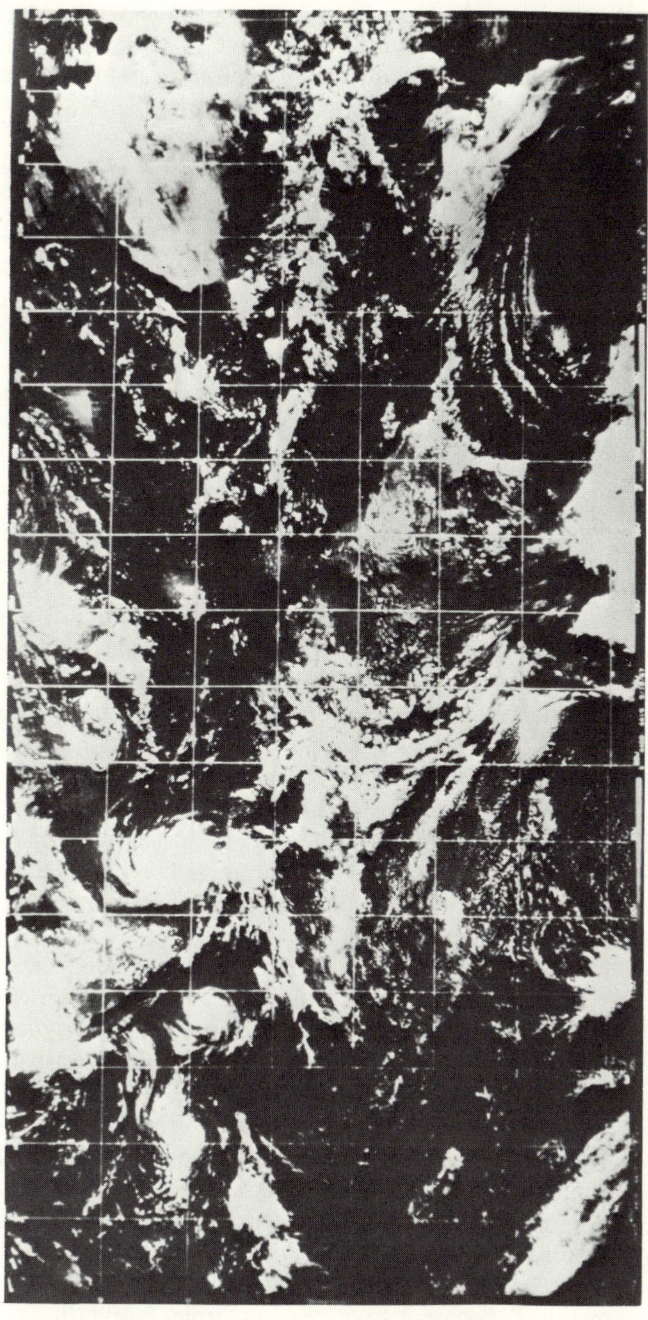

Fig. 7. World satellite mosiac for September 18, 1967.

the close of World War II. This open frontier is the widest gap in the scientific basis for the modification of convective precipitation.

With single isolated cumuli, operational precipitation augmentation would rarely be economic. Even isolated cumuli respond differently to seeding depending upon their interaction with other clouds and systems (Simpson and Woodley, 1971). In some regions cumulus mergers, as mentioned, are the main summer rain producers. In other regions, winter cyclonic storms, with convective bands as the modification unit, are the useful contexts for experimentation. How clouds interact with each other and with their surroundings and boundaries undoubtedly determines whether, to what degree, and by what means they may be treatable in any target, from single cloud to extensive area. Effects beyond the specified target, in both space and time, also form a wide open frontier in cloud modification. Positive downwind effects on rainfall now appear firmly related to seeding in at least two well-documented precipitation augmentation experiments on winter storms, one in southern California (Brown, 1971), the other in Israel (Brier *et al.*, 1974). As yet no physical hypotheses have been advanced and tested for these positive downwind effects, which must involve the direct or indirect interaction of clouds.

As in these downwind effect cases and that of merger, cloud interaction can apparently lead to rain increases. In other situations, the impacts of clouds upon each other can decrease precipitation, as for example when compensating subsidence suppresses the neighbors of a large cloud or when an extensive anvil shadow wipes out convection over many square miles of land (Woodley *et al.*, 1974). These effects need not be confined to a downwind direction. Neither must clouds coexist in time to affect each other. Wetting and/or cooling the ground, changing the air stability or humidity, or washing out aerosols are just a few among many possible means by which predecessor clouds can influence successors for hours or perhaps longer periods of time.

Clearly, the validity of randomized crossover or any dual target modification experiment designs must be carefully reassessed in each case in the light of the interaction problem, which has been called "dynamic contamination" (Simpson and Dennis, 1974) and warrants extensive research effort.

Another of the principal controversies in weather modification relates to cloud interaction, namely, the famous "robbing Peter to pay Paul" dilemma. Now the question is: If we show we can augment precipitation deliberately in a given cloud or target area, does precipitation elsewhere undergo a compensating decrease? If so, where, to what extent, and how is the magnitude and distribution of the compensation controlled? The dilemma has not been resolved because first, the scientific foundations in cloud interaction do not yet exist to formulate specific questions tractably and second, measurement tools are just now becoming evolved for accurate documentation of convective precipitation over large inaccessible areas (see Section 3.2).

A closely related question concerns the degree to which the total precipitation within a region and time period can be specified by large-scale variables. This question is, of course, basic to numerical forecasting as well as to weather modification. For the former reason, it was one of the main foci of the 1974 GARP (Global Atmospheric Research Program) Atlantic Tropical Experiment (GATE). The extent to which rainfall can be predicted from large-scale parameters would appear at first glance to bear an inverse relation to its overall modification potential. Even ignoring valid reservations concerning open thermodynamic systems and feedback effects of cloud processes upon large-scale parameters, it is still possible that 10–15% variability may be within the indeterminacy of large-scale assessment, yet of vital importance to rain modifiers and those served by them (Gagin and Neumann, 1974). The scientific foundations upon which to build toward answering this vital question might be greatly strengthened by the late 1970s as the GATE analyses become available, although the probability of land-sea differences must be kept in mind in their application.

Very crude attempts to simulate cumulus interaction by adjacent or successive thermals have been made in the one- and two-dimensional frameworks, discussed earlier in Section 2.2. For our purposes, these efforts are less than useful for two reasons: first, cumulus interaction probably depends on entrainment and turbulent mixing processes that are not yet well modeled and second, they surely depend on the lower boundary, upon wind and wind shear, and other meso- and larger scale features such as convergence.

To date, the only meaningful simulation of cumulus interaction over land is the fine effort by Hill (1974) which is, nevertheless, only a bare beginning to include the important physics. He treats a uniformly heated lower boundary with random temperature perturbations, uses variable viscosity, good parameterization of precipitation growth and fallout, and has cyclic (open) boundaries. The model is still restricted by two-dimensionality, a coarse grid, and absence of a large-scale wind field. Even so, model clouds interact with each other and the subcloud layer in a manner that is qualitatively realistic. Improved stages of this model are ready to guide and evaluate modification experiments when computer time and capacity become available.

With the 1970s, cumulus interactions with the subcloud layer are just becoming part of the scientific basis for modification experiments. Long ago glider pilots' experience (Scorer and Ludlam, 1953; Ludlam and

Scorer, 1953) suggested that on fair days over land small cumuli are often found at the tops of buoyant columnar thermals. In calm air or light winds, these may remain rooted over warm spots, while in moderate flow they can travel approximately with the wind. With stronger flow, oceanic and laboratory studies (Woodcock, 1940; Malkus and Riehl, 1964) suggest that thermals give way to rolls oriented along the shear vector between the lower boundary and the convecting layer. Oceanic cloud lines are commonly found where the ascending portions of these rolls can be inferred. Over flat land, Plank's (1969) and other observations suggest that the same mechanisms may be responsible for the closely spaced rows of small cumuli seen in the morning hours of fine days.

For large clouds and precipitation, however, evidence has been accruing over the past several years that convergence lines must be set up. We hypothesize but have not yet demonstrated that these are necessary for merger also.

The development of these convergence lines and their relationships to precipitating cumuli has been studied observationally over the tropical oceans by Janota (1971), over a tropical island by Bhumralkar (1973a, b) and over a subtropical peninsula by Pielke (1973, 1974) and Fernandez-Partagas (1973). In each case, the convergence line preceded attainment of the precipitation stage by the clouds. Over Florida, the sea breeze influence is apparently so strong that the locations of the major convergence lines, which observationally coincide with radar echo distributions throughout the afternoons, are not altered even by the growth therein of rows and clusters of 15–18 km tall cumulonimbi (Fig. 8). The model is discussed more at the end of this section.

In contrast, over the small island of Grand Bahama (130 km long by 10 km across, with prevailing winds normal to the long axis), the cumulus rainfall reduces the island heating directly below the clouds, causing the convergence line and its associated showers to migrate toward the lee shore, where they die out near sunset.

Over the oceans, two kinds of boundary layer control on penetrative cumuli appear to exist that are not necessarily mutually exclusive. One is heating, either uniform or localized in warm spots, which have been related to cloud groups observationally (Malkus, 1957). The other is frictional pumping in the boundary layer in the presence of large-scale cyclonic vorticity. Charney and Eliassen (1964) postulated a positive feedback loop between cumuli and synoptic-scale disturbance under these conditions, which they have called CISK, or Conditional Instability of the Second Kind. This mechanism has been extensively exploited in models of tropical disturbances and hurricanes, although proper observational testing awaits the GATE data analyses.

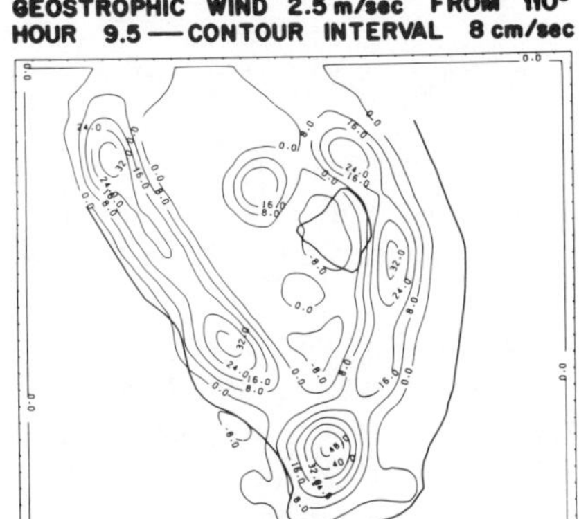

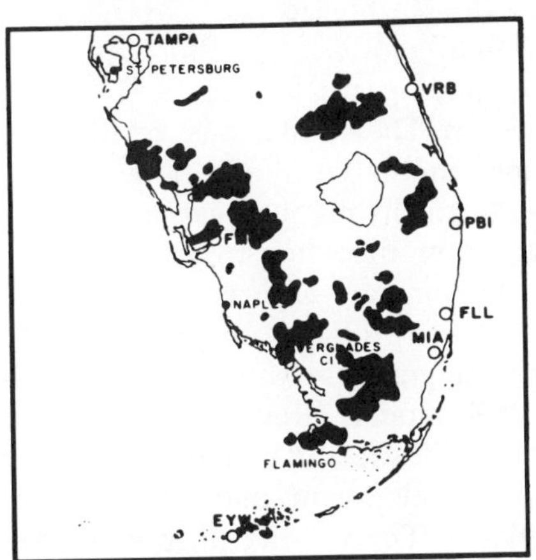

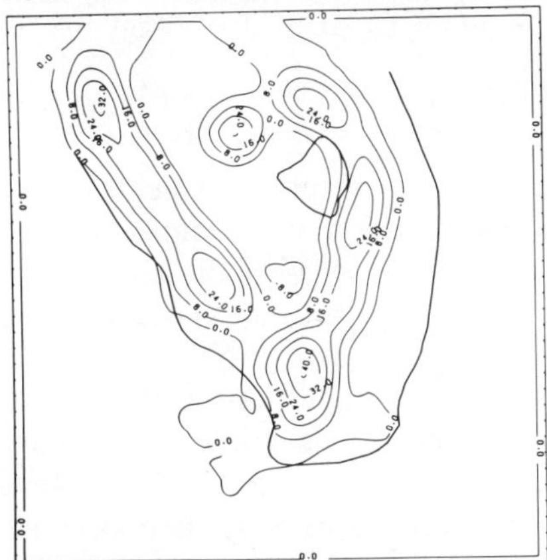

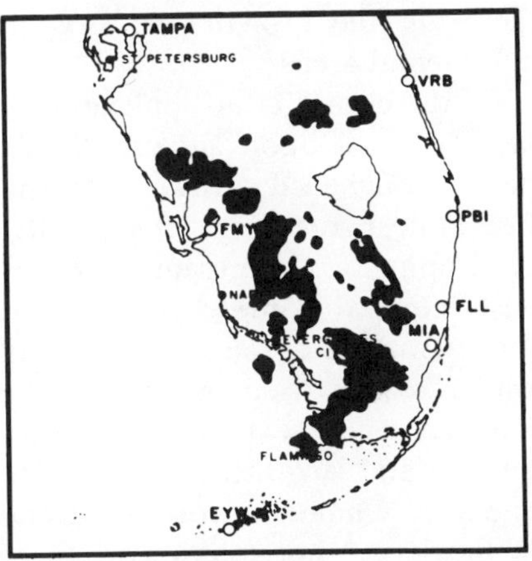

FIG. 8. Comparison of Pielke model-predicted vertical motion field at 1.22 km and the radar echo map at equivalent times for June 29, 1971. (After Pielke, 1973.)

The control upon cloud growth and merger exerted by the boundary layer and its flexibility or rigidity must be investigated further to develop the necessary scientific basis for rain augmentation. In cases where boundary layer control is as rigid as it appears to be on sea-breeze days in Florida, only a rather small "window" in space and time might be available for modification. Attempting to target modified clouds over some desired areas might be working against insuperable natural forces that are concentrating the merger potential in convergence lines established elsewhere by the sea breeze.

In all modification efforts, man must be seeking cooperation with nature by finding weak links and Achilles' heels in a precariously balanced system. These links must be discovered through physical hypotheses developed in analytical and numerical models, which can then be validated by strategically planned measurement programs, often those involving the test of the modification hypothesis itself.

Fortunately, aspects of boundary layer coupling with clouds are becoming tractable problems subject to rapidly advancing numerical simulation and meaningful observational tests. The literature abounds with tropical oceanic CISK models (Rodenhuis, 1971; Bates, 1973) in which the interaction of frictionally forced boundary layer pumping, parameterized cloud growth, and the development and propagation of a synoptic-scale disturbance are related. The validity of these models will be scrutinized using the GATE data. Their applicability to precipitation modification over land has not been considered, much less explored, although the exploration might prove valuable. Frictionally produced convergence in low levels could be a factor in the growth and organization of convective rain systems in continental areas.

Of direct and clear application to cloud organization and to boundary layer control upon shower cloud systems, is the Pielke model of the heated roughened surface layer and its effects upon the wind fields of the free atmosphere (Figs. 9 and 10). The substitution of specified surface orography for the present flat lower boundary is now virtually completed. Also required is improved representation of surface heating and allowance for interaction of the sea breeze with the prevailing synoptic flow offshore, which is held fixed in the 1974 version of the model. Another missing link, namely, the parameterization of cumulus processes, will be more difficult and lengthier the greater the degree of realism and sophistication intended. Cumulus processes cannot be introduced directly, first because of the large[4] horizontal grid spacing and second, because the cumulus and meso-scales of motion differ sufficiently that the same simplifying assumptions cannot be applied to both. On the other hand, the cumulus scale and meso-scale may not differ sufficiently to allow the simplifying equilibrium assumptions basic to the Arakawa and Shubert (1974) and Betts (1973) closed parameterization models of cumulus ensembles. Nor should the common mistake be made which includes explicit microphysics without actual introduction of the convective-scale motions.

[4] Presently 11 km.

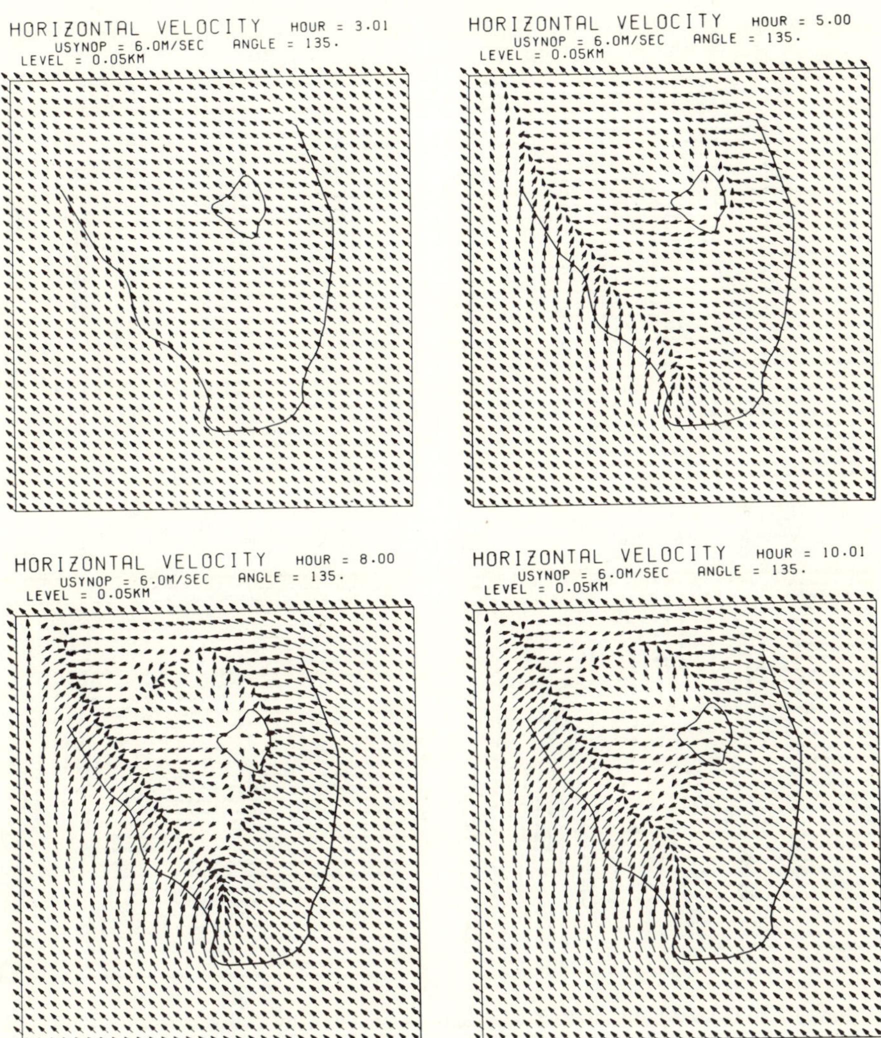

FIG. 9. Low-level horizontal wind velocity as predicted by Pielke model with southeasterly synoptic flow. (After Pielke, 1973.)

One approach to treating the variety of motion scales and their interaction is that of "nested models." For example, one could nest a Hill-type model within the Pielke model. Perhaps then one could even go further, nesting one of several individual cumulus models within the Hill multicloud model. In addition to the vast effort and expense entailed, there are many physical and mathematical difficulties involved in joining

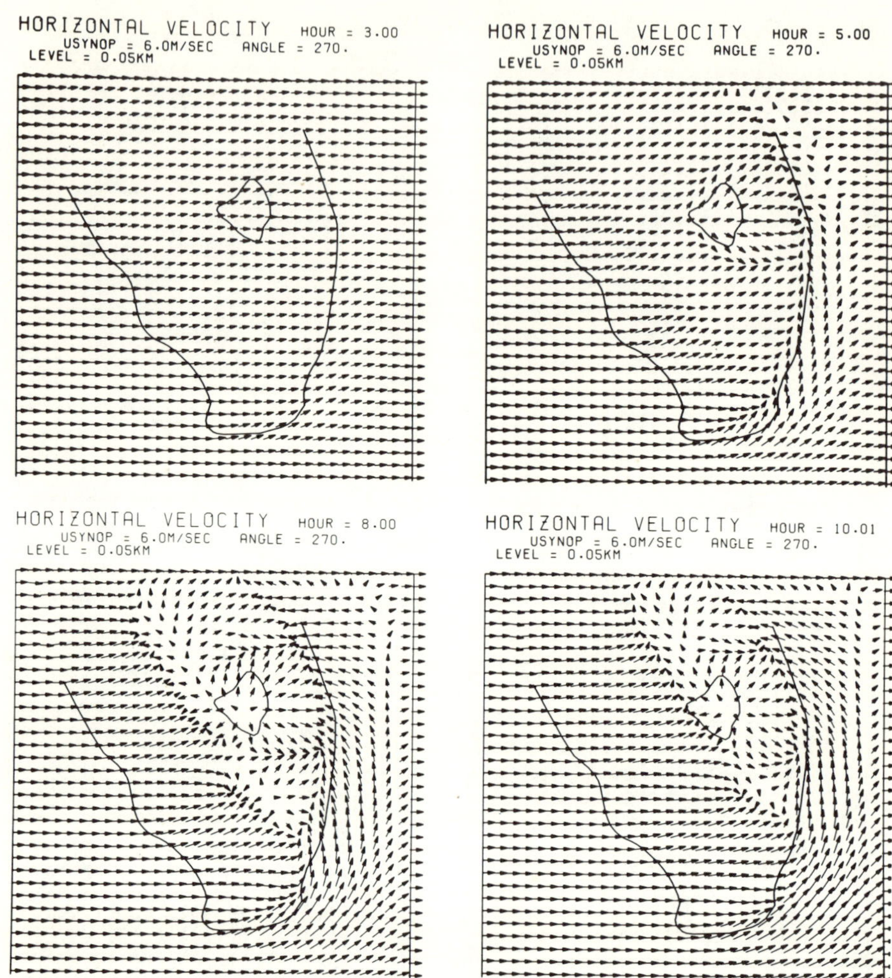

FIG. 10. Low-level horizontal wind velocity as predicted by Pielke model with westerly synoptic flow. (After Pielke, 1973.)

the domains. Consequently, successful multiple nesting is unlikely to be achieved within the coming decade or computer generation. An alternative, less ambitious, but still demanding approach is recommended as an interim measure, namely, parameterization of key processes belonging to one scale of motion in a model of a larger or smaller scale. A particularly needed example for modification is parameterization of cumulus processes in meso-scale models. In the framework of the Pielke model, this task is off to a promising start by Cotton[5] and collaborators at EML. The sea breeze circulations simulated by Pielke (Fig. 10) modify the cumulus environment, providing different stratification of temperature and humidity in the convergence zones than away from them. The investigation so far consists of running the best available one-dimensional model first in the unmodified early morning sounding, then in several "model soundings" modified by the simulated sea breeze. Preliminary results show considerably enhanced cumulus growth in the convergence zones; so far only thermodynamic effects of the sea breeze upon the clouds have been included in the models. A complementary ongoing program involves using the Pielke model to prescribe the locations and structure of the convergence lines and an observational study relating these lines to growth and merger of precipitation echoes.

A later phase, after cumulus parameterization in the Pielke model, may permit it actually to simulate seeding experiments. With an all-out effort, the trial state of this simulation could be reached in 2–3 years. Introduction of shearing and convergent wind fields in the Hill and Pielke models is an urgent need in developing the scientific foundation for precipitation modification, particularly when areas and cloud systems become the targets of modification; however, it is difficult at present to assess the time frame or effort necessary to achieve useful results.

3. Assessment of the Technological Basis for the Modification of Cumulus Clouds and Systems

3.1. Modification Agents, Delivery Systems, and Targeting

The most advanced techniques for cumulus modification are means of ice nucleation of supercooled clouds. In the past decade, most progress has been concentrated in the area of the generation of silver iodide smoke by means of ground and airborne burners, in the form of droppable pyrotechnics and in the form of artillery shells and rockets, the latter fired from aircraft as well as from the ground.

Much has been learned about how silver iodide induces supercooled clouds to glaciate, but very much more remains to be learned than is known today. Recently there has been evidence (Weickmann *et al.*, 1970; Parungo, 1973a) that in relatively warm supercooled clouds, nucleation is either by direct contact or by condensation and freezing, while at colder temperatures (below about $-17.5°C$) the more classic process of vapor diffusive growth upon the ice-forming nucleus becomes important.

---

[5] Personal communication.

The old-fashioned simplified explanation of the nucleation effectiveness of silver iodide in terms of the resemblance of its crystal lattice structure to that of ice has been demonstrated relevant, in that pure single silver iodide crystals readily nucleate laboratory and natural clouds. However, the nucleation activity of silver iodide is much more complex than that explanation covers and the complexities impact seriously upon the scientific and technological basis of cumulus modification. Most silver iodide generators in common use produce particles that are not single perfect crystals but are agglomerates containing impurities and other chemical substances in addition to AgI. Evidence has been presented, in fact (Edwards and Evans, 1968, and references), that fractures and cracks on the crystal surfaces favor nucleation by providing preferred spots for enhanced ice growth.

There has been controversy regarding the role of impurities and other chemicals present. The sum of evidence suggests that impurities may aid or retard nucleation depending on their nature and the nature of the cloud. For example, most burners in operational use generate silver iodide in combination either with sodium iodide or with ammonium iodide. Sodium iodide is hygroscopic, which can be advantageous if the smoke is introduced near or below freezing temperatures in the subject cloud. However, if the seeding is conducted from aircraft flying at the base of a warm cloud, with several thousand feet of vertical depth through which the smoke must rise to reach the $-4°C$ level, nucleation can be inhibited through wetting and dissolution of the particles or by means of detrimental changes to their surface structure.

Generators producing silver iodide–ammonium iodide smoke provide particles which are nonhygroscopic; these smokes may retain their nucleation ability and survive through a wet cloud but would not have the possible assist of hygroscopicity where it could be advantageous.

A major advance in generators has been made by Patten *et al.* (1971) whose device uses the aircraft engine itself to burn the mixture. Their burners use either sodium or ammonium iodide. Laboratory tests indicate that the Patten burners are about an order of magnitude more efficient in terms of output of active nuclei per gram of AgI than other generators currently in common use. The augmented efficiency is probably related to the smaller particle size produced. The small size is in turn caused by the higher burning temperature. Another potentially very practical improvement in the Patten burner is that it has the capability of injecting the salts directly into the flame without the necessity of first dissolving them in a solution such as acetone.

A major advance in cumulus modification technology was the development of pyrotechnic generators at the Naval Weapons Center (NWC) in

China Lake, California, which reached the first practical use in the early 1960s. The early pyrotechnics (St. Amand and Donnan, 1963; Simpson *et al.*, 1963) were droppable rockets with propellants and therefore hard cases, so that use only over oceanic areas was possible. The Cyclops and Alecto units produced massive quantities of silver iodide (each Alecto produced about 1.18 kg of AgI), thus revolutionizing hurricane experimentation and enabling dynamic seeding of cumulus clouds to become a reality (Simpson *et al.*, 1965).

Subsequently, pyrotechnic flares burning to complete consumption at ambient pressures were developed at NWC and elsewhere (Simpson *et al.*, 1970) so that overland experimentation was possible. Pyrotechnics are in use in hail suppression and rain augmentation efforts on a worldwide basis today. The difference is that pyrotechnics, while generally less efficient per gram of material, can inject much more AgI into a cloud in a short time period, and a more complete vertical distribution can be obtained. The vertically falling plumes can be placed as close together as desired to get an adequate horizontal distribution of the material. Pyrotechnics, however, are expensive and most apparently emit a significant fraction of their particles in the relatively inefficient size range of 0.1–1.0 μm, where the particles cannot be captured readily by cloud drops. Particles smaller than 0.1 μm can be captured diffusively, while those larger than 1 μm are subject to hydrodynamic capture. Lack of direct capture nucleation is a handicap at warm supercooled temperatures.

Promising technological developments with pyrotechnics are continuing, which will contribute to more effective experimentation. Fusees which can either be attached to aircraft or burned on the ground, for example, within convective bands as they pass over the mountain tops near Santa Barbara (Elliott *et al.*, 1971), have been available for some years. Recently pyrotechnics have been developed which explode and scatter their fragments widely after a specified time from ejection, permitting a better distribution of material in the horizontal. Flares will soon be available which emit combined or successive mixtures, such as hygroscopic substances and silver iodide, timed for emission in the appropriate parts of the cloud.

Concerning delivery systems of ice nucleants, the most interesting recent development has been that of rockets that are fired from aircraft, for the purposes of the National Hail Research Experiment. This type of delivery system has the dual advantage of the mobility characteristic of aircraft, without the hazard of entering the dangerous portions of huge cumuli. Radar targeting of missiles from aircraft in turbulent conditions is difficult, however, and demonstration of accuracy is awaited.

Meanwhile, most rain augmentation programs are still conducted by generators mounted on aircraft flying either upwind of a target area, as in Israel (Gagin and Neumann, 1974) or circling at the base of the subject clouds and relying on updrafts to convey the material into the cold (below $-4°C$) portion. Both of these approaches require confirmation that the seeding material arrives at the proper location in the clouds. An even more shaky situation is posed by ground generators, except in those noteworthy experiments where pains have been taken to demonstrate by tracing techniques and laboratory models that the material in fact reaches the intended clouds. These successful situations have been confined to mountains, where the generators are placed near to or within cloud base.

Three serious gaps exist in the technological basis for artificial ice nucleation. First, the only existing measures of nucleation efficiencies have been obtained from laboratory experiments, which may or may not be a meaningful representation of conditions existing in real clouds. Nucleation efficiency is defined as the number of active ice nuclei produced per gram of material, such as silver iodide. Laboratory tests involve producing the smoke under as realistic conditions as possible, capturing some and injecting it into an artificial laboratory cloud, then counting the ice crystals falling out in a specified space and time. Lack of realism can creep in at each step, from burning the material to sampling the smoke, and in the laboratory cloud and cloud chamber. The particle size produced by generators is sensitive to the airspeed and to the burning and quenching temperatures. The laboratory cloud is not generating new droplets, does not resemble the real cloud in turbulence and other features, and is confined within walls. Particularly troublesome is the question of time dependence of nucleation and the effects of laboratory simulation versus real cloud conditions. The experiments may miss nucleation mechanisms that occur in real clouds or introduce spurious ones. The solution to this problem does not promise to be simple. Meaningful tests in real cumuli would appear virtually prohibitive in view of natural fluctuations, while test results in stratus, although valuable, would not necessarily apply to cumuli.

The second gap in knowledge is compounded by the first but would exist even if we knew generator efficiencies. The fact is that we do not know enough yet of natural or artificial ice nucleation mechanisms to prescribe the optimum amounts of material to inject into clouds under given circumstances nor, for the same reason, do we know the most effective size spectra or chemical constituents. It has generally been accepted that more massive seeding is required if dynamic effects are desired than if we are aiming just to increase the precipitation efficiency via the microphysics. In neither case, however, has it yet been properly specified under what conditions "overseeding" sets in, leaving much of the potential precipitation suspended in small ice particles. A notable and puzzling recent development has been the apparent release of dynamic effects using relatively small concentrations of silver iodide in South Dakota cumuli (Miller *et al.*, 1974). Rainfall increases related to dynamic seedability have apparently been measured when only about 10–15 gm per cumulus tower are introduced. Numerical simulation both suggested and apparently clarified this result, in that the Hirsch one-dimensional model leads one to believe that with that amount of material, all of the rain hydrometeors and a fraction of the cloud hydrometeors could be frozen in 2–3 min.[6] Direct observational tests of whether dynamic effects are actually released in these experiments were attempted in the summer of 1974, with subsequent analyses in progress.

The third and perhaps most troublesome of the serious gaps in basic ice nucleation technology concerns the distribution of material. We do not know what distribution is required, for example, relative to draft and hydrometeor structure and their life cycles. We do not know how the cloud turbulence spreads silver iodide smoke. Although tests have been made in clear air, it is unlikely that they can be extrapolated to cloud interiors. Rough calculations made years ago by McCready (1959) with assumptions regarding eddy coefficients are only slowly being improved.

An important new area of activity which will help to narrow the gaps cited is the use of tracers and the chemical sampling of precipitation. Tracers such as indium oxide are being used in one natural rain study in Illinois and in several modification programs. The tracer material is both being detected in unexpected places and often not being found where "plume theory" would predict its presence.

Neutron activation and atomic absorption techniques have recently permitted very sensitive sampling of precipitation for silver and other trace metals. With the latter, silver can now be detected in concentrations as low as 10^{-13} gm ml^{-1}. In a western winter snow project, seeded snowfalls contained silver in concentrations up to 70 times "background" (about 2–4×10^{-12} gm ml^{-1}). In a Florida massive seeding project, no seed-control differences were detected, owing probably to the great depth of cloud (~ 3000 m) below the burnout level of the pyrotechnics, although strangely the silver counts were higher than in the west, namely about 5×10^{-11} gm ml^{-1} at the ground and more than 10^{-9} gm ml^{-1} at cloud base. The cause of the high values, particularly at cloud base level is being investigated further.[7] In the Great Lakes Winter Snowstorm Program (Warburton and Owens, 1972), seeded precipitation not only showed silver

---

[6] Dr. Arnett S. Dennis, personal communication.

[7] A paper on this work appears in the Preprint Volume of the Fourth Conference on Weather Modification of the American Meteorological Society, Ft. Lauderdale, November, 1974. The work is entitled "Tracing Silver Iodide in the South Florida Area" by J. Wisniewski, W. R. Cotton, and R. I. Sax.

concentrations enhanced 100 times above background ($\sim 2 \times 10^{-11}$ gm ml^{-1}) but the silver-enhanced particles showed changes from rimed to diffusional growth crystals, helping to confirm the modification hypothesis.

Electron microscopes together with X-ray analyzers are being used to locate the ice nucleus within three-dimensional images of snow crystals (Parungo, 1973b). These investigations determine the chemical composition of the nucleus, helping to understand better the role of natural and artificial particles in the nucleation process.

In addition to vital aids to seeding technology, these tracer and chemical analysis procedures are essential for investigating possible effects of the seeding upon the ecology and upon water supplies, matters of concern in the interface between weather modification and society.

Another advancing technological frontier in supercooled cloud modification involves the use of organic compounds to induce freezing. Some of these have the ability to form hydrogen bonds with water and probably also have a molecular structure that is favorable to ice formation. Among those substances found most effective in the laboratory are phloroglucinol, metaldehyde, and urea, which is also used to enhance coalescence in warm clouds. These chemicals offer the potential advantage of nucleating at warm supercooled temperatures. However, their use still is in the early experimental stages and delivery systems present a problem (see Simpson and Dennis, 1974).

In the case of warm clouds, the technology of coalescence enhancement is newer and therefore less advanced than are the procedures of artificial ice nucleation. Nevertheless, significant progress has been made in the past five years, particularly in the Dakota programs and at the Air Force Cambridge Research Laboratory (AFCRL). With hygroscopic substances such as salt, it is now possible to control particle sizes and even to obtain monodisperse distributions. Delivery systems have also advanced to obviate clumping, sticking, and other problems, mainly through the technique of microencapsulation. Salt-generating pyrotechnics have been invented, but so far the means of controlling their particle size output has not been found.

Despite this fine technological progress, however, the basic scientific problems related to the technology have not been solved. This gap is at least as large in regard to coalescence enhancement as it is for ice nucleation. A particularly important unsolved question concerns the optimum sizes of the particles and the place in the cloud to introduce them. Some arguments favor placing large particles fairly high in the cloud and allowing them to grow by coalescence as they fall. The opposing and more popular view advocates smaller sizes near cloud base which are intended to be lifted by the updraft.

3.2. Measurement Systems

Measurement systems to evaluate the modification of precipitation from cumulus clouds and systems have two related purposes: first, the observation of the rain or snow itself to compare seeded and control targets and second, measurements of treated and untreated clouds and their surroundings to test each link in the modification hypothesis. In both cases, the tools may involve direct or remote probing, or for best results, both together. For an excellent review of weather modification instrumentation, the reader is referred to Ruskin and Scott (1974).

The greatest progress in measurement technology in weather modification has been in its philosophy, advancing from almost blind statistics applied to seed-control raingage records with little else, to an overall systems approach with multifaceted measurements used to design experiments, to determine the number of cases needed, to test each chain in a compound modification hypothesis usually involving models, and to stratify experimental situations into favorable and unfavorable categories.

The technology of precipitation determination has advanced on all fronts. As demonstrated by Bergeron's pioneer Project Pluvius (1960) gaging requirements to analyze natural precipitation in high latitudes are stringent. Tropical air masses pose an even more severe challenge, owing to the huge space and time variations in convective rainfall (Fig. 11). Only recently have rain-measuring requirements been examined to determine detectability of modification effects. Detailed criteria have been obtained from dense gage networks in Illinois (Huff, 1971) and in Florida. In a 93-day study of a 655 km^2 mesonetwork in Florida, it was found that a requirement of 99% detection of 24-hour rain amounts greater than 0.02 mm necessitates a gage density of approximately 52 km^2/gage. For the measurement of areal convective rainfall greater than 0.02 mm within a factor of two on 90, 70, and 50% of the days, gage densities of 31, 91, and 208 km^2/gage, respectively, are required. These requirements are apparently more stringent than those for Illinois convective showers (Fig. 12), with the reservation that the test area in Illinois was 1037 km^2, and gaging requirements for areal rainfall decrease with area size.

It is important to summarize the methods of these gaging studies. Definition of the "true rainfall" was that obtained from 1 gage/2.6 km^2. Subnetworks are then selected at the hierarchy of densities desired and the area mean rainfall calculated for these and compared to the "true" rainfall.

In the Florida study, only 22 gages out of the 220 in the micronet were recording gages, owing to limited funds. Telemetering gages are, however, available. For example, they have been used in the Santa Barbara project where they had to be placed on virtually inaccessible mountain tops. Within the coming decade, gages interrogated by satellites should become

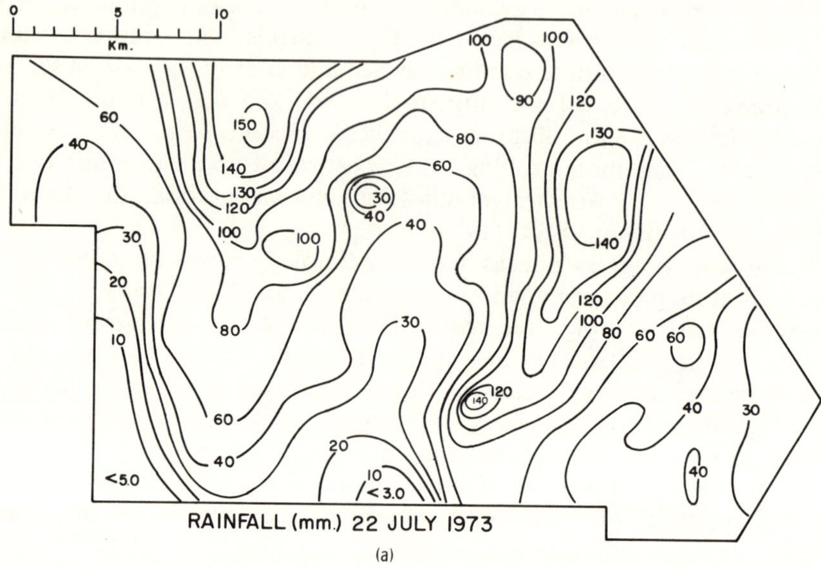

(a)

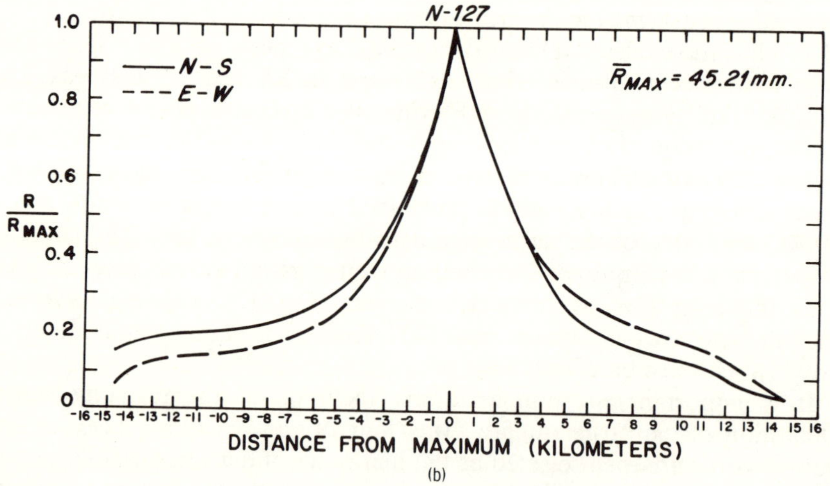

(b)

FIG. 11. (a) Typical summer 24-hr rainfall in mm in south Florida micronetwork. This University of Virginia micronet was 655 km² in area, with one raingage per 2.6 km². (After Woodley *et al.*, 1975.) (b) Average rain gradients in micronet of (a), for 127 days. (After Woodley *et al.*, 1975.)

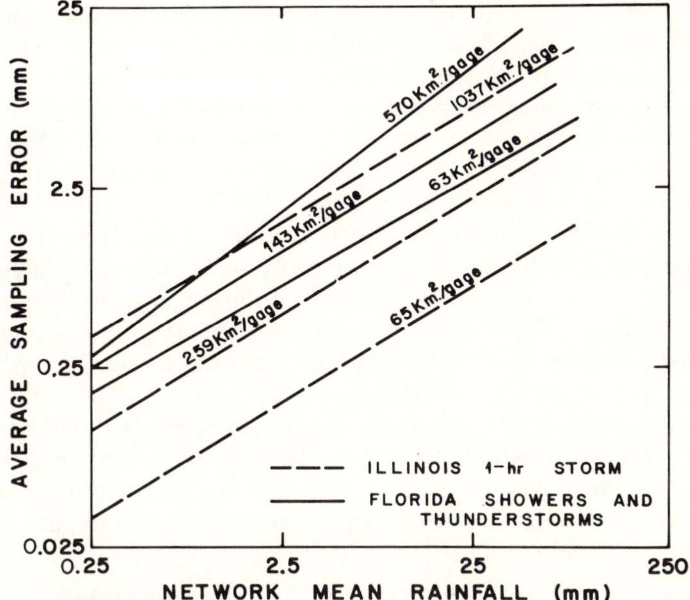

Fig. 12. Comparison of gaging requirements and sampling errors for Illinois and south Florida. (After Woodley *et al.*, 1975.)

available, reaching a still higher level in areal rain determination for modification programs concerned with fixed land targets.

For moving or inaccessible targets, over-water determinations and details of rain development, radar offers an important supplement to gages, although it is unlikely to replace them entirely. In principle, radar should be a near-perfect tool for convective precipitation measurement because it approximates an infinitely dense network by detecting all of the rainfall within range. Unfortunately, this expectation is not realized, because the magnitudes of the radar precipitation estimates are never without errors.

Radar calibration carries always an uncertainty to some extent because the radar beam is usually not uniformly filled with precipitation. Further, the relationship of radar reflectivity (Z) to rainfall rate (R) is variable between storms and within storms, even in the same geographic location and season (Stout and Mueller, 1968). One cannot count on "normal" refractive conditions; with anomalous propagation, there is false echo and uncertainty as to what is being measured. Rather than attempt a quantitative correction for all these error sources, which does not appear feasible in the foreseeable future, it now appears more practical to calibrate the radars against a few rain gages in clusters. The radar defines the spatial variability and provides a first estimate of the magnitude of the precipitation and the calibrating gages allow for adjustment (Wilson, 1970; Woodley and Herndon, 1970; Woodley, *et al.*, 1975).

The results presented by Woodley *et al.* (1975) demonstrate that radar is a reasonably adequate tool for rainfall comparisons in tropical cumulus modification experiments. Salient results are summarized here. Radar performance in estimating convective rainfall over south Florida was determined during the summers of 1972 and 1973. Two collocated calibrated 10-cm radars (UM/10-cm of the University of Miami and WSR-57 of the National Hurricane Center) were operated during 1972. Only the WSR-57 was used during 1973. In all cases, the radar estimates were compared with the rainfall as determined by rain gages (densities 2.6 to 7.8 km^2/gage) in cluster arrays as shown in Fig. 14 on p. 42.

On a daily basis in 1972, the mean absolute percentage difference between gage and radar rainfalls for the periods of operation of the two radars ranged between 35 and 40%. The radars were within a factor of two of the cluster standard 70% of the time. The correlations between gage and radar

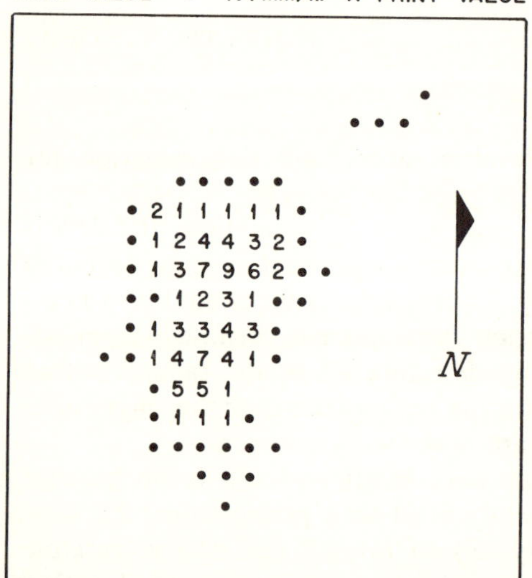

FIG. 13. Illustration (left) of a digitized radar echo on the WSR-57 10-cm radar of the National Weather Service at Miami looking at the experimental cloud shown in the photograph on the right. (After Wiggert and Ostlung, 1975.)

rainfalls were 0.87 and 0.84 for the UM/10-cm and WSR-57 radars, respectively. The correlation between radar rainfalls for 46 showers common to UM/10-cm and WSR-57 was 0.94.

In 1973, WSR-57 radar-derived rainfalls were calculated by hand, as in 1972, and by computer using taped radar observations. Upon comparison, no systematic differences between the rainfalls generated manually and by computer were noted. The mean gage and radar correspondence improves with heavier rain, with a larger time frame for the radar-rain estimates and with an increase in the area size over which the estimates are made. On a daily basis, 80% of the radar estimates were within a factor of two of the cluster standard. The mean factor of difference was 1.51. In 1972 and 1973 combined, the accuracy of the WSR-57 radar in estimating convective rainfall approximated that which one would obtain with a gage density of 65 km^2/gage over an area the size of the mesonet (see Fig. 11). The radar digitation is illustrated in Fig. 13.

The daily representation of rainfall by the radar improves if one adjusts it using gages. For the mesonet, the radar estimates of rainfall were compared to gage measurements before and after the radar representation of rainfall had been adjusted by the ratio of the summed gage to radar rainfalls obtained from peripheral gage clusters. In the mean, this adjustment produced a statistically significant 15% improvement (better than 1% significance with a two-tailed "t" test) in radar accuracy. The adjusted radar measurement then had an approximate gage density equivalence of 26 km^2/gage.

For the 1973 Florida program (FACE 1973, Fig. 14), a digitizer was constructed and linked to the Miami WSR-57 (Wiggert and Andrews, 1974) with the guidance of NOAA's National Severe Storms Laboratory. The tape-recorded output from the digitizer is processed by a sequence of computer programs written at EML (Wiggert and Ostlund, 1975). One program assesses radar-derived rainfall rates and total rain volumes over selectable areas for selectable time periods; another isolates and tracks radar echoes and, while so doing, calculates the rainfall from each echo as it grows, moves, merges, splits, or dies.

In subtropical areas like Florida with low cloud base and fairly constant high humidity in the subcloud layer, radar looking at cloud base has been shown to give an adequate estimate of rain reaching the ground and measured there by gages. With high cloud bases and intervening dry air of low and variable humidity, as in the High Plains region, the problem may be more difficult to resolve.

In this context, still a third method of rainfall sensing is proving helpful, namely the so-called "flying rain gage." These instruments all operate on aircraft using the same principle, namely, counting and sizing raindrops. Three ways of doing this have been seriously attempted, namely, the old-fashioned foil impactor, a microphone, and the Knollenberg set of optical arrays (Knollenberg, 1972). The foil impactor, requiring calibration and arduous labor in reduction, has proved fairly reliable except perhaps when drops come close to breakup size, or about 6 mm, when double impressions

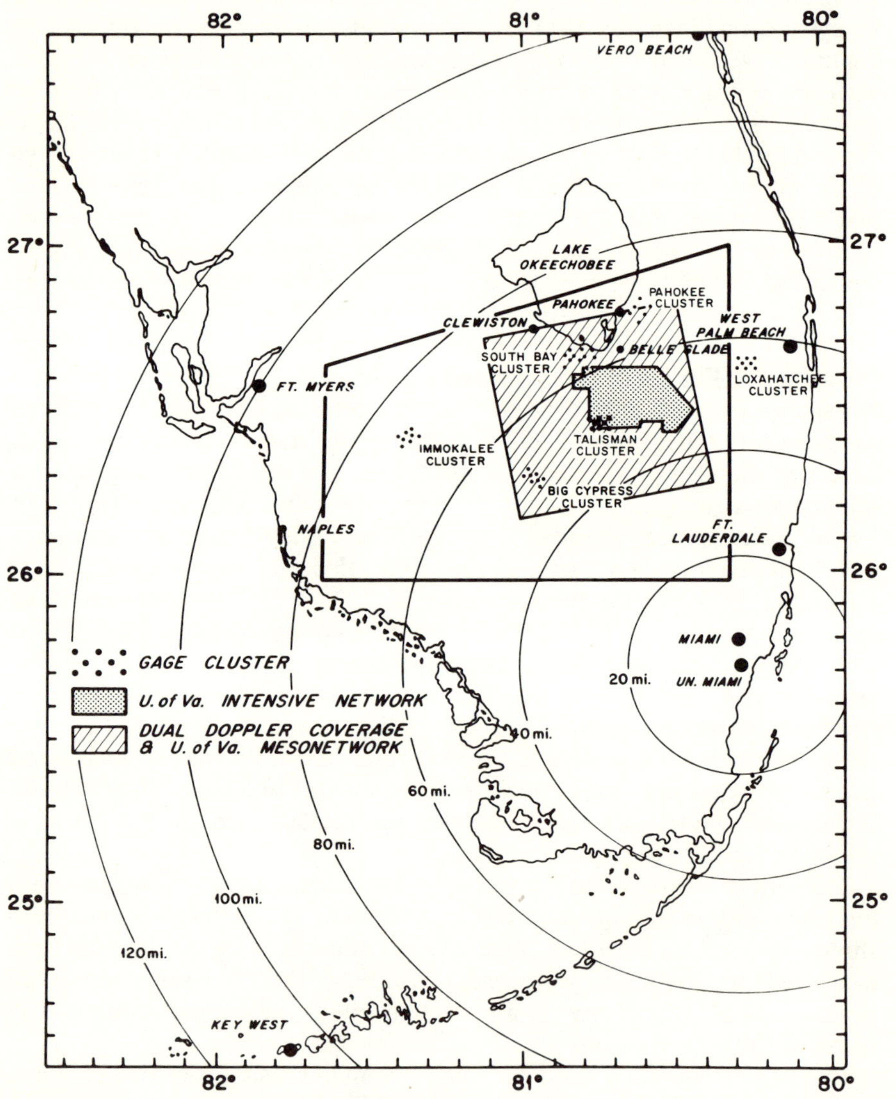

FIG. 14. South Florida seeding target of EML (large polygon of area 1.3×10^4 km^2 enclosed by solid line) showing raingage clusters. Range circles from Miami radar are spaced at 20 NM (37 km) intervals.

have sometimes been detected. Doubts about the microphone method at least on large, vibrating aircraft are still prevalent. The Knollenberg probe appears to be a major advance, although it is expensive and requires considerable experience to operate reliably.

All the "flying rain gages" suffer a severe sampling problem and so can only be used in conjunction with other rain-measuring techniques. They scoop out a very narrow tunnel through a cloud, not exceeding about 1 m^3 in volume and hence could fail to properly sample the large drop sizes, even if the aircraft is correctly guided through the main rainshaft. Aside from this difficulty, horizontal variations across a cumulus require an integrating technique to measure rain volume accurately enough to compare treated and untreated cases.

Methods of probing clouds to test models and modification hypotheses have also advanced significantly. Aircraft measurement capability has progressed on several fronts. Probably the most potentially valuable innovation is the inertial platform-gust probe system for measuring drafts, vertical motions, and fluxes (Fig. 15). The previous lack of these measurements was one of the main factors holding back progress on cumulus dynamics

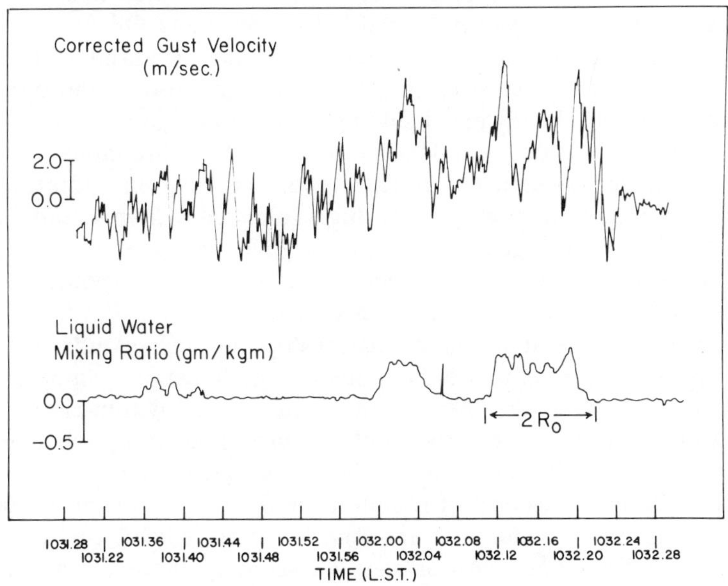

Fig. 15. Cloud penetration from aircraft carrying gust-probe inertial platform system near cloud base. Top trace is corrected gust velocity in m sec^{-1}. Lower trace is liquid water mixing ratio as measured with a hot wire instrument. R_0 is the characteristic radius of the ascending region. The abscissa is time. (After Cotton, 1975b.)

and entrainment studies. Fluxes through cloud base are essential to improve single cloud models and to begin modeling cloud interaction.

In the areas of thermodynamics and microphysics, measurement capability has also moved forward since 1970. It is now possible to obtain better in-cloud temperature records by means of correcting the traces for instrumental lag (McCarthy, 1974). Much better determination of the total water substance and its breakdown into water and ice, and large and small particles is now possible. The use of Lyman-α devices with evaporator (Ruskin, 1965, 1967) permits fast-response assessment of the total H_2O; to get that in liquid plus solid form, the water vapor content must be separately measured and subtracted so that often a fairly small difference between large quantities is being sought in tropical clouds with high vapor contents. For measurement of total liquid water content, the NHRL hot wire nimbiometer (Merceret and Schricker, 1974) and the Levine probe (Levine, 1974) have recently been improved and tested, while the Johnson-Williams reliably measures that fraction of the water content in the cloud-size droplets, below about 40 μm radius. Drop spectra in any desired size range can now be obtained by an appropriately designed Knollenberg probe.

The most important microphysical advances in aircraft probing, however, are the optical devices to measure continuously ice crystal and ice nucleus concentrations by means of polarized light (Sheets and Odencrantz, 1974). While the ice nucleus counter is still subject to the uncertainties mentioned earlier in regard to all existing ice nucleus measurements, the optical ice crystal counter has already provided important growth in the scientific basis of cumulus modification from its use in FACE 1973.[8] Preliminary evidence strongly suggests the expected direct relationship between updraft age and natural ice content. Strong updrafts, high Johnson-Williams readings, and low ice crystal counts generally coincide, while the presence of high ice crystal counts usually can be detected on later penetrations, with low Johnson-Williams readings and diminished updrafts.

Experimental pursuit of these relationships in such clouds and other types is essential in order to identify potential modification "windows." This type of data set will also permit an evaluation of within-cloud spatial relationships of water-ice together with updrafts, a necessity for model and concept testing that has never before been done adequately.

Several important aspects of remote probing are just reaching practical results in cumulus modeling and modification efforts. Aerial photogrammetry, although still laborious, is becoming more exact (Herrera-

---

[8] Described by R. I. Sax in a paper entitled "On the Microphysical Differences Between Populations of Seeded vs. Non-seeded Florida Cumuli," Preprint Vol., Fourth Conference on Weather Modification of the American Meteorological Society, Ft. Lauderdale, November, 1974.

Cantilo, 1969). Visual depiction is essential in relating the dynamic and structural aspects of the clouds to their images from satellites and radars.

Outstanding among the latter developments are Dopplers radars. With single Dopplers, it is possible to obtain useful information on the time history of particle spectra, terminal velocities, and, with some assumptions, vertical air motions inside clouds. With dual Doppler, the history of the three-dimensional particle and air motion field in cumulus cells at a hierarchy of levels can be documented (Fig. 16). Results are now available contemporaneously using a special processing and display system developed by Lhermitte[9] and used in FACE 1973 and 1975. Motions around clouds and below cloud base are beginning to be obtained by chaff dispersal, a potentially powerful tool for resolving the entrainment problem.

Application of this dual Doppler system for seed-control comparisons opens up a whole new dimension in the evaluation of modification experiments and in augmenting their scientific foundations. Other types of remote probes are now reaching productivity. Examples include infrared radiometry to look at cloud temperatures and the impacts of clouds upon the temperature of the lower boundary, microwave radiometry to measure integrated water content, and IR lidar to obtain information on small-scale variability in droplet spectra (this method requires precautions because of high sensitivity to drop size). Acoustic sounders and radars are starting to provide data on temperature patterns in clouds, and other advanced radars, such as polarized beam dual wavelength, broad-band noise pulse, FM-CW, and high power, and narrow beam pulse radars will aid but not replace direct probing by aircraft.

The other key remote probing area offering promise to modification experiments is the satellite. Methods to measure rainfall using enhanced satellite imagery are under development at several laboratories, including EML and NESS in NOAA and Space Sciences at the University of Wisconsin (Woodley *et al.*, 1972; Martin and Scherer, 1973). The satellite rainfall evaluation is being evolved and tested against calibrated radar and gages in south Florida. A solid relationship between cloud height and cloud brightness has been established (Griffith and Woodley, 1973) as has a relation between echo intensity and brightness. This approach offers one of the few hopes of detecting extended area effects of modification experiments over ocean or inaccessible landmasses. Other satellite developments such as infrared radiometry to map cloud top heights and ERTS-type photography

---

[9] Described by R. M. Lhermitte and R. I. Sax in a paper entitled "Use of Dual Doppler Radar in the Florida Area Cumulus Experiment," Preprint Vol., Fourth Conference on Weather Modification of the American Meteorological Society, Ft. Lauderdale, November, 1974, pp. 89–94.

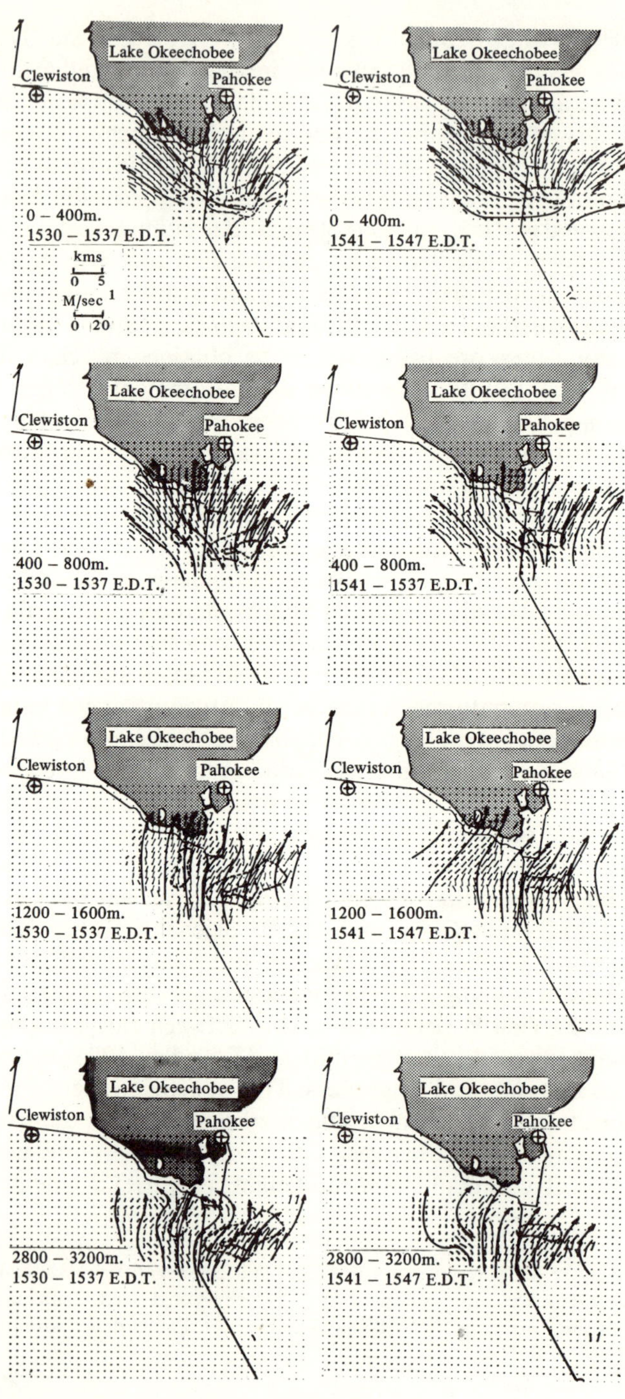

Lake Okeechobee
Clewiston
Pahokee
0 – 400m.
1530 – 1537 E.D.T.
kms
0   5
M/sec
0   20
Lake Okeechobee
Clewiston
Pahokee
0 – 400m.
1541 – 1547 E.D.T.
Lake Okeechobee
Clewiston
Pahokee
400 – 800m.
1530 – 1537 E.D.T.
Lake Okeechobee
Clewiston
Pahokee
400 – 800m.
1541 – 1537 E.D.T.
Lake Okeechobee
Clewiston
Pahokee
1200 – 1600m.
1530 – 1537 E.D.T.
Lake Okeechobee
Clewiston
Pahokee
1200 – 1600m.
1541 – 1547 E.D.T.
Lake Okeechobee
Clewiston
Pahokee
2800 – 3200m.
1530 – 1537 E.D.T.
Lake Okeechobee
Clewiston
Pahokee
2800 – 3200m.
1541 – 1547 E.D.T.

to relate cloud fields to surface moisture and other soil characteristics will contribute to modification technology and also to the scientific basis and usefulness of rain augmentation.

3.3. Computational Tools

A significant fraction of the progress in weather modification has been made possible by the development and use of high-speed computers. As mentioned previously, even the one-dimensional and first field-of-motion thermal models were prohibitive prior to the advent of the early computers, while three-dimensional single cloud simulations or models of cumulus interactions and meso-scale processes strain the biggest and fastest computers now available and demand the ultimate in finite differencing techniques and programming skills available (Cotton, 1975b). Major advances in these models probably must await the next generation of computers beyond the CDC 7600s and IBM 360-195s.

Equally important for weather modification has been the automation of data recording and processing. When recorded by pens on analog charts in the 1950s, two or three cloud penetrations could require months to reduce and analyze. The same results can now be made available contemporaneously.

Valuable contributions have been the development of remote linkage and minicomputers, so that models can be run in advance of launching expensive experiments and during them to guide possible changes in plan, and so that data can be processed in real time or nearly real time to be fed back into the experimental decision process. Nowadays many research aircraft have onboard computers which display data graphically, such as instantaneous wind fields.

Of value to several aspects of modification, including modeling, decision making, and evaluation, are the several means of computer-generated graphics, such as the cathode ray tubes, the automatic plotter, and time-lapse and/or animated movies. The latter have been particularly helpful in relating complex, time-dependent combinations of variables more meaningfully, as for example, the radar presentation in the Dakota program and the film of the Pielke three-dimensional model of the evolution of the Florida sea breeze. The more complex three-dimensional models will constitute a real

---

FIG. 16. Dual Doppler radar (3.2 cm wavelength) presentation of the motion fields in horizontal planes at selected altitudes. Case study of August 4, 1973 (during EML experimental period of FACE 1973) in seeding target area of Fig. 14. Grid point spacing is 1.2 km. Streamlines and regions of maximum echo intensity. Contours for $Z = 10^4$ and $Z = 3 \times 10^4$ are indicated. (After R. Lhermitte and R. Sax, unpublished, see footnote 9, p. 45).

challenge to computer experts in their presentation, to depict the results in an intelligible and useful fashion.

An increasingly useful mode of computer use is the time-share system, extensively employed at EML. Programming is usually carried out in BASIC at these terminals, permitting an average user to acquire facility in a day or two. More important for modification, the interactive mode permits instant reply by the computer to the input of initial conditions or data, or both, by the programmer, who can then choose a different chain of sub-programs depending on the computer's response to each input. The more sophisticated portable calculators have some of these advantages but, of course, are limited in storage. Already the decision process in many weather modification experiments is evolving as a "decision ladder" with interactive computations forming many of the rungs.

3.4. Statistical Tools and Approaches

In the early 1960s statistics, if used at all in weather modification, was generally confined to designing a randomization scheme and to testing the null hypothesis on the resulting data, where the null hypothesis stated that the treated and untreated samples were drawn from the same population. If the null hypothesis could be rejected at the 5% level, it was generally accepted that the populations were different and that the difference was related to the treatment.

The advance has been in the entire methodology and attitude as well as in numerous specific tools and their applications. Statistics today is an integral part of the systems approach to weather modification. Statistical methods of a wide variety are used in every stage of experimentation, from preliminary design through analysis, evaluation, and the decision processes involved in applications.

The application of statistical tools must be inextricably interwoven with meteorological insight. It begins with the earliest consideration of possible experiment designs and optimally is an interactive process with the modeling and measuring aspects at all stages. It never terminates because a given data set can be reevaluated as many times as new ideas present themselves.

Fortunately, the time is past when a modification experiment consisted of blind statistics applied to whatever rain gage data happened to be available. Hopefully, the time is passing when statisticians working in isolation, without knowledge of atmospheric processes, make pronouncements about weather modification. A new generation of statisticians is learning meteorology. Many are rolling up their sleeves to fly in aircraft, building on the knowledge gained of cloud behavior to design improved tests of hypotheses. Others are active participants in experimental design, using their statistical tools to specify gaging and radar accuracy requirements, subsequently serving their turn in manning the radars. Actual exposure to the atmosphere's complexities is a brutal but effective cure for the tendency to oversimplify or to daydream. Conversely, a growing population of meteorologists have recognized that they must learn and apply statistics, which should be a firm requirement for a graduate degree in atmospheric science. Best of all, there are several active weather modification groups in the United States which have one or more meteorological statisticians as integral members of their staffs, and several others where there is close collaboration with a nearby statistics department or laboratory.

A most important advance in statistical applications is the growing recognition of the importance of calculating early in an experiment, or even before it, how many cases are needed to establish a postulated range of seeding effects to a specified significance level. This calculation requires knowing or assuming the distribution of the variable in question. Results have been reached in several different ways depending on the data available. One method involves use of historical records, if such exist, as exemplified by Gabriel's (1970) calculations for the experiment in Israel. There the dual target design used a double ratio statistic. Historical data were permuted in Monte Carlo experiments to find out in what fraction of the cases various values of the double ratio would be obtained by chance.

Another approach is exemplified in the Florida cumulus experiments (Simpson *et al.*, 1973). There it was found that a gamma function was a good fit to the rainfall distributions, so that the gamma properties could be used to calculate the number of cases required to resolve a range of seeding factors (Fig. 17). Still a third method of estimation is to take whatever data set is available, normalize it by a root transformation, and use the properties of normal distributions.

These efforts have contributed to the recognition of the importance of determining the natural distributions involved and their impact upon deductions from experiments. In particular, the "heavy tailed" rainfall distributions from cumulus have come under the joint scrutiny of meteorologists and statisticians, leading to identification of their pitfalls and development of means to cope with them which require both meteorological insights and sharp statistical tools (Simpson *et al.*, 1975).

A coming important advance in cumulus modification may be the combination of model simulation and statistical methods in evolving predictors and regressions. This approach will be particularly valuable in those situations where the likelihood of contamination precludes the use of dual target design to mitigate the effects of natural variability. In short range forecasting, a technique called "model output statistics," or MOS, has been developed (Klein, 1965; Klein *et al.*, 1959) empirically relating parameters predicted by the large-scale circulation models to precipitation,

JOANNE SIMPSON

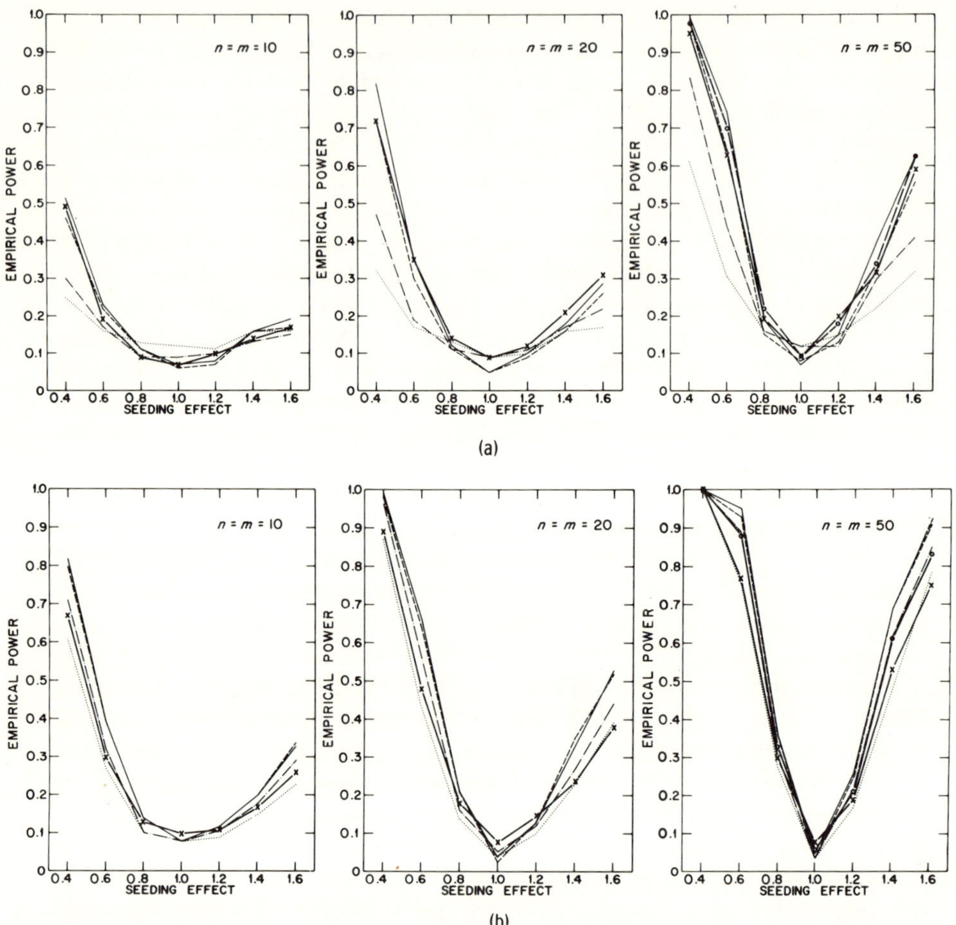

FIG. 17. Power of likelihood ratio test to resolve a postulated range of seeding factors to the 5% significance level, assuming the actual rainfall obeys a gamma distribution with known shape parameter, characteristic of 24-hr rainfall in south Florida, two sizes of areas. The solid curve shows the test power with the true gamma distribution. The other five curves show degradation in test power from empirically determined errors caused by reducing the gage spacing below 2.6 km² per gage and by unadjusted and adjusted radar (using special gage clusters). Note the comparatively small degradation owing to use of unadjusted and adjusted radar. Particularly important is increase of test power with sample size (from left to right). The number of seeded cases is n, unseeded m. (After Olsen and Woodley, 1975.) (a) EML mesonet area 655 km². ——— Error dist. 62 km²/gage; — · — 286 km²/gage; ····· 572 km²/gage; — × — unadjusted radar; (— ○ —) adjusted radar. Even with 50 pairs of cases (right diagram), there is discouragingly low probability of resolving seeding factors in the range 0.6–1.6. (b) EML seeding target (area 1.3×10^4 km²). ——— Error dist. 73 km²/gage; — · — 216 km²/gage; ····· 432 km²/gage; — × — unadjusted radar; — ○ — adjusted radar. Note improvement in test power for comparable measuring errors, relative to smaller area.

fog, and other local weather occurrences. The equations are developed by stepwise regression. It is hoped, for example, that parameters from the neso-scale simulations such as the Pielke sea-breeze model of south Florida (Pielke, 1974) can be selected, which will serve as predictors for one or more of the key variables in experiments on cumulus modification. A promising approach would be to combine one or more model outputs with one or more observational predictors, as for example, the rain in or upwind of the target prior to the experiment (see Woodley *et al.*, 1974, Part II). Unfortunately, a reasonably large data sample, of the order of 100 cases, is necessary to develop a significant regression involving several independent variables.

Bayesian statistics is increasingly used in meteorological problems. This approach requires the assignment of prior probabilities to one or more variables, apparently the nub of the controversy which rages between its adherents and strict classicists.

In weather modification, we see two rather different uses of Bayesian statistics. One is strictly scientific; it is aimed at testing hypotheses or at determining probability distributions of seeding effects, using experimental data. In this usage, some degree of knowledge regarding the natural distributions of the key variables is required (Olsen *et al.*, 1975).

The prior probability assignment may either incorporate prior knowledge if such exists, or may be "diffuse" or unprejudiced. In the latter case, Olsen (1975) has shown, using the Florida single cumulus experiment as an example, that when the physical assumptions are the same, the same results are obtained as with classical statistics. Choice of the Bayesian method is then a matter of convenience for some workers, with its advantages increasing as informative prior probabilities become available.

The other application of Bayesian statistics to cumulus modification is in the decision analysis framework, bringing in factors such as economics, environmental, and societal impacts, when attempting to decide whether and how a modification technique should be applied operationally. In urgent situations this type of analysis has been undertaken far in advance of resolution of the scientific aspects of the modification experiment, as in the case of the hurricane (Howard *et al.*, 1972). Many convective cloud experiments are sufficiently advanced scientifically relative to hurricane experimentation that various types of decision analysis could be warranted, particularly in the winter cyclonic situations in Israel and southern California.

An encouraging development from the joint use of Bayesian and classical approaches has been the concept of a "semirandomized" experiment. These experiments obtain an adequate number of random control cases to avoid bias or its allegation. At the same time, extensive studies of the unmodified clouds or cloud systems are made to document the natural distributions and possibly to stratify them in terms of large-scale flow, cloud top temperature, or some other key parameters. Once the natural distributions are documented, it is of course necessary to relate the random control cases to them and to determine the probability that they are from the same population.

4. Summary and Conclusions

The foundations for modifying cumulus clouds and systems have advanced to the point where they are strong enough to undertake meaningful experiments with good chances for success. These advances have been made despite limited resources.

The major advance has been the growing concept of an overall systems approach, with simulation, cloud physics-dynamics, measurements of all sorts, and several types of statistical tools, building interactively upon each other to develop, test, and improve modification hypotheses and to design and execute complex long-lasting experiments.

The strongest specific advances have been in four areas: (1) precipitation measurement, (2) cloud physics-dynamics observations, (3) application of statistical tools, and (4) numerical simulation.

The measurement of convective precipitation by calibrated digitized radars adjusted by gage clusters is now on an established footing. For example, the method can be used to resolve seeding effects of 1.5 or more if about 50 pairs of cases are obtained or smaller seeding effects with a correspondingly longer experiment. The point is that the requirements can be firmly documented in terms of the natural rain distribution and the accuracy of the system. The frontier lies in satellite adaptation, which may prove indispensable in the crucial unsolved question of extended area effects.

In cloud physics-dynamics, documentation of water-ice budgets is now feasible, so that, for example, already glaciated clouds can be eliminated as static seeding targets. However, the reason for the large concentrations of natural ice at fairly high supercooled temperatures remains an open gap in the scientific foundations of modification.

Advances in the application of statistical tools have been particularly useful in calculating the number of cases required in an experiment to establish the treatment effect. These calculations have brought home the long effort that is usually required to resolve modification effects that are necessarily much smaller than the natural variability. Statistical tools have also become sharper and easier to apply in hypothesis testing and in decision making in the face of uncertainty, concerning whether to continue an experiment or to apply its results operationally. A frontier in application of statistics, requiring the combination with better meteorological measurements and knowledge, is in the area of predictors and covariates. In cumulus modification, this frontier is a particularly important hope of mitigating the effects of natural variability since dual target designs are often suspect because of possible "dynamic contamination."

The model simulation foundation of convective cloud modification exists in varying degrees of development on three scales of motion: the single cumulus, the interaction of several cumuli, and the meso-scale.

Concerning single cumuli, the one-dimensional models have been widely used but their severe limitations require that they be applied with caution. Multidimensional models may be just approaching the useful stage as foundation for modification, but they still have a high ratio of complexity to usefulness and a number of major unresolved problems, in particular the mixing between cloud and environment.

Meso-scale models have sprung into existence in the past 5 years, and are already useful in guiding modification experiments and will hopefully soon be providing covariates. However, a major effort lies ahead to parameterize cloud processes in them. Meaningful cloud interaction models have just begun to appear. These should very soon provide insight on the merging processes and on cloud-boundary layer interaction, of vital concern to modification.

Two frontier areas that are just beginning to add a whole spectrum to the foundations of cumulus modification are: (1) Doppler radars and (2) tracers and chemical sampling of precipitation.

Doppler radars will permit, without penetration, histories of the motions in clouds at several levels and with the addition of chaff, the motions around, below, and above cloud—an inestimable gain.

Tracers and chemical sampling can be used to test and improve the targeting of seeding materials, to study nucleation mechanisms, and to learn more on the unclear subject of photolysis of silver iodide. The techniques can also provide information on cloud-environment interaction, contamination between targets and extra-area, and persistent effects of deliberate and inadvertent seeding.

APPENDIX. SUMMARY OF CUMULUS MODEL SIMULATIONS

This is a brief, therefore necessarily incomplete and nonrigorous summary. Its purpose is to clarify for the average reader the basic mathematical and physical differences between 3-D, 2-D, and 1-D cumulus models, indicating the problems and advantages special to each class. It should be clear that each type of model has its value to cumulus prediction and modification, and also to other aspects of atmospheric science upon which cumulus processes impact. Hence, the development of these modeling approaches interact, overlap, reinforce, test each other, and therefore should be carried out simultaneously and in close collaboration. For a much more complete and rigorous exposition of cumulus models, the reader is referred to a review by Cotton (1975a).

A. Fully Three-Dimensional Cumulus Simulations

Multidimensional cumulus models have often been called "field-of-motion" models because the clouds are part of an overall motion field in a domain. "Cloud" is identified as the region or regions in which solid or liquid H_2O is present.

The hydrodynamic and thermodynamic equations are usually solved by finite differences in time steps on a spatial grid, the size and resolution of which are limited by computer space and cost. Another approach, outside the scope of this discussion (Fox, 1972) involves expansion of the velocity field in an orthogonal eigenfunction series.

In fully 3-D simulations, all three velocity components and their spatial derivatives in x, y, and z are considered, as are all derivatives of the thermodynamic and water substance variables. The convection can consist of one or many interacting clouds, depending on initial and boundary conditions. Arbitrary assumption about the scale of the convection is not required; the motions may be initiated, for example, by uniform heating or by random or regular perturbations in temperature or humidity with or without orography. In principle, the convection can develop in a current varying in space and time, although in practice the lateral boundary conditions pose difficulties if varying flow across them is desired.

For illustration here, a reference state (denoted by subscript o) in steady hydrostatic motion is assumed. The effects of radiation, viscosity, and heat conduction are assumed unimportant in the reference state, so that the Boussinesq approximation is appropriate to both deep and shallow convection (Dutton and Fichtl, 1969). The Boussinesq approximation means physically that variations in density or specific volume are negligible except when multiplied by gravity; mathematically it leads to considerable simplification in the equations to be solved. With it, the approximate equations of motion are

$$(A.1) \quad \frac{d(u_i)}{dt} = -\alpha_o \frac{\partial p'}{\partial x_i} + g \frac{\alpha'}{\alpha_o} \delta_{i3} - Q_T g \, \delta_{i3} + \gamma_o \left[\frac{\partial^2 u_i}{\partial x_j \, \partial x_j} + \frac{1}{3} \frac{\partial}{\partial x_i} \left(\frac{\partial u_j}{\partial x_j} \right) \right]$$

The primes denote departures from the basic state. The dummy subscripts i and j run through 1, 2, and 3 corresponding to x, y, and z components, respectively. Q_T is the mixing ratio of total condensate; γ_o is μ/ρ_o, the molecular kinematic viscosity; δ is the Kronecker delta, bringing in the gravitational terms in the z direction only. The remaining symbols have their usual meanings.

When an approximate linearized equation of state is used, the bouyancy term becomes

$$(A.2) \qquad g\frac{\alpha'}{\alpha_o} = g\left[\frac{\Theta'_v}{\Theta_{vo}} - \frac{c_v}{c_p}\frac{p'}{p_o}\right]$$

where Θ_v stands for virtual potential temperature. The pressure perturbation is negligible only for shallow convection.

An expansion of Poisson's equation for potential temperature and the use of the hydrostatic relation for the reference state lead to the thermodynamic energy equation in the form

$$(A.3) \qquad \frac{d}{dt}\left(\frac{\Theta'}{\Theta_o}\right) + \frac{w}{\Theta_o}\frac{\partial\Theta_o}{\partial z} = \frac{1}{c_p}\frac{ds}{dt}$$

where s is entropy. The evaluation of ds/dt in moist deep convection results in an implicit relation among the thermodynamic variables. Usually the pressure perturbation is ignored in determining saturation vapor pressure in the cloud.

The preceding equations apply to nonturbulent as well as turbulent motion. An outstanding feature of atmospheric convection is its turbulence, which poses the most serious difficulty faced by modelers. In practice, most workers have equated subgrid-scale motions and fluxes to the subconvection or turbulent scale. In order to solve the turbulent equations numerically, they must be averaged over a time interval and geometric volume, the latter corresponding to the grid intervals. The dependent variables are usually therefore decomposed as follows:

$$(A.4) \qquad u_i = \bar{u}_i + u''_i$$

where double-primed quantities denote subgrid or subconvective scale variables.

Then the equations of motion to be solved become

$$(A.5) \qquad \frac{\partial u_i}{\partial t} = -\frac{1}{\rho_o}\frac{\partial(\rho_o\bar{u}_j\bar{u}_i)}{\partial x_j} - \frac{1}{\rho_o}\frac{\partial p'}{\partial x_i} - g\frac{\rho'}{\rho_o}\delta_{i3} - Q_T g\,\delta_{i3} - \frac{1}{\rho_o}\frac{\partial(\rho_o\overline{u''_j u''_i})}{\partial x_j}$$

where molecular stresses have now been neglected relative to Reynolds or turbulent stresses, represented by the last term on the right. Analogous terms appear in the equations for potential temperature, mixing ratio, and hydrometeor content. Treatment of these turbulent flux terms is the most difficult challenge in field-of-motion models, which has not yet been satisfactorily met.

Physically, these "eddy" correlation terms comprise the vital interactions and exchanges between cloud and environment. They may contain a significant portion of the key fluxes and transports by cumuli, and they may also significantly regulate or control the microphysical spectra and precipitation development.

A complete equation for each of the Reynolds stresses involves nine components. Two are triple correlation terms, requiring some form of closure, a subject of sophisticated research programs in turbulence theory. There are also terms involving pressure-velocity, temperature-velocity, and liquid water–velocity correlations, the latter of which can add to the drag exerted by the condensate.

No existing cumulus models explicitly or rigorously, or both, include all these terms. Some models include or parameterize some of them (for details see Cotton, 1975a). Even when and if adequate turbulent flux formulations are established, they will almost surely present computational demands beyond those facilities readily accessible today to most modelers, especially in the moist version where H_2O phase changes and complex microphysical processes must also be simulated.

The earliest 3-D models bypassed these difficulties by using the pseudo-viscosity concept, namely,

$$(A.6) \qquad -\overline{u_j'' A''} = K_A \frac{\partial \overline{A}}{\partial x_j}$$

where the A's stand for the dependent variables u_i, Θ, q, etc. In the simplest treatments, the eddy exchange coefficients are assumed identical for all variables and constant in space and time. Magnitudes are often adjusted to minimize truncation errors or computational instabilities in order to obtain "realistic looking" results—an unsatisfying procedure.

Various attempts at use of "nonlinear viscosities" are underway, as discussed in the review by Cotton (1975a) and its references. In these efforts, the stress terms are usually related to the mean rate of deformation tensor and sometimes also to the average turbulence energy. Adequate internal and external tests have not yet been possible for determining whether these formulations represent sufficiently improved approximations to reality to warrant their general use.

When precipitating and/or glaciating cumuli are to be simulated, cloud microphysical processes must be included or parameterized. In the multidimensional models, air parcels must be traced in order to perform hydrometeor phase change, fallout, and water budget calculations at frequent time intervals. Even with highly parameterized treatments of particle growth and fallout, the additional demand upon computer time and space is usually so great that sacrifices are required in rigor, grid spacing, domain, dimension, or some compromise between these desiderata is needed.

At present, few if any 3-D models attempt formulation of detailed microphysical water particle growth beginning with condensation on nuclei, and none have attempted to simulate the details of ice particle growth, collection, riming, etc. Ironically, the simplification of model dynamics, particularly restrictions in dimensionality and treatment of turbulence, is currently the price paid for increased detail and elegance of the microphysical simulations.

B. Two-Dimensional Cumulus Simulations

In 2-D simulations, the independent space variables are reduced to two, with immense reduction in both conceptual complexity and in computational demands. Treatment of the triple correlation terms in the Reynold's stress equations is avoided. However, most current working 2-D models use pseudoviscosity, either linear or nonlinear. The dynamic equations are solved in a two-dimensional domain, so that in Equation (A.1), i and j take up values only 1 and 3. Alternatively, two-dimensionality permits definition of a velocity stream function, ψ, so that the two dynamic equations may be combined into one vorticity equation, namely,

$$(\text{A.1a}) \qquad \frac{\partial}{\partial t}(\nabla^2\psi) = -J(\psi, \nabla^2\psi) + \frac{g}{\Theta_{vo}}\frac{\partial\Theta'_v}{\partial x} - g\frac{\partial Q_T}{\partial x} + v_M \nabla^4\psi$$

where the vorticity is $\nabla^2\psi$, defined as positive counterclockwise, J is the Jacobian, and v_M is eddy kinematic viscosity for momentum. Boundary conditions and solution procedures are sometimes simpler or more economical with (A.1a) than with the primitive equation (A.1).

Two-dimensional cumulus models exist in two geometries, namely, slab and axisymmetric.

B.1. Slab Symmetric. The dynamic equations are solved in an x–z domain with no variations in one horizontal direction (y), or alternatively a vorticity equation is solved for that component about the y-axis.

Initial and boundary conditions, as well as the degree of microphysical sophistication depend on the objectives of the modeler. The lateral boundaries most often consist of distant walls or are cyclic. In a few slab simulations, the effects of vertical wind shear or convergence upon convection are the direct objectives so that the more difficult case of open boundaries is attacked (Asai, 1964).

One cloud or several interacting clouds may be treated. In multiple cloud models arbitrary specification of horizontal scale can be avoided by random heating from below superposed on uniform warming (Hill, 1974).

The lower boundary may be flat or mountainous, heated, cooled, or thermally inert. With the latter, convection is often initiated by one or several density perturbations at "cloud base" level. Several 2-D slab models treat the entire life cycles of one or more cumuli with sophisticated microphysics leading to precipitation. At least two have reached the stage where their authors believe that simulated modification experiments upon them may be meaningful in guiding field experiments. One is the heated mountain model by Orville and collaborators (Orville and Sloan, 1970; Liu and Orville, 1969) which now includes parameterized glaciation processes and may be ready to explore the consequences of ice phase seeding. Another is the shallow convection model by Arnason and collaborators (Arnason *et al.*, 1968; Arnason and Greenfield, 1972) which treats "warm" clouds beginning with CCNs and the details of condensation, so that it may be a vehicle to simulate hygroscopic seeding.

However commendable their progress and insights, predictions of these slab simulations must be viewed with caution owing to their inherent dynamic and geometrical oversimplifications. Experts have contended (see Cotton, 1975a) that the impossibility of vortex stretching in two dimensions leads to an unrealistic energy transfer by the inertial terms to increasingly larger scales of motion. The extent of the damaging effects of this shortcoming for the purpose of model-guided field experiments is not known.

To cope with this difficulty, work is underway to formulate eddy viscosity equations, coefficients, and differencing schemes that will remove or compensate the errors. For example, it is found that the transfer coefficients for the scalar variables must be much larger than that for momentum to obtain realistic temperature and velocity spectra, a situation unhappily reminiscent of the "coefficient juggling" criticized in the 1-D models.

Furthermore, the geometric constraint to meet continuity in a slab requires unrealistically large downdrafts relative to updrafts, hence the deductions regarding preferred convective scale from these simulations must not be taken too literally.

B.2. Axisymmetric Models. In these 2-D simulations, cylindrical coordinates are used, so that the independent spatial variables are r and z. Variations in the azimuthal (Θ) direction are ignored. Primitive equation or vorticity formulation may be used. The only advantage over the slab geometries is that more realistic down-versus-updraft relations are obtained (Ogura, 1963; Murray, 1970). Disadvantages are that only one isolated cumulus can be simulated and it is difficult if not impossible to avoid some degree of horizontal scale selection.

Nevertheless, axisymmetric models played a pioneering role in field-of-motion simulation (e.g., Murray, 1971) and lately have been effectively used by Murray and Koenig (1972) in illuminating important physical-dynamical interactions caused by evaporation at the cloud edges and as a vehicle for improved ice process simulation (Koenig, 1972).

C. One-Dimensional Simulations

These are the pioneering cumulus models, the first convective simulations permitting dynamical-physical interaction, and the cloud models most extensively used to guide and evaluate cumulus modification experiments.

C.1. Classic 1-D Models. Their dynamics is based on the third equation of motion, which may be written

$$(A.7) \qquad \frac{dw}{dt} = -\frac{1}{\rho}\frac{\partial p}{\partial z} - g - \text{drag}$$

The drag in (A.7) may consist of three parts: the transfer of vertical momentum from the ascending cloudy air to the air entrained therein from the environment, μw^2, a drag due to the weight of liquid or solid H_2O present, gQ_T, which is commonly subtracted from the bouyancy, and an "aerodynamic" or "form" drag omitted below and in most current models (see Simpson *et al.*, 1965 for discussion).

It is usually assumed that the environment is in hydrostatic equilibrium, that the ideal gas law holds, and that horizontal pressure gradients between elements in the cloud and environment can be neglected. Under these conditions (A.7) becomes

$$(A.8) \qquad \frac{dw}{dt} = \frac{\partial w}{\partial t} + w\frac{\partial w}{\partial z} = \left(\frac{T_v - T_{ve}}{T_{ve}} - Q_T\right)g - \mu w^2$$

where the subscript v refers to virtual temperature and e to the cloud environment. The dynamics now diverges sharply depending upon whether or not $\partial w/\partial t$ is included, allowing for time dependence.

The classic 1-D models are based on an entrainment concept where it is usually assumed that

$$(A.9) \qquad \mu = \frac{1}{M}\frac{dM}{dz}$$

and further that

$$(A.10) \qquad \mu = \frac{1}{M}\frac{dM}{dz} \approx \frac{2\alpha}{R}$$

where the fractional entrainment rate with height is postulated to depend inversely on a horizontal dimension. This postulate may be derived either empirically, from dimensional arguments, or from similarity theory or by analogy, or both, with laboratory plumes, thermals, and jets.

It is the use of this postulate that has led to calling this class of simulations "entity" models of a cumulus, since the cloud is envisaged as a discrete recognizable circulation unit, often with a laboratory analog. Both the entrainment concept and the $1/R$ postulate have been subject to controversy (Warner 1970b, 1972) with recent observational verification to first order under carefully selected conditions (McCarthy, 1974).

Once the $1/R$ entrainment postulate is made, R must be specified to integrate the equation for w and other cloud variables. The scale of convection is an imperfectly understood frontier, as described in the main text and elsewhere. Therefore, the specification of R must usually be made from observations or empirical predictions. There are two other environmental and initial conditions required to solve 1-D model equations. First, an ambient thermodynamic sounding is required for the entrainment calculation, and second in-cloud properties at cloud base are necessary initial conditions.

In some of these models (e.g., Simpson and Wiggert, 1969, 1971) a further reduction in buoyancy due to a "virtual mass" effect is introduced by means of a coefficient less than one which multiplies the buoyancy term. The "extra" mass is envisaged as due to acceleration of outside air around the element's top and its internal circulation. Its use attempts to make some allowance for the perturbation pressure effect (see equation A.2) omitted in equation (A.8) and most 1-D models. In earlier versions of the Simpson-Wiggert model an aerodynamic drag was sometimes also included. Both of these effects were introduced owing to apparent over-prediction of top heights and rise rates; their justification was by analogy with laboratory experiments.

Recently, Holton (1973) has developed a 1-D model including the pressure perturbation term for shallow nonprecipitating cumuli. In this framework he found growth was inhibited in increasing proportion to cloud radius, which he suggested would provide a scale selection. Cotton (1975a) cautions that the pressure perturbation effect may be reduced if a significant fraction of cumulus kinetic energy and buoyancy are on the fluctuating, subcumulus scale, a still unresolved problem probably outside the scope of 1-D treatments.

The basic thermodynamics of 1-D models involves two processes: the reduction of buoyancy by entrainment (at low levels in tropical air, drying is a more drastic brake than the direct momentum attrition) and the change in phase of water substance. If precipitating clouds are treated, equations or assumptions regarding hydrometeor fallout must be introduced. These range from crude "calibrations" in the earliest models (Simpson *et al.*, 1965) to evolution of sophisticated equations prescribing conversion, accretion, crystal growth, riming, etc. in later developments (see Cotton, 1972a, b).

The simplest and most often used 1-D models are the Lagrangian tower model by Simpson and collaborators (1965; Simpson and Wiggert, 1969, 1971) and the steady-state model by Weinstein and Davis (1968). In both of these $\partial w/\partial t$ is omitted and the basic dynamic equation is written

$$(A.11) \qquad \frac{1}{2}\left(\frac{dw^2}{dz}\right) = g\left(\frac{T_v - T_{ve}}{T_{ve}} - Q_T\right) - \mu w^2$$

when entrainment is the sole braking effect and neither virtual mass nor aerodynamic drag are included.

The two approaches, however, differ conceptually. The origin of coordinates in the Simpson model follows an active tower, where w is the rise rate. R in equation (A.10) denotes tower radius, usually measured from photographs. It may vary as desired. The continuity equation is not explicitly required, since the tower is envisaged as simultaneously entraining and shedding mass.

The early Simpson-Wiggert models considered only one tower. Later Mason and Jonas (1974) developed a "successive thermal" model mainly for investigating the effects of vertically interacting thermals on cloud microstructure. The failure of the predictions to meet Warner's simultaneous water and top height data has been noted by Cotton (1975a). More seriously, dynamic tower interactions remain beyond the scope of this class of simulation.

In the steady-state 1-D model, the entire vertical profile of a steady state cumulus is treated with the origin of the z coordinate at cloud base. The continuity equation specifically requires constant upward mass flux, so that the cloud has an hourglass shape, being narrowest at the level of maximum w, which represents the updraft speed in this conceptualization. R is the updraft radius, with variations prescribed by continuity. Commonly the updraft radius at cloud base is either approximated from observations or assumed.

The first complete time-dependent 1-D entraining cumulus model was published by Weinstein (1970). The origin of z was at cloud base, precipitation growth was treated by Kessler's (1969) parameterizations, and the freezing process was empirically assumed to occur in specified temperature ranges, as in the early Lagrangian and steady-state 1-D models. Dynamic seeding was simulated and partial observational comparisons with Arizona experimental data were encouraging. Weinstein chose to maintain a constant radius in his model, while at the same time limiting horizontal transport to the entrainment hypothesis; hence his model was not fully closed. Wisner and co-workers (1972) met continuity with variable radius and, like Weinstein, neglected vertical eddy transport. This neglect was somewhat compensated numerically by the implicit diffusional characteristics of forward upstream differencing. However, the nature of vertical eddy transport in a cumulus convective environment is complex. Adequate representation by any differencing scheme would be strictly fortuitous and would lend little understanding to cloud processes.

The classic 1-D models assume that their dependent variables either represent mean or peak active cloud properties; no attempts are made to treat gradients within cloud or at its edges, or cloud alterations of the environment sounding.

C.2. "One and One Half" Dimensional Models. In these attempts, the basic dynamic equation to be integrated remains equation (A.7), but efforts are made to improve the formulations of turbulent exchanges with inclusion of some effects of the cloud upon its surroundings which could influence its further development. The reason these models are informally called "one and one half" dimensional is therefore apparent, since some of the extradimensional effects, such as those due to horizontal gradients (within clouds or at their edges) or environmental subsidence are parameterized to be incorporated in the basic one-dimensional mathematics.

So far, these models fall into two classes. One is the ingenious "two shell" multiple parcel model of Lopez (1973). This model parameterizes the tendency of thermals to form toroidal circulations by ejecting mass from a central convective region to an outer shell whenever a negative vertical velocity gradient exists. Both regions are treated independently in a one-dimensional Lagrangian manner. Unfortunately, this promising model has not yet been subjected to observational tests.

The other class of "one and one half dimensional" models was pioneered by Asai and Kasahara (1967). These focus on improved formulation of the exchanges across cloud boundaries. Following early ideas by Houghton and Cramer (1951), entrainment is decomposed into that component required to satisfy continuity ("dynamic" or, better, "kinematic" entrainment) and turbulent exchange, which depends on gradients at the cloud boundary. The equation system is closed by diagnosing radial velocity at the cloud edge with the radial average continuity equation.

Another innovative feature of this model is that it can take account of compensating motions in the cloud environment. Microphysical extensions have been made for warm clouds (Ogura and Takahashi, 1973;

Silverman and Glass, 1973) and by Ogura and Takahashi (1971) including the ice phase. Attempts to improve the dynamic aspects by means of non-linear viscosities are underway by Cotton (1975b), who is guiding this work by continual observational comparisons.

C.3. "Tuned" or "Calibrated" Models. All atmospheric models contain choices of empirical equations, coefficients, and selections between differencing or integration schemes. "Tuning" or "calibrating" a model means adjusting the various constants, equations, and procedures (within the range of observational limits, internal consistency, and common sense) to achieve joint predictions of the dependent variables in optimal agreement with one or more data sets.

Since 1-D cumulus models are highly simplified, their "tuning" is most necessary, apparent, and exposed to criticism. For the simplest illustration, in the early version of the Lagrangian tower model (Simpson *et al.*, 1965) the entrainment coefficient was adjusted downward slightly relative to the laboratory plume and a fraction of the condensate was dropped out so that predicted top heights, water contents, and buoyancies would agree with a tropical oceanic data set. When it was found that cloud tops and rise rates were overpredicted when water contents and buoyancies roughly agreed with observations, form drag or virtual mass (guided by laboratory results) was called upon to reduce momentum without further reduction in temperature or water content.

In the next stage of the 1-D models, precipitation growth and fallout were simulated by means of parameterized formulas for conversion and accretion developed by Kessler (1969) and Berry (see Simpson and Wiggert, 1971). Although a step toward reality was clearly made, "tuning" the parameters of the growth equations, particle spectra, etc. is still required, particularly when modeling less well known ice processes is attempted. The eventual aim is to replace empiricisms by sound physical laws whenever and wherever possible.

When 1-D models are extended to calculate rain amounts, or hail presence or fallout, in the complex cloud systems corresponding to many experimental situations, they undergo extensive tuning which is justified when honestly explained. A model tuned for a convective storm situation in South Africa may not, for example, give good results for Alberta and vice versa. No 1-D model has been able, so far, to predict correct detailed variable relationships in the very small nonprecipitating clouds off the coast of Australia (Warner 1970b, 1972; Cotton 1975b). This failure may be related to the most serious oversimplification of 1-D models which lies in the entrainment formulation, with concomitant neglect of cloud fine structure and internal gradients. Warner and others have contended that this formulation is beyond oversimplification and is, in fact, erroneous. In their view, 1-D models are adjusted empiricisms and not meaningful approximations to physical reality.

Oversimplification becomes erroneous or unmeaningful depending on the purpose for which a model is used. While for improved detailed understanding of cloud dynamical-physical interaction, 1-D models have probably outlived their initial great usefulness, they have contributed and will continue to contribute major advances in controlled weather modification (Simpson, 1970; Simpson and Dennis, 1974).

ACKNOWLEDGMENTS

The preparation of this paper was supported by the Division of Atmospheric Water Resources Management, of the Bureau of Reclamation, U.S. Department of the Interior and by Grant No. GI-43764 from the RANN (Research Applied to National Needs) Division of the National Science Foundation.

The author would like to thank her colleagues Roger Pielke, William Woodley, and Joseph Wisniewski for comments and help, and for provision of several figures and calculations for this article.

Particular gratitude is owed Robert Sax who helped extensively with the material on aerosols, nucleation, cloud microphysics, and electrification; to William Cotton who provided indispensable aid throughout; and to Wallace Howell who offered constructive criticism and advice during the first two drafts.

Broadening, gap filling, and improved writing were instigated by the editor, Helmut Landsberg, whose wide knowledge was generously made available during the final revisions.

Fine work on manuscript preparation and correction was done by Mary Morris and Bobby Cassidy.

In preparing the Appendix, my colleagues William Cotton, Roger Pielke, and Francis Murray have been very helpful. In particular, Cotton's preparation of a review of cumulus models has been used extensively as a guide, which I adapted to the modification context.

REFERENCES

Arakawa, A., and Schubert, W. H. (1974). Interaction of a cumulus cloud ensemble with the large-scale environment. *J. Atmos. Sci.* **31**, 674–701.

Arnason, G., and Greenfield, R. S. (1972). Micro- and macro-structures of numerically simulated convective clouds. *J. Atmos. Sci.* **29**, 342–367.

Arnason, G., Greenfield, R. S., and Newberg, E. A. (1968). A numerical experiment in dry and moist convection including the rain stage. *J. Atmos. Sci.* **25**, 404–415.

Asai, T. (1964). Cumulus convection in the atmosphere with vertical wind shear. *J. Meteorol. Soc. Jpn., Ser. 2* pp. 245–259.

Asai, T., and Kasahara, A. (1967). A theoretical study of the compensating downdrafts associated with cumulus clouds. *J. Atmos. Sci.* **24**, 487–496.

Austin, J. M., and Fleisher, A. (1948). A thermodynamic analysis of cumulus convection. *J. Meteorol.* **4**, 91–94.

Austin, P. M., and Houze, R. A. (1972). Analysis of the structure of precipitation patterns in New England. *J. Appl. Meteorol.* **11**, 926–935.

Bates, J. R. (1973). A generalization of the CISK theory. *J. Atmos. Sci.* **30**, 1509–1519.

Battan, L. J. (1963). Relationship between cloud base and initial radar echo. *J. Appl. Meteorol.* **2**, 333–336.

Battan, L. J. (1974). "Doppler Radar Observations of a Hailstorm," Sci. Rep. No. 28, 25 pp., Inst. Atmos. Phys., University of Arizona, Tucson.

Battan, L. J., and Theiss, J. B. (1970). Measurement of vertical velocities in convective clouds by means of pulsed-Doppler radar. *J. Atmos. Sci.* **27**, 293–298.

Bergeron, T. (1960). Operation and results of "Project Pluvius." *In* "Physics of Precipitation," Am. Geophys. Union Publ. **746**, 152–157, Washington, D.C.

Berry, E. X. (1967). Cloud droplet growth by collection. *J. Atmos. Sci.* **24**, 688–701.

Betts, A. K. (1973). Non-precipitating cumulus convection and its parameterization. *Q. J. R. Meteorol. Soc.* **99**, 178–196.

Bhumralkar, C. (1973a). An observational and theoretical study of atmospheric flow over a heated island: Part I. *Mon. Weather Rev.* **101**, 719–730.

Bhumralkar, C. (1973b). An observational and theoretical study of atmospheric flow over a heated island: Part II. *Mon. Weather Rev.* **101**, 731–745.

Braham, R. R. (1958). Cumulus cloud precipitation as revealed by radar, Arizona, 1955. *J. Meteorol.* **10**, 311–324.

Braham, R. R. (1964). What is the role of ice in summer rainshowers? *J. Atmos. Sci.* **21**, 640–645.

Braham, R. R., and Squires, P. (1974). Cloud Physics—1974. *Bull. Am. Meteorol. Soc.* **55**, 543–556.

Brier, G. W., Grant, L. O., and Mielke, P. W., Jr. (1974). An evaluation of extended area effects from attempts to modify local clouds and cloud systems. *Proc. W.M.O./I.A.M.A.P. Sci. Conf. Weather Modif. Tashkent U.S.S.R., Oct., 1973* W.M.O. No. 399, pp. 439–447.

Brown, K. J. (1971). Evidence of large scale effects of cloud seeding. *Trans. Semin. Extended Area Eff. Cloud Seeding*, Santa Barbara, California, Feb., 1971, pp. 58–90.

Byers, H. R., and Braham, R. R. (1949). " The Thunderstorm," 287 pp., Final Rep. US Govt. Printing Office, Washington, D.C.

Charlton, R. B., and List, R. (1972a). Hail size distributions and accumulation zones. *J. Atmos. Sci.* **29**, 1182–1193.

Charlton, R. B., and List, R. (1972b). Cloud modelling with spheroidal hailstones. *J. Rech. Atmos.* **6**, 55–62.

Charney, J. G., and Eliassen, A. (1964). On the growth of the hurricane depression. *J. Atmos. Sci.* **21**, 68–75.

Cotton, W. R. (1972a). Numerical simulation of precipitation development in supercooled cumuli. Part I. *Mon. Weather Rev.* **100**, 757–763.

Cotton, W. R. (1972b). Numerical simulation of precipitation in supercooled cumuli. Part II. *Mon. Weather Rev.* **100**, 764–784.

Cotton, W. R. (1975a). Theoretical cumulus dynamics. *Rev. Geophys.* **13**, 419–448.

Cotton, W. R. (1975b). On parameterization of turbulent transport in clouds. *J. Atmos. Sci.* **32**, 548–564.

Danielsen, E. F., Bleck, R. and Morris, D. A. (1972). Hail growth by stochastic collection in a cumulus model cloud. *J. Atmos. Sci.* **29**, 135–155.

Dutton, J. A., and Fichtl, G. H. (1969). Approximate equations of motion for gases and liquids. *J. Atmos. Sci.* **26**, 241–254.

Edwards, G. R., and Evans, L. F. (1968). Ice nucleation by silver iodide:III. The nature of the nucleating site. *J. Atmos. Sci.* **25**, 249–256.

Elliott, R. D., St. Amand, P., and Thompson, J. R. (1971). Santa Barbara pyrotechnic cloud seeding test results 1967–1970. *J. Appl. Meteorol.* **10**, 785–795.

Fernandez-Partagas, J. J. (1973). "Subsynoptic Convergence—Rainfall Relationships Based upon 1971 South Florida Data," 76 pp., NOAA Tech. Memo. ERL WMPO-9.

Fitzgerald, J. W. (1972). "A Study of the Initial Phase of Cloud Droplet Growth by Condensation. Comparison between Theory and Observation," Tech. Rep. No. 44. Cloud Phys. Lab., Dept. of Geophys. Sci., University of Chicago, Illinois. 144 pp.

Fitzgerald, J. W., and Spyers-Duran, P. A. (1973). Changes in cloud nucleus concentration and cloud droplet size distribution associated with pollution from St. Louis. *J. Appl. Meteorol.* **12**, 511–516.

Fox, D. G. (1972). Numerical simulation of three-dimensional, shape-preserving convective elements. *J. Atmos. Sci.* **29**, 322–341.

Gabriel, K. R. (1970). "The Israeli rainmaking experiment 1961–1967. Final statistical tables and evaluation," 47 pp., Rep., Hebrew University of Jerusalem, Israel.

Gagin, A., and Neumann, J. (1974). Rain stimulation and cloud physics in Israel. *In* "Weather Modification" (W. N. Hess, ed.), Chap. 13, pp. 454–494. Wiley, New York.

Griffith, C. G., and Woodley, W. L. (1973). On the variation with height of the top brightness of precipitating convective clouds. *J. Appl. Meteorol.* **12**, 1086–1089.

Hallet, J., and Mossop, S. C. (1974). Production of secondary ice particles during the riming process. *Nature (London)* **249**, 26–28.

Herrera-Cantilo, L. M. (1969). "Aerial Cloud Photogrammetry Based on Doppler Navigation," Final Rep. NOAA Grant 22-2-69 (N), 65 pp., Rosenstiel School of Marine and Atmospheric Science, University of Miami, Coral Gables, Florida.

Hocking, L. M., and Jonas, P. R. (1970). The collision efficiency of small drops. *Q. J. R. Meteorol. Soc.* **96**, 722–729.

Hill, G. E. (1974). Factors controlling the size and spacing of cumulus clouds as revealed by numerical experiments. *J. Atmos. Sci.* **31**, 646–673.

Holle, R. L. (1974). Populations of parameters related to dynamic cumulus seeding over Florida. *J. Appl. Meteorol.* **13**, 364–373.

Howard, R. A., Matheson, J. E., and North, D. W. (1972). The decision to seed hurricanes. *Science* **176**, 1191–1202.

Holton, J. R. (1973). A one-dimensional model including pressure perturbation. *Mon. Weather Rev.* **101**, 201–205.

Houghton, H. G., and Cramer, H. E. (1951). A theory of entrainment in convective currents. *J. Meteorol.,* **8**, 95–102.

Huff, F. A. (1971). Evaluation of precipitation records in weather modification experiments. *Adv. Geophys.* **15**, 59–135.

Huff, F. A., ed. (1973). Summary report of Metromex studies 1971–1972. Rep. Invest. 74, 169 pp., Illinois State Water Survey, Urbana, Illinois.

Huff, F. A., and Changnon, S. A. (1972). Climatological assessment of urban effects on precipitation at St. Louis. *J. Appl. Meteorol.* **11**, 823–842.

Janota, P. (1971). An empirical study of the planetary boundary layer in the vicinity of the intertropical convergence zone. Ph.D. Dissertation. Dept. Meteor., MIT, Cambridge, Massachusetts.

Kessler, E., III (1969). On the distribution and continuity of water substance in atmospheric circulation. 84 pp., *Meteorol. Monogr.* **10**.

Klein, W. H. (1965). "Application of Synoptic Climatology and Short-range Numerical Prediction to Five-Day Forecasting," Res. Pap. No. 46, 109 pp., U.S. Weather Bureau, Washington, D. C.

Klein, W. H., Lewis, B. M., and Enger, I. (1959). Objective prediction of five-day mean temperatures during winter. *J. Meteorol.,* **16**, 672–682.

Knollenberg, R. G. (1972). Comparative liquid water content measurements of conventional instruments with an optical array spectrometer. *J. Appl. Meteorol.* **11**, 501–508.

Koenig, L. R. (1963). The glaciating behavior of small cumulonimbus clouds. *J. Atmos. Sci.* **20**, 29–47.

Koenig, L. R. (1972). Parameterization of ice growth for numerical calculations of cloud dynamics. *Mon. Weather Rev.* **100**, 417–423.

Kovetz, A., and Olund, B. (1969). The effects of coalescence and condensation on rain formation in a cloud of finite vertical extent. *J. Atmos. Sci.* **26**, 1060–1065.

Langmuir, I. (1948). The production of rain by a chain reaction in cumulus clouds at temperatures above freezing. *J. Meteorol.* **5**, 175–192.

Levin, Z., and Ziv, A. (1974). The electrification of thunderclouds and the rain gush. *J. Geophys. Res.* **79**, 2699–2704.

Levin, Z., Neiburger, M., and Rodriguez, L. (1973). Experimental evaluation of collection and coalescence efficiencies. *J. Atmos. Sci.* **30**, 944–946.

Levine, J. (1974). "A Simple Aircraft Hot Wire Instrument System for Measuring Cloud and Rain Water. NOAA Tech. Memo. ERL WMPO-16, 30 pp.

List, R., and Schemenauer, R. S. (1971). Free-fall behavior of planar snow crystals, conical graupel and small hail. *J. Atmos. Sci.* **28**, 110–115.

List, R., Cantin, J. G., and Ferland, M. G. (1970). Structural properties of two hailstone samples. *J. Atmos. Sci.* **27**, 1080–1090.

List, R., Murray, W. A., and Dyck, C. (1972). Air bubbles in hailstones. *J. Atmos. Sci.* **29**, 916–290.

Liu, J. V., and Orville, H. D. (1969). Numerical modelling of precipitation and cloud shadow effects on mountain-induced cumuli. *J. Atmos. Sci.* **26**, 1283–1298.

Lopez, M. E., and Vickers, W. W. (1973). "Modelling and field measurement of the Saturn rocket cloud," Mitre Tech. Rep. MTR-2543. 124 pp. Mitre Corp., Bedford, Massachusetts.

Lopez, R. E. (1973). A parametric model of cumulus convection. *J. Atmos. Sci.* **30**, 1354–1373.

Ludlam, F. H. (1951). The production of showers by the coalescence of cloud droplets. *Q. J. R. Meteorol. Soc.* **86**, 107–113.

Ludlam, F. H., and Scorer, R. S. (1953). Reviews of modern meteorology 10: Convection in the atmosphere. *Q. J. R. Meteorol. Soc.* **79**, 317–341.

McCarthy, J. (1974). Field verification of the relationship between entrainment rate and cumulus cloud diameter. *J. Atmos. Sci.* **31**, 1028–1039.

McCready, P. B., Jr. (1959). The lightning mechanism and its relation to natural and artificial freezing nuclei. *Rec. Adv. Atmos. Electri.* **25**, 369–381.

Malkus, J. S. (1949). Effects of wind shear on some aspects of convection. *Trans. Am. Geophys. Union* **30**, 19–25.

Malkus, J. S. (1952). Recent advances in the study of convective clouds and their interaction with the environment. *Tellus* **4**, 71–87.

Malkus, J. S. (1954). Some results of a trade cumulus cloud investigation. *J. Meteorol.* **11**, 220–237.

Malkus, J. S. (1957). Trade cumulus cloud groups: Some observations suggesting a mechanism of their origin. *Tellus* **9**, 33–44.

Malkus, J. S. (1960). Recent developments in studies of penetrative convection and an application to hurricane cumulonimbus towers. *In* "Cumulus Dynamics" (C. E. Anderson, ed.), pp. 65–84. Pergamon, Oxford.

Malkus, J. S. (1962). Large-scale interactions. *In* "The Sea: Ideas and Observations" (M. N. Hill, ed.), Vol. 1, pp. 88–294. Wiley (Interscience), New York.

Malkus, J. S., and Riehl, H. (1964). "Cloud Structure and Distributions Over the Tropical Pacific Ocean." 229 pp., Univ. of California Press, Berkeley.

Malkus, J. S., and Scorer, R. S. (1955). The erosion of cumulus towers. *J. Meteorol.* **12**, 43–57.

Martin, D. W., and Scherer, W. D. (1973). Review of satellite rainfall estimations. *Bull. Am. Meteorol. Soc.* **54**, 661–674.

Mason, B. J. (1971). "The Physics of Clouds." 671 pp., Oxford Univ. Press (Clarendon), London and New York.

Mason, B. J. (1974). Review of "Weather and Climate Modification" by The U.S. National Academy of Science Panel. *WMO Bull.* **23**(2), 136–137.

Mason, B. J., and Jonas, P. R. (1974). The evolution of droplet spectra and large droplets by condensation in cumulus clouds. *Q. J. R. Meteorol. Soc.* **100**, 23–38.

Mee, T. R., and Takeuchi, D. M. (1968). "Natural Glaciation and Particle Size Distribution in Marine Tropical Cumuli," Meteorol. Res., Inc. Final Rep. Exp. Meteorol. Branch, ESSA, Coral Gables, Florida, under Contract E 22-30-68 (N), 71 pp.

Merceret, F. J., and Schricker, T. L. (1974). A new hot-wire liquid cloud water meter. *J. Appl. Meteorol.* **14**, 319–326.

Miller, J. R., Dennis, A. S., Boyd, E. I., Smith, P. L., and Cain, D. E. (1974). "The North Dakota Pilot Project: Annual Report for 1973, Inst. of Atmos Sci., South Dakota Sch. of Mines and Tech., Rapid City, 46 pp.

Moore, C. B., and Vonnegut, B. (1973). Reply to P. R. Brazier-Smith, S. G. Jennings and J. Latham on "Increased rates of rainfall production in electrified clouds." *Q. J. R. Meteorol. Soc.* **99**, 779–786.

Moore, C. B., Vonnegut, B., Vrablik, E. A., and McCaig, D. A. (1964). Gushes of rain and hail after lightning. *J. Atmos. Sci.* **21**, 646–665.

Morton, B. R. (1968). "Non-similar Turbulent Plumes," GFDL Pap. No. 7, 50 pp., Geophys. Fluid Dynamics Lab., Monash Univ., Clayton, Victoria, Australia.

Mossop, S. C., Ruskin, R. E., and Heffernan, K. J. (1968). Glaciation of a cumulus at approximately −4°C. *J. Atmos Sci.* **25**, 889–899.

Murray, F. W. (1970). Numerical models of tropical cumulus cloud with bilateral and axial symmetry. *Mon. Weather Rev.* **98**, 14–28.

Murray, F. W. (1971). Humidity augmentation as an initial impulse in a numerical cloud model. *Mon. Weather Rev.* **99**, 37–48.

Murray, F. W., and Koenig, L. R. (1972). Numerical experiments on the relations between microphysics and dynamics in cumulus convection. *Mon. Weather Rev.* **100**, 717–732.

National Academy of Sciences (1973). "Weather and Climate Modification." 258 pp., Washington, D.C.

Neiburger, M., Lee, I. Y., Lobl, E., and Rodriguez, L. (1974). Computed collision efficiencies and experimental collection efficiencies of cloud drops. (Prepr.) *Conf. Cloud Phys. Tucson, Arizona, Oct., 1974*, pp. 73–78.

Ogura, Y. (1963). The evaluation of a moist convection element in a shallow, conditionally unstable atmosphere, a numerical calculation. *J. Atmos. Sci.* **20**, 407–424.

Ogura, Y., and Takahashi, T. (1971). Numerical simulation of the life cycle of a thunderstorm cell. *Mon. Weather Rev.* **99**, 895–911.

Ogura, Y., and Takahashi, T. (1973). The development of warm rain in a cumulus model. *J. Atmos. Sci.* **30**, 262–277.

Olsen, A. R. (1975). Bayesian and Classical Statistical Methods Applied to Randomized Weather Modification Experiments. *J. Appl. Meteorol.* **14**, 970–973.

Olsen, A. R., and Woodley, W. L. (1975). On the effect of natural rainfall variability and measurement errors in the detection of seeding effect. *J. Appl. Meteorol.* **14**, 929–938.

Olsen, A. R., Simpson, J., and Eden, J. C. (1975). A Bayesian analysis of a multiplicative treatment effect in weather modification. *Technometrics* **17**, 161–166.

Orville, H. D., and Sloan L. J. (1970). A numerical simulation of the life history of a rainstorm. *J. Atmos. Sci.* **8**, 1148–1159.

Parungo, F. P. (1973a). Electron-microscopic study of silver iodide as contact or sublimation nuclei. *J. Appl. Meteorol.* **12**, 517–521.

Parungo, F. P. (1973b). Ice nucleation: Elemental identification of particles in snow crystals. *Science* **180**, 1057–1058.

Patten, B. T., McFadden, J. D., and Friedman, H. A. (1971). NOAA Research Flight Facility capabilities for weather modification research. Airborne cloud seeding systems developed and utilized by the Research Flight Facility. (Prepr.) *Int. Conf. Weather Modif. Canberra, Australia, Sept., 1971,* pp. 358–360.

Pielke, R. (1973). "A Three-dimensional Numerical Model of the Sea Breezes over South Florida," NOAA Tech. Memo. ERL WMPO-2. 136 pp.

Pielke, R. (1974). A three-dimensional numerical model of the sea breezes over south Florida. *Mon. Weather Rev.* **102**, 115–139.

Plank, V. G. (1969). The size distribution of cumulus clouds in representative Florida populations. *J. Appl. Meteorol.* **8**, 46–67.

Riehl, H., and Malkus, J. S. (1958). On the heat balance in the equatorial through zone. *Geophysica* **6**, 503–538.

Rodenhuis, D. (1971). A note concerning the effect of gravitational stability upon the CISK model of tropical disturbances. *J. Atmos. Sci.* **28**, 126–129.

Ruskin, R. E., ed. (1965). Principles and methods of measuring humidity in gases. *In* "Humidity and Moisture Measurement and Control in Science," Vol. 1. Van Nostrand-Reinhold, Princeton, New Jersey. 704 pp.

Ruskin, R. E. (1967). Measurements of water-ice budget changes at $-5°C$ in AgI-seeded tropical cumulus. *J. Appl. Meteorol.* **6**, 72–81.

Ruskin, R. E., and Scott, W. D. (1974). Weather modification instruments and their use. *In* "Weather Modification" (W. N. Hess, ed.) Chap. 4. Wiley, New York.

St. Amand, P., and Donnan, J. A. (1963). Investigations in atmospheric environmental control. *Nav. Res. Rev.* **16**, 10–13.

Sartor, J. D. (1973). The electrification of thunderstorms and the formation of precipitation. *Naturewissenschaften* **60**, 19–31.

Saunders, P. M. (1961). An observational study of cumulus. *J. Meteorol.* **18**, 451–467.

Saunders, P. M. (1965). Some characteristics of tropical marine showers. *J. Atmos. Sci.* **22**, 167–175.

Sax, R. I., and Goldsmith, P. (1972). Nucleation of water drops by Brownian contact with AgI and other aerosols. *Q. J. R. Meteorol. Soc.* **98**, 60–72.

Scorer, R. S., and Ludlam, F. H. (1953). Bubble theory of penetrative convective. *Q. J. R. Meteorol. Soc.* **79**, 94–103.

Scorer, R. S., and Ronne, C. (1956). Experiments with convection bubbles. *Weather* **11**, 151–154.

Scott, W. D., and Levin, Z. (1974). Stochastic rain development in an electrified cloud. (*Prepr.*) *Conf. Cloud Phys., Tucson, Arizona, Oct., 1974* pp. 377–382.

Sheets, R. C., and Odencrantz, F. K. (1974). Response characteristics of two automatic ice particle counters. *J. Appl. Meteorol.* **13**, 148–155.

Silverman, B., and Glass, M. (1973). A numerical simulation of warm cumulus clouds: Part I, Parameterized vs. non-parameterized microphysics. *J. Atmos. Sci.* **30**, 1620–1637.

Simpson, J. (1970). Cumulus cloud modification: Progress and prospects. *In* "A Century of Weather Progress," Amer. Meteor. Soc., pp. 143–155.

Simpson, J. (1971). On cumulus entrainment and one-dimensional models. *J. Atmos. Sci.* **28**, 449–455.

Simpson, J. (1972a). Reply to Warner. *J. Atmos. Sci.* **29**, 220–225.

Simpson, J. (1972b). Use of the gamma distribution in single cloud rainfall analysis. *Mon. Weather Rev.* **100**, 309–312.

Simpson, J. (1973). "The Global Energy Budget and the Role of Cumulus Clouds." NOAA Tech. Memo. ERL WMPO-8. 175 pp.

Simpson, J., and Dennis, A. S. (1974). Cumulus clouds and their modification. *In* "Weather Modification" (W. N. Hess, ed.) Chap. 6, pp. 229–280. Wiley, New York.

Simpson, J., Eden, J. C., and Olsen, A. R. (1975). On the design and evaluation of cumulus modification experiments. *J. Appl. Meteorol.* **14**, 946–958.

Simpson, J., and Wiggert, V. (1969). Models of precipitating cumulus towers. *Mon. Weather Rev.* **97**, 471–489.

Simpson, J., and Wiggert, V. (1971). 1968 Florida cumulus seeding experiment: Numerical model results. *Mon. Weather Rev.* **99**, 87–118.

Simpson, J., and Woodley, W. L. (1971). Seeding cumulus in Florida: New 1970 results. *Science* **172**, 117–126.

Simpson, J., Simpson, R. H., Andrews, D. A., and Eaton, M. A. (1965). Experimental cumulus dynamics. *Rev. Geophys.* **3**, 387–431.

Simpson, J., Brier, G. W., and Simpson, R. H. (1967). Stormfury cumulus seeding experiment 1965: Statistical analysis and main results. *J. Atmos. Sci.* **24**, 508–521.

Simpson, J., Woodley, W. L., Friedman, H. A., Slusher, T. W., Scheffee, R. S., and Steele, R. L. (1970). An airborne pyrotechnic and seeding system and its use. *J. Appl. Meteorol.* **9**, 109–122.

Simpson, J., Woodley, W. L., Miller, A. H., and Cotton, G. F. (1971). Precipitation results of two randomized pyrotechnic cumulus seeding experiments. *J. Appl. Meteorol.* **10**, 526–544.

Simpson, J., Woodley, W. L., Olsen, A. R., and Eden, J. C. (1973). Bayesian statistics applied to dynamic modification experiments on Florida cumulus clouds. *J. Atmos. Sci.* **30**, 1178–1190.

Simpson, J., Ackerman, B., Betts, A., McCarthy, J., Newton, C., Orville, H., Ruskin, R., Sax, R., Warner, J., and Weickmann, H. (1974). Achievements and expectations in cloud physics. *Bull. Am. Meteorol. Soc.* **55**, 556–572.

Simpson, R. H., Ahrens, M. R., and Decker, R. D. (1963). "A Cloud Seeding Experiment in Hurricane Esther, 1961," Nat. Hurricane Res. Proj. Rep. No. 60, 30 pp., U.S. Weather Bureau, Washington, D.C.

Sloss, P. W. (1967). An empirical examination of cumulus entrainment. *J. Appl. Meteorol.* **6**, 878–881.

Stommel, H. (1947). Entrainment of air into a cumulus cloud. *J. Meteorol.* **4**, 91–96.

Stout, G. E., and Mueller, E. A. (1968). Survey of relationship between rainfall rate and radar reflectivity in the measurement of precipitation. *J. Appl. Meteorol.* **7**, 465–473.

Telford, J. W. (1955). A new aspect of coalescence theory. *J. Meteorol.* **12**, 436–444.

Tennekes, H., and Woods, J. D. (1973). Coalescence in a weakly turbulent cloud. *Q. J. R. Meteorol. Soc.* **99**, 758–763.

Twomey, S. (1964). Statistical effects in the evolution of a distribution of cloud droplets by coalescence. *J. Atmos. Sci.* **21**, 553–557.

Twomey, S. (1966). Computations of rain formation by coalescence. *J. Amos. Sci.* **23**, 405–411.

Twomey, S., and Warner, J. (1967). Comparison of measurements of cloud droplets and cloud nuclei. *J. Atmos. Sci.* **24**, 702–703.

Warburton, J. A., and Owens, M. S. (1972). Diffusional deposition of ice on silver iodide in seeded lake effect storms. *J. Rech. Atmos.* **6**, 679–692.

Warner, J. (1969a). The microstructure of cumulus cloud. Part I. General features of the droplet spectrum. *J. Atmos. Sci.* **26**, 1049–1059.

Warner, J. (1969b). The microstructure of cumulus cloud. Part II. The effect on droplet size distribution of the cloud nucleus spectrum and updraft velocity. *J. Atmos. Sci.* **26**, 1272–1282.

Warner, J. (1970a). The microstructure of cumulus cloud. Part III. The nature of the updraft. *J. Atmos. Sci.* **27**, 682–688.

Warner, J. (1970b). On steady-state one-dimensional models of cumulus convection. *J. Atmos. Sci.* **27**, 1035–1040.

Warner, J. (1972). Comments "On cumulus entrainment and one-dimensional models." *J. Atmos. Sci.* **29**, 218–219.

Warner, J. (1973a). The microstructure of cumulus cloud. Part IV. The effect on the droplet spectrum of mixing between cloud and environment. *J. Atmos. Sci.* **30**, 256–261.

Warner, J. (1973b). The microstructure of cumulus cloud. Part V. Changes in droplet size with cloud age. *J. Atmos. Sci.* **30**, 1724–1726.

Weickmann, H. K., Katz, U., and Steele, R. (1970). AgI—Sublimation or contact nucleus? (*Prepr.*) *2nd Nat. Conf. Weather Modif.* Santa Barbara, California, April, 1970, pp. 332–336.

Weinstein, A. I. (1970). A numerical model of cumulus dynamics and microphysics. *J. Atmos. Sci.* **27**, 246–255.

Weinstein, A. I. (1972). Ice-phase seeding potential for cumulus cloud modification in the western United States. *J. Appl. Meteorol.* **11**, 202–210.

Weinstein, A. I., and Davis, L. G. (1968). "A Parameterized Numerical Model of Cumulus Convection," Report II, N.S.F. GA-777, 43 pp., National Science Foundation, Washington, D.C.

Whelpdale, D. M., and List, R. (1971). The coalescence process in raindrop growth. *J. Geophys. Res.* **76**, 2836–2856.

Wiggert, V. (1972). "Cumulus Simulations by a Modified Axisymmetric Model, with Comparisons to Four Observed Tropical Clouds," NOAA Tech. Memo. ERL OD-12. 96 pp.

Wiggert, V., and Andrews, G. (1974). "Digitizing, Recording and Computer Processing Weather Radar Data at EML, NOAA," 65 pp., NOAA Tech. Memo. ERL WMPO-17.

Wiggert, V., and Ostlund, S. (1975). Computerized rain assessment and tracking of south Florida WSR-57 weather radar echoes. *Bull. Am. Meteorol. Soc.* **56**, 17–26.

Wilson, J. W. (1970). Integration of radar and raingage data for improved rainfall measurement. *J. Appl. Meteorol.* **9**, 489–497.

Wisner, C., Orville, H. D., and Myers, C. (1972). A numerical model of a hail-bearing cloud. *J. Atmos. Sci.* **29**, 1160–1181.

Woodcock, A. H. (1940). Convection and soaring over the open sea. *J. Mar. Res.* **3**, 248–253.

Woodley, W. L. (1970a). Precipitation results from a pyrotechnic cumulus seeding experiment. *J. Appl. Meteorol.* **9**, 242–257.

Woodley, W. L. (1970b). Rainfall enhancement by dynamic cloud modification. *Science* **9**, 127–132.

Woodley, W. L., and Herndon, A. (1970). A raingage evaluation of the Miami-reflectivity rainfall rate relation. *J. Appl. Meteorol.* **9**, 258–264.

Woodley, W. L., Sancho, B., and Norwood, J. (1971). Some precipitation aspects of Florida showers and thunderstorms. *Weatherwise* **24**, 106–119.

Woodley, W. L., Sancho, B., and Miller, A. H. (1972). "Rainfall Estimation from Satellite Photographs," NOAA Tech. Memo. ERL OD-11, 43 pp.

Woodley, W. L., Donaldson, J., Simpson, J., Olsen, A. R., and Eden, J. C. (1974). "On the Design and Evaluation of Cumulus Modification Experiments," NOAA Tech. Memo. ERL WMPO-13, 65 pp.

Woodley, W. L., Olsen, A. R., Herndon, A., and Wiggert, V. (1975). Comparison of gage and radar methods of convective rain measurement. *J. Appl. Meteorol.* **14**, 909–928.

Woodward, B. (1959). The motion in and around isolated thermals. *Q. J. R. Meteorol. Soc.* **85**, 144–151.

Ziv, A., and Levin, Z. (1974). Thundercloud electrification: cloud growth and electrical development. *J. Atmos. Sci.* **31**, 1652–1661.

THE PROPERTIES OF ATMOSPHERIC AEROSOL PARTICLES AS FUNCTIONS OF THE RELATIVE HUMIDITY AT THERMODYNAMIC EQUILIBRIUM WITH THE SURROUNDING MOIST AIR

GOTTFRIED HÄNEL

Institut für Meteorologie
Johannes Gutenberg-Universität, Mainz, Germany

1. INTRODUCTION

1.1. Problem

The atmosphere is a colloidal system in which the gaseous phase is the dispersion medium and the suspended particles constitute the dispersed phase; such a system is called an aerosol. Therefore, atmospheric suspensoids will be called aerosol particles or atmospheric aerosol particles.

Condensation of water vapor takes place on the suspensoids with increasing relative humidity, whereas evaporation of the water retained in the suspensoids takes place with decreasing relative humidity. This implies that the physical parameters of the suspensoids this review deals with, namely, their mass, volume or size, mean density, mean refractive index, and their radiative properties, depend on the relative humidity. The knowledge of all these parameters is necessary for the solution of problems involving meteorology and aerosol measuring techniques.

For instance, one of the parameters that has an influence on the lifetime of suspensoids in the atmosphere is their gravitational settling velocity. The computation of this lifetime requires knowing not only the size and the dynamic shape factor but also the mean density of the particles. If the latter two properties of the particles are known, it is possible to measure their size with impactors or centrifuges (cf. Stöber *et al.*, 1970; Hänel, 1970b; Hochrainer, 1971). Size distributions of suspensoids obtained in this way are prerequisites for the computation of scattering and absorption processes of electromagnetic radiation in the atmosphere.

Fog and cloud droplets are formed from atmospheric suspensoids primarily caused by condensation of water vapor. The computational treatment of the condensation process (Howell, 1949; Keith and Arons, 1954; Mordy, 1959; Neiburger and Chien, 1960; Kornfeld, 1970; Paluch, 1971; Chen, 1971) has to be based upon experimental investigations that require measurements of the water vapor condensation on the particles at very high relative humidities.

A quantitative visibility forecast (Kasten, 1969; Hänel, 1971) for very high relative humidities, especially for fog, is essential for the traffic system. Besides the particle concentration, relative humidity, and the wind field, this forecasting requires the knowledge of how the extinction coefficient of the particles depends on the relative humidity.

The divergence of the electromagnetic radiation in the atmosphere controls the cooling or heating rates of the air. Its computation (Grassl,

1973b, 1974; Eiden and Eschelbach, 1973; Yamamoto and Tanaka, 1972) requires—among other quantities—the knowledge of the refractive index of the particles and the particle size distribution as well as their changes with the relative humidity.

These few examples are proof enough for the need to know about the changes in size, density, and refractive index of aerosol particles with relative humidity. Fortunately, many measurements and computations in this field are available (Junge, 1952a, b; Volz, 1954; Eiden, 1966; Hänel, 1966, 1968, 1969; Winkler, 1969, 1970; Fischer, 1971, 1973; Winkler and Junge, 1971, 1972; Meszaros, 1971; Grassl, 1973a; Volz, 1972 and 1973). The integration of all these results, however, into one harmonious concept is open to objection and in most cases even impracticable. This is because of the heterogeneous ways and durations of sampling. Moreover, the samples considered have been taken at different locations and during different seasons.

It is therefore our objective to base each experimental investigation of a sample of aerosol particles on the evaluation of the mass, volume or size, mean density, and the mean refractive index as functions of the relative humidity. Together with measured size distributions these data shall allow model calculations on the topics mentioned above.

1.2. Selection of Sampling and Measuring Methods

Atmospheric suspensoids come from various origins. They are results of man's activity, plant life, photochemical processes, chemical reactions, and wind erosion. Furthermore, the so-called sea spray originates at the sea surface. Aerosol particles of different origin mix while they travel through the atmosphere. Mixed nuclei are formed by coagulation or coalescence of cloud droplets which later on evaporate. During their way through the atmosphere the individual aerosol particles are not expected to be chemically stable. Therefore, it must be assumed that each individual particle may have a chemical composition and structure of its own. Thus, at first, it might be concluded that only individual particles should be subject to measurements of physical quantities.

If so, the evaluation of all aerosol particles within 1 cm^3 would require from a hundred up to a hundred thousand measurements for each parameter. For interpreting this type of measurements of individual particles, e.g., with regard to the formation of clouds, it would be necessary to track the development and the path of each particle. Such requirements in the fields of measuring technique and numerical capacity of computers, however, cannot yet be met. Furthermore, no proof has been furnished for the prerequisite that numerical models of the atmosphere must be based upon detailed data on individual particles. Therefore, it is deemed appropriate to begin with measurements of mean values of physical properties of aerosol particle collectives.

There are two ways of measuring mean values: either investigating aerosol particle collectives in an airborne state by means of scattered light or collecting the particles in an impactor or on a filter for taking the measurements later in the laboratory. The interpretation of results obtained from measurements of scattered light (e.g., Frankenberger, 1964; Eiden, 1966; Pueschel *et al.*, 1969) has not yet been feasible without making assumptions on the structure and the shape of the particles (see e.g., Bullrich, 1964). Therefore, measurements on samples of aerosol particles obtained with an impactor are made in order to compare with and supplement scattered light measurements.

2. THEORETIC EVALUATION OF MASS, SIZE, MEAN DENSITY, AND MEAN REFRACTIVE INDEX AS FUNCTIONS OF THE RELATIVE HUMIDITY

The individual airborne aerosol particles as well as the samples of aerosol particles taken with an impactor or on a filter are mixtures of various substances. Therefore, the same theoretic approaches for computing the mean density and the mean refractive index apply to both. The difference between the growth with relative humidity of an individual particle and of a sample of aerosol particles can be described analytically, because the sample of aerosol particles reacts like an individual, very large particle. Thus, the values of the mass, the mean density, the volume or size, and the mean refractive index of the airborne particles and samples versus relative humidity can be treated with one and the same analytic procedure.

In Section 2.1 a single aerosol particle with a liquid cover is considered, which is in thermodynamic equilibrium with the surrounding moist air. First, the basic equation describing this case with the help of the water activity and the curvature correction is discussed. Then, the water activity, which is the equilibrium humidity over a plane surface, is described in terms of the linear and the exponential mass increase coefficients. Both these coefficients relate quantities that can be measured even on unknown substances without any assumptions. They are discussed for various cases. Second, the curvature correction term is described in terms of measurable quantities. In Section 2.2 the thermodynamic equilibrium between an aerosol particle and the surrounding moist air is considered for the case of an incomplete liquid cover. It is shown that with the concept of the linear mass increase coefficient this case can also be treated. Third (Section 2.3), the thermodynamic equilibrium between a sample of atmospheric aerosol particles with the surrounding moist air is described. Finally (Sections 2.4

and 2.5), the dependence of the mean density and of the real part and the imaginary part of the mean refractive index upon relative humidity are derived for both single aerosol particles and samples of aerosol particles.

2.1. The Basic Equations for Thermodynamic Equilibrium between a Spherical Particle with Liquid Cover and the Surrounding Moist Air

The condensation process of water vapor has been studied theoretically for the case of thermodynamic equilibrium between uncharged aerosol particles and the surrounding moist air by a large variety of authors (i.e., Koehler, 1936; Wall, 1942; Junge, 1952a, b; Zebel, 1956; Mason, 1963; Dufour and Defay, 1963; Low, 1969; Herbert, 1975). All these models are valid only for specific cases and are not applicable to a sufficiently wide range of relative humidity, covering both cloud physics as well as aerosol physics application. Moreover, these theories neglect a phenomenologic coefficient that can be measured directly on real atmospheric aerosol particles. Therefore, the outlines of a model with such an empirical coefficient have been developed (Hänel, 1968, 1970a).

The new model is based on an equation that describes quite generally the influence of curvature and dissolved material on the equilibrium partial vapor pressure p_r over an uncharged particle. Following the considerations of Harrison (1965) it is necessary to relate p_r to the equilibrium partial vapor pressure p_w over a plane surface of water being saturated with dry air. p_w has to be calculated for the total pressure p of the moist air surrounding the particle and for the particle's absolute temperature T. This can be done easily with the data compiled by Goff and Gratch (1945). From rigorous thermodynamic considerations similar to those of Dufour and Defay (1963), but regarding imperfect gas relationships with the formulation of Goff and Gratch (1945), the compressibility of pure water after Tait (1898), and the compressibility of the solution as well as the deviation of the volume of the solution from the sum of volumes of pure water and pure solute after Gibson (1934, 1935), the final equation reads

$$\frac{p_r}{p_w} + X = \frac{a_{wp}}{a_{wa}} \exp\left(\frac{2\sigma v_w}{R_w Tr}[1 + Y]\right) \qquad \text{with} \qquad \frac{a_{wp}}{a_{wa}} = a_w + Z$$

In this equation σ is the surface tension on the particle's surface, v_w the specific volume of pure water at the standard pressure 1 bar, r the radius of curvature of the particle's surface, and R_w the specific gas constant of pure water.

a_{wp} is the water activity of the particle's liquid cover at the pressure p of the surrounding moist air and the absolute temperature T of the particle.

It is well known that the pressure inside the particle increases with de-creasing radius r and that for radii between 10^{-5} and 10^{-6} μm the internal pressure lies in between about 15 and 150 bars. This pressure increase due to curvature of the particles has been regarded within the property Y.

Thus, a_{wp} is defined for a hypothetical plane surface of the particle's liquid cover. a_{wa} is the water activity for a plane surface of water saturated with dry air at the pressure p and the absolute temperature T. a_w is the water activity of the particle's liquid cover with a hypothetical plane surface in absence of dry air in the gaseous as well as the liquid state. a_w can be measured within a closed laboratory system at well-defined equilibrium vapor pressure of water and temperature.

X is a correction function that regards moist air as an imperfect gas. Its value is in the order of 0.002 to $0.008 \cdot p_r/p_w(1 - p_r/p_w)$ within the temperature range from $+30°C$ to $-30°C$ and thus negligible in the whole range of values that p_r/p_w can attain.

Y is a correction function needed to account for compressibility of pure water and of the liquid cover of the particle, and moreover it contains the deviation of the volume of this cover from the sum of volumes of pure water and pure solvent. The contributions to Y resulting from compressi-bility effects are negligible in each case, whereas the contributions caused by the deviation from volume additivity is of the same sign and the same magnitude as the percentage deviation of radius due to deviation from volume additivity. Thus, assuming volume additivity for the calculation of r, the correction Y can be generally neglected. This will be done in further considerations (cf. Section 2.1.2.2).

The function Z governs the transition from a laboratory system without dry air to the true atmospheric conditions. Z is in the order of 0.004 to $0.01 \cdot (1 - a_w)$ within the temperature range $+30°C$ to $-30°C$ and thus negligible. It must be emphasized that the corrections X and Z balance out each other to a high degree. Thus, X, Y, and Z can be omitted and with high accuracy it can be written even for further theoretic considerations.

$$(2.1) \qquad \frac{p_r}{p_w} = a_w \exp\left(\frac{2\sigma v_w}{rR_w T}\right)$$

This equation is the basis for the following considerations. The same equation has been used by Hänel (1968, 1970a, 1971) but without considera-tion of the effect on p_r and p_w of the dissolved dry air. From the fore-going discussion it is evident that this effect can be neglected if the pressure ratio p_r/p_w is considered, and that it cannot be neglected if p_r or p_w are regarded alone. Because only pressure ratios were used in these papers their results are still valid.

In the following sections the abbreviation

$$(2.2) \qquad \delta = \exp\left(\frac{2\sigma v_{\mathrm{w}}}{R_{\mathrm{w}}\, Tr}\right)$$

for the curvature correction term will be used.

We shall assume that the particle is in thermodynamic equilibrium with the surrounding moist air. Under this condition the relative humidity f of the moist air is given by

$$(2.3) \qquad f = p_{\mathrm{r}}/p_{\mathrm{w}}$$

Then Eq. (2.1) reads

$$(2.4) \qquad f = \delta a_{\mathrm{w}}$$

which, in the case of a plane surface, reduces to

$$(2.5) \qquad f = a_{\mathrm{w}}$$

Thus the water activity is the equilibrium humidity over a plane surface. In the following considerations, the curvature correction and the water activity are expressed in terms of quantities which can either be measured directly on atmospheric aerosol particles or calculated when the chemical composition of the water-soluble substance of the particles is known.

2.1.1. Concept of the Mass Increase Coefficient for the Description of the Water Activity. In the following sections, the water activity a_{w} will be discussed for a particle with a liquid cover. For this case the water activity describes primarily the reduction of water vapor pressure due to the dissolved material and the influence of the interaction between the undissolved material and the liquid cover. The discussions in the next sections are related to the exponential and the linear mass increase coefficients for four cases: a pure solute particle, a mixed solute particle, an aerosol particle under idealized conditions of water uptake, and an aerosol particle under real conditions of water uptake.

2.1.1.1. The case of a pure solute. First, a particle is considered which is an aqueous solution of a pure solute. For this unrealistic but simple case, both the linear and exponential mass increase coefficients can be defined and discussed very easily. Moreover, some main features of these mass increase coefficients can be described.

For the case of a pure solute there are four equivalent definitions for a_w, each containing a different empirical coefficient.

(2.6a)

$$a_w = \left(\frac{n_w}{n_w + n_s}\right)^{\phi_a} \qquad a_w = \gamma_w \frac{n_w}{n_w + n_s}$$

$$a_w = \exp\left(-\phi v \frac{n_s}{n_w}\right) \qquad a_w = \frac{n_w}{n_w + i n_s}$$

n_w is the mole number of water, n_s the number of moles of the dry solute, ϕ_a the apparent osmotic coefficient, γ_w the activity coefficient of water, ϕ the practical osmotic coefficient and i the van't Hoff factor. v denotes the number of moles of ions formed from 1 mole of solute, v being equal to one if there is no dissociation of the solute molecules. ϕ_a, γ_w, ϕ, and i describe empirically the deviation of a_w from the ideal case denoted by $a_w = n_w/(n_w + n_s)$. Thus they are quantities that have to be measured.

Considering that $n_w = m_w/M_w$ and $n_s = m_s/M_s$, Eqs. (2.6a) read

(2.6b)

$$a_w = \left(1 + \frac{M_w}{M_s} \cdot \frac{m_s}{m_w}\right)^{-\phi_a} \qquad a_w = \gamma_w\left(1 + \frac{M_w}{M_s} \cdot \frac{m_s}{m_w}\right)^{-1}$$

$$a_w = \exp\left(-\phi v \frac{M_w}{M_s} \cdot \frac{m_s}{m_w}\right) \qquad a_w = \left(1 + i \frac{M_w}{M_s} \cdot \frac{m_s}{m_w}\right)^{-1}$$

m_w is the mass of water, M_w its molar mass, m_s the mass of the solute, and M_s its molar mass. In the case of totally dissolved solute, the ratio m_s/m_w is the mass concentration of the solute within water. It is evident that even for a solute with unknown chemical composition the relationship between the water activity a_w and the mass concentration m_s/m_w can be described empirically by the third or fourth equation given in (2.6b) with two new empirical coefficients:

(2.7)
$$\eta_s = \phi v M_w/M_s \qquad \text{and} \qquad \mu_s = i M_w/M_s$$

where η_s is the exponential mass increase coefficient and μ_s the linear mass increase coefficient of the solute. The linear mass increase coefficient has already been used earlier (Hänel, 1968, 1970a). Both the practical osmotic coefficient ϕ and the van't Hoff factor i are tabulated (i.e., by Robinson and Stokes, 1959; Low, 1969) as a function of solute concentration.

Introducing definitions (2.7) into (2.6b) yields

(2.8a) $\qquad a_w = \exp(-\eta_s m_s/m_w) \qquad$ or $\qquad a_w = (1 + \mu_s m_s/m_w)^{-1}$

and

(2.8b) $\qquad m_w/m_s = -\eta_s/\ln a_w \qquad$ or $\qquad m_w/m_s = \mu_s a_w/(1 - a_w)$

Equation (2.8a) describes the water activity a_w as a function of the solute concentration m_s/m_w and Eq. (2.8b), the water uptake per unit mass of solute m_w/m_s as a function of the water activity a_w. When a_w is computed from m_s/m_w the coefficients η_s and μ_s must be known as functions of m_s/m_w. In the reverse case, when m_w/m_s is computed as a function of a_w, the properties η_s and μ_s must be known as functions of a_w. Both problems can be treated, because a_w and m_s/m_w are the properties that are usually measured (cf. Robinson and Stokes, 1959; Section 4.1).

Like any other thermodynamic property, both mass increase coefficients depend on temperature as well as on pressure. For a given water activity they are directly proportional to the water uptake per unit mass of dry solute. Thus, at a specific water activity the ratio between the exponential or linear mass increase coefficients of different solutes directly gives the ratio of the relevant water uptakes per unit mass of the solutes.

At infinite dilution, i.e., $a_w = 1$, the practical osmotic coefficient is $\phi = 1$ and the van't Hoff factor is $i = v$. Denoting $\eta_s^0 = \eta_s(a_w = 1)$ and $\mu_s^0 = \mu_s(a_w = 1)$, Eq. (2.7) gives

$$(2.9) \qquad \eta_s^0 = \mu_s^0 = vM_w/M_s$$

In Fig. 1 the dependence of the linear mass increase coefficient μ_s upon water activity a_w is plotted for some selected salts. It can be seen that μ_s varies characteristically for each salt; of particular interest is the increase of μ_s shortly before a_w reaches one, because this increase has to be considered for cloud physics application (cf. McDonald, 1953). Winkler (1973) has used the approximation $a_w \cong 1 - \alpha m_s/m_w$ stating that within this formula α could be regarded as a constant, irrespective of the fact that α is defined by $\alpha = iM_w/M_s$ like the linear coefficient of mass increase μ_s in Eq. (2.7). A proof of this approximation showed that for aqueous NaCl solutions at $a_w = 1, 0.9836, 0.9011$, and 0.760 the corresponding α-values are 0.616, $0.562, 0.604$, and 0.684, whereas for aqueous $(NH_4)_2SO_4$ solutions at $a_w = 1, 0.9819, 0.9021$, and 0.812 the corresponding α-values are 0.409, $0.274, 0.247$, and 0.259. In comparison with Fig. 1 it should be noted that α can be regarded as a constant only when this is true for μ_s as well. Consequently the conclusions derived from the assumption of a constant α-value must be regarded with caution.

2.1.1.2. The case of a mixed solute. The water-soluble material within atmospheric aerosol particles is expected to be a mixture of different chemicals. This has been demonstrated by a large number of chemical analyses. In general the water-soluble material within atmospheric aerosol particles is regarded to be a mixture of electrolytes together with any other water-soluble material. Thus, a realistic consideration upon the thermodynamics of aerosol particles has to treat mixed solutes.

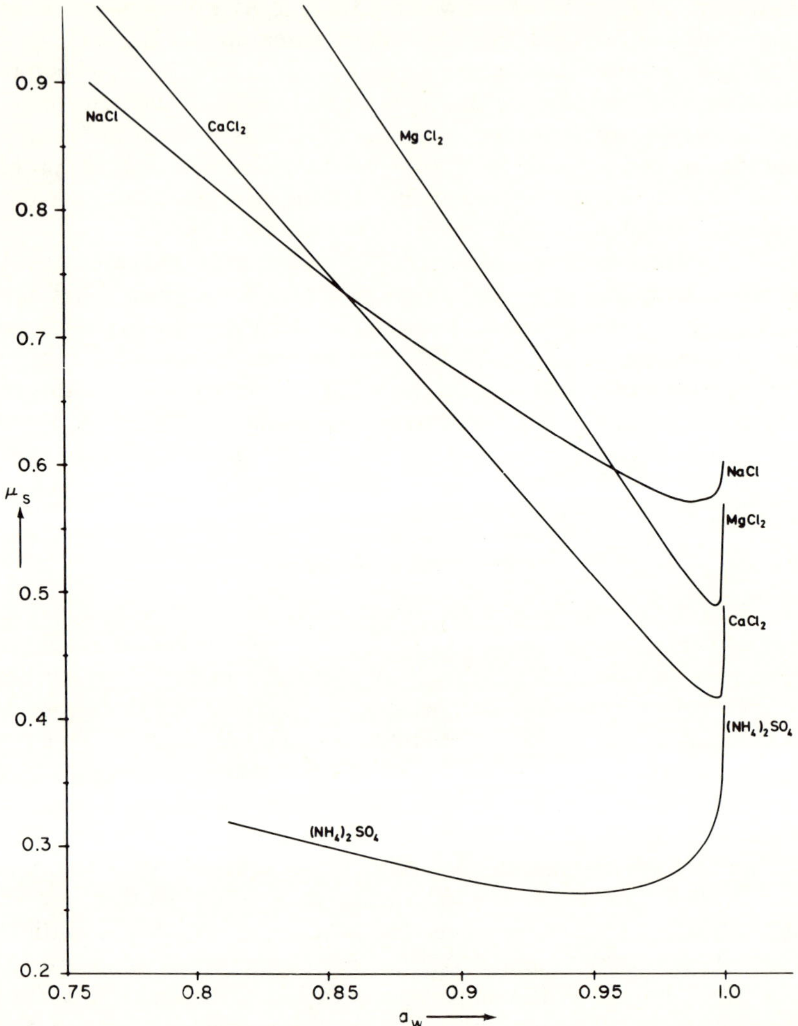

FIG. 1. Linear mass increase coefficient μ_s versus water activity a_w at 25°C for selected salts calculated from the experimental data presented by Robinson and Stokes (1959).

Mass m_s and total mole number n_s in dry state for a mixture are given by

$$(2.10) \qquad m_s = \sum_i m_{si} \qquad \text{and} \qquad n_s = \sum_i n_{si} = \sum_i \frac{m_{si}}{M_{si}}$$

where m_{si}, n_{si}, and M_{si} denote the masses in dry state, the mole numbers in dry state, and the molar weights of the components, respectively. The mean molar weight of the mixture is defined by

$$(2.11) \qquad 1/\overline{M}_s = n_s/m_s$$

and the mean number of moles of ions and undissociated molecules per mole of the mixture by

$$(2.12) \qquad \bar{v} = \sum_i v_i \frac{n_{si}}{n_s}$$

v_i is the number of moles of ions formed from 1 mole of the component i. Thus, in agreement with Harned and Owen (1958) the exponential and the linear mass increase coefficients are

$$(2.13) \qquad \bar{\eta}_s = \bar{v}\bar{\phi} M_w/\overline{M}_s \qquad \text{and} \qquad \bar{\mu}_s = \bar{i} M_w/\overline{M}_s$$

$\bar{\phi}$ is the observed practical osmotic coefficient and $\bar{i}$ the observed van't Hoff factor of the mixture, if the value of its mean molar weight $\overline{M}_s$ is known. Both mass increase coefficients for mixtures can be used in Eqs. (2.8a and b) instead of those for pure substances. Therefore a mixed solute can be treated like a pure solute, and both the exponential and the linear mass increase coefficients can be obtained by measurement without knowledge of the chemical composition. The following considerations deal with the problem of computing both mass increase coefficients when the chemical composition of the solute mixture is known.

Again it is $\bar{\phi} = 1$ and $\bar{i} = \bar{v}$ for infinite dilution at $a_w = 1$. Thus both the mass increase coefficients are identical at $a_w = 1$ and equal to

$$(2.14a) \qquad \bar{\eta}_s^0 = \bar{\mu}_s^0 = \bar{v} M_w/\overline{M}_s$$

Considering definitions (2.11) and (2.12) and Eq. (2.9) the exponential and the linear mass increase coefficients of a mixture at infinite dilution, $a_w = 1$, are

$$(2.14b) \qquad \bar{\eta}_s^0 = \sum_i \eta_{si}^0 \frac{m_{si}}{m_s} \qquad \text{and} \qquad \bar{\mu}_s^0 = \sum_i \mu_{si}^0 \frac{m_{si}}{m_s}$$

The η_{si}^0 and the μ_{si}^0 are the exponential and linear mass increase coefficients, respectively, of the components at infinite dilution. The ratios m_{si}/m_s are the mass concentrations of the components within the dry solute mixture.

The idealized exponential and linear mass increase coefficients are defined as in Eq. (2.14b), assuming

$$(2.15) \qquad \bar{v}\bar{\phi} n_s = \sum_i v_i \phi_i n_{si} \qquad \text{and} \qquad \bar{i} n_s = \sum_i i_i n_{si}$$

where ϕ_i and i_i are the practical osmotic coefficients and the van't Hoff factors of the pure components if they were alone. In these equations only the interaction between the ions of each chemical component is considered, but there is no interaction between those ions the chemical components do not have in common. For example, in a mixture of KCl and $MgCl_2$ the interaction between K^+ and Mg^{2+} is not considered. With Eq. (2.15) the idealized exponential mass increase coefficient $\bar{\eta}_s^{id}$ and the idealized linear mass increase coefficient $\bar{\mu}_s^{id}$ can be written as

$$(2.16) \qquad \bar{\eta}_s^{id} = \sum_i \eta_{si} \frac{m_{si}}{m_s} \qquad \text{and} \qquad \bar{\mu}_s^{id} = \sum_i \mu_{si} \frac{m_{si}}{m_s}$$

as in Eq. (2.14b). At high dilution the idealized coefficients should give the most reliable results. Both idealized coefficients can be regarded as functions of the concentration m_s/m_w or the water activity a_w corresponding to the specific problem to be solved.

  Like all mass increase coefficients defined above, the idealized ones are related by an equation of the type

$$\bar{\eta}_s^{id} = \bar{\mu}_s^{id} \frac{a_w}{1 - a_w} \ln(1/a_w)$$

when they are known as a function of the water activity a_w. Thus, at a given water activity the percentage errors are the same for both coefficients, and the reliability of only one of these two coefficients has to be tested. For this, the idealized mass increase coefficient has been chosen.

The additivity rule (2.16) gives reliable results even for concentrated solutions at a_w smaller than 0.95. This is shown in Fig. 2 for the idealized linear mass increase coefficient. In Table I a summary of maximum errors for the idealized mass increase coefficients is given for a variety of mixed electrolytes. The largest errors have been found to be smaller than 10 to 15% at water activities smaller than 0.85 to 0.90.

Equation (2.16) has the same accuracy as the rule presented by Robinson and Stokes (1959, p. 449) for the molar vapor pressure lowering of a mixed electrolyte. Equations (2.16) mean that for a given water activity both idealized mass increase coefficients lie between the mass increase coefficients of the pure components. This may not be so, when the mass increase coefficients of the components differ only by little and at the same time the error of the idealized one exceeds this difference. For example, Robinson and Stokes (1945) give measurements on a mixture of KCl and $CuCl_2$ in aqueous solution. In this case at the water activity $a_w = 0.852$ the measured linear mass increase coefficient is $\bar{\mu}_s = 0.438$, whereas the calculated one is $\bar{\mu}_s^{id} = 0.500$. For the pure components, $\mu_s = 0.514$ for KCl and $\mu_s = 0.484$ for $CuCl_2$ at the same water activity. It is striking that the calculated coefficient is only 12.4% too large. For the same electrolyte mixture, the

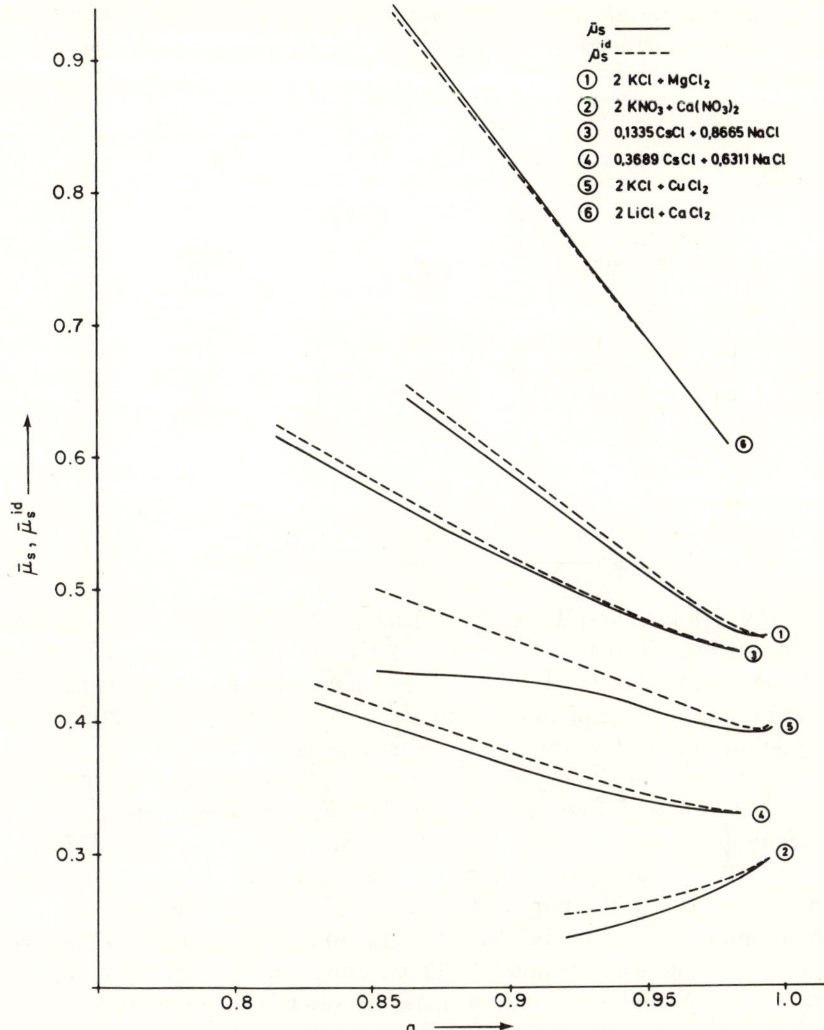

FIG. 2. Measured and idealized linear mass increase coefficients $\bar{\mu}_s$ and $\bar{\mu}_s^{id}$ versus water activity a_w at 25°C. Calculations from the experimental data from Robinson and Stokes (1959, 1945) and Robinson (1945, 1952).

situation changes at larger water activities: At $a_w = 0.9447$ it is $\bar{\mu}_s^{id} = 0.427$, $\bar{\mu}_s = 0.413$, $\mu_s = 0.451$ (KCl), $\mu_s = 0.400$ (CuCl₂), and at $a_w = 0.98904$ it is $\bar{\mu}_s^{id} = 0.393$, $\bar{\mu}_s = 0.391$, $\mu_s = 0.440$ (KCl), $\mu_s = 0.340$ (CuCl₂). Here the idealized linear mass increase coefficients lie between those of the pure components, the errors being 3.3 % at $a_w = 0.9447$ and 0.5 % at $a_w = 0.98904$.

TABLE I. Maximum errors for the idealized mass increase coefficient $\bar{\mu}_s^{id}$ of mixtures of salts together with the range of applicability in the water activity a_w

System	a_w—range of applicability	Maximum percentage error of $\bar{\mu}_s^{id}$
Sea salt	0.75 —1	8.7
$2KCl + MgCl_2$	0.865—1	1.7
$2KNO_3 + Ca(NO_3)_2$	0.92 —1	7.1
$2LiCl + CaCl_2$	0.23 —1	0.6
$2KCl + CuCl_2$	0.85 —1	12.4
$0.8665\ NaCl + 0.1335\ CsCl$	0.816—1	1.8
$0.6311\ NaCl + 0.3689\ CsCl$	0.83 —1	3.3
$x\ NaCl + (1 - x)\ KCl$ $0 \leqq x \leqq 1$	0.8515	0.9
$2KCl + BaCl_2$ $2KCl + MnCl_2$ $2KCl + CoCl_2$ $2KCl + NiCl_2$	0.9 —1	5

Recently Winkler and Junge (1972) have reported their measurements on some mixed electrolytes at water activities smaller than 0.95. They found one example for which the value of the measured linear mass increase coefficient was more and more reduced with chemical complexity of the mixed solute. In contradiction to this finding is the behavior of sea salt, which is a mixture of six major water-soluble salts (Arons and Keith, 1954). In Fig. 3 it is shown that for sea salt the idealized linear mass increase coefficient agrees with the measured one within 10 % in the range 0.75 to 1 of water activity. Moreover, the measured linear mass increase coefficient lies between those of the pure components at the same water activities.

To summarize, it can be stated that both idealized mass increase coefficients can be used for model calculations in cloud physics or for fog situations. Only small errors in the mass of water being condensed on the mixed solute, corresponding to small errors in the heat released during the condensation process, are introduced using Eq. (2.16). Even in the case of water activities smaller than 0.95 Eq. (2.16) gives reliable results.

2.1.1.3. Aerosol particles under idealized conditions of water uptake. Up to now only pure or mixed solutes have been considered. But it is well known (Junge, 1952a; Winkler and Junge, 1972) that the atmospheric aerosol particles contain insoluble substances too. Therefore, let us consider a particle that in its dry state is composed of a mixed solute and insoluble substances and has at large relative humidity, i.e., large water activity, a liquid cover. This means that no portion of its surface contains water-

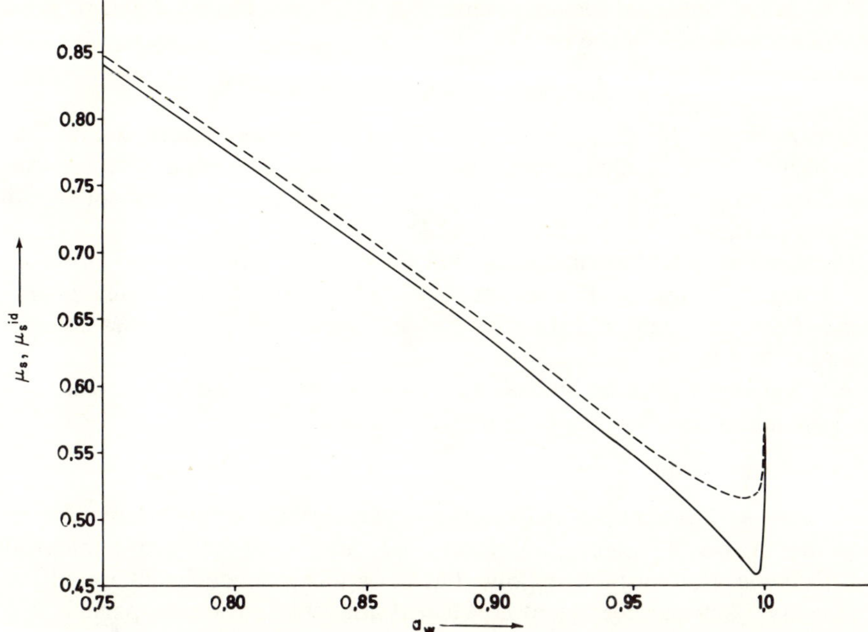

FIG. 3. Measured and idealized linear mass increase coefficients $\bar{\mu}_s$ and $\bar{\mu}_s^{id}$ versus water activity a_w at 25°C for sea salt. Experimental data from Arons and Keith (1954).

insoluble material. Moreover, it is assumed that the insoluble fraction has no influence on the mass of water taken up by the solute, thus neglecting interphase phenomena between solution and insoluble material as well as the possible presence of surface-active substances. Then the mass of water taken up by the mixed solute is given by means of an equation like (2.8b) in which the mass increase coefficients of the mixed solute have been introduced:

$$m_w = -m_s \cdot \bar{\eta}_s / \ln(a_w) \qquad m_w = m_s \cdot \bar{\mu}_s a_w / (1 - a_w)$$

where m_s denotes the mass of the dry mixed solute. It must be emphasized that the water activity is related to a plane surface of the solution.

The total mass of the dry particle is defined by

$$(2.17) \qquad m_0 = m_u + m_s$$

m_u denoting the mass of the insoluble material. Dividing the equations above by m_0, the water uptake per unit mass of the dry material is obtained:

$$(2.18) \qquad m_w/m_0 = -\eta^{id}/\ln(a_w) \qquad \text{or} \qquad m_w/m_0 = \mu^{id} a_w / (1 - a_w)$$

η^{id} is called the idealized exponential and μ^{id} the idealized linear mass increase coefficient of the particle. They are given by

$$(2.19) \qquad \eta^{id} = \bar{\eta}_s \, m_s/m_0 \qquad \text{and} \qquad \mu^{id} = \bar{\mu}_s \, m_s/m_0$$

where m_s/m_0 is the mass concentration of the mixed solute within the dry particle. The idealized mass increase coefficients should give reliable results at large water activities. Thus they should be applicable for cloud physics considerations (cf. Section 6.2.2).

However it must be considered that an accurate experimental proof of this concept is missing. The results of Winkler (1970) and of Winkler and Junge (1972) only allow distinctive conclusions for water activities smaller than 0.7 to 0.8.

Solving Eqs. (2.18) for the water activity, we have formulations that can be introduced directly into (2.1) or (2.4):

$$(2.20) \qquad a_w = \exp(-\eta^{id} m_0/m_w) \qquad \text{or} \qquad a_w = (1 + \mu^{id} m_0/m_w)^{-1}$$

2.1.1.4. Aerosol particles under real conditions of water uptake. Considering now the effects of water insoluble material and of surface-active material (cf. Defay *et al.*, 1966), the exponential mass increase coefficient η and the linear mass increase coefficient μ of a real atmospheric aerosol particle can be described by

$$(2.21) \qquad \eta = \bar{\eta}_s \, \gamma \, m_s/m_0 \qquad \text{and} \qquad \mu = \bar{\mu}_s \, \gamma \, m_s/m_0$$

where γ is an empirical correction factor, which describes the deviation from the idealized conditions (2.19) for both coefficients.

As in (2.18) and (2.20), the water uptake per unit mass of dry matter and the water activity related to a plane surface can be written as

$$(2.22) \qquad m_w/m_0 = -\eta/\ln(a_w) \qquad \text{or} \qquad m_w/m_0 = \mu \, a_w/(1 - a_w)$$

and

$$(2.23) \qquad a_w = \exp(-\eta \, m_0/m_w) \qquad \text{or} \qquad a_w = (1 + \mu \, m_0/m_w)^{-1}$$

We wish to emphasize that each of the equations (2.23) for the water activity can be used directly to describe the chemical potential of water being condensed on an atmospheric aerosol particle. Again the exponential and the linear mass increase coefficients of a real atmospheric aerosol particle relate measurable properties (cf. Section 2.3).

In the following discussions, mostly the linear mass increase coefficient will be used. Moreover, only the case of a real aerosol particle will be considered. All the other cases can be derived regarding the definitions given above.

2.1.2 Discussion of the Curvature Correction. The curvature correction term relates relative humidity and water activity. Thus it is necessary to describe the curvature correction in terms of properties that can either be measured or be calculated.

2.1.2.1. The surface tension. From experimental and theoretic results (compare Harned and Owen, 1958; Defay *et al.*, 1966; Low, 1969; Dufour and Defay, 1963) the surface tension σ_s of an aqueous electrolyte solution with totally dissolved solute can be approximated by

$$(2.24) \qquad \sigma_s = \sigma_w(T) + b\, m_s/m_w$$

$\sigma_w(T)$ is the surface tension of pure water at the absolute temperature T. b is constant for each solute, $b = 27.6$ dynes/cm (NaCl) and $b = 16.4$ dynes/cm $[(NH_4)_2SO_4]$. These two values approximate the range of b expected for atmospheric aerosol particles.

The temperature dependence of the surface tension of water can be approximated by the linear relationship

$$(2.25) \qquad \sigma_w(T) = \sigma_w(T_0) + a(T_0 - T)$$

holding within the temperature range $-10°C$ to $+30°C$ with $a = 0.153$ dynes/cm °K and $\sigma_w(T_0 = 273.16°K) = 75.6$ dynes/cm.

In considering water uptake by aerosol particles under real conditions, one has to replace the concentration m_s/m_w of the solution by the water uptake per unit mass of the dry particle m_w/m_0. Moreover, the possible influence of surface-active material has to be considered. The first can be done regarding idealized conditions of water uptake since even errors of 10 or 20% in m_s/m_w do not influence the value of the surface tension by more than 2 to 4%, even in the worst case of highly concentrated solutions. For the case of water uptake under idealized conditions it is

$$(2.26) \qquad \frac{m_s}{m_w} = \frac{m_s}{m_0}\bigg/\frac{m_w}{m_0}$$

The influence of surface-active material is introduced by an empirical factor γ_σ.

Combining the last three equations, the surface tension of the liquid cover of a real aerosol particle can be written as

$$(2.27) \qquad \sigma = \gamma_\sigma\left[\sigma_w(T_0) + a(T_0 - T) + b\,\frac{m_s}{m_0}\bigg/\frac{m_w}{m_0}\right]$$

The latter equation is valid under the condition that the solute is completely dissolved. The constant b can be determined from knowledge of the chemical composition of the solute matter (cf. Defay *et al.*, 1966), a good value for practical purposes being $b = 20$ dynes/cm.

2.1.2.2. The equivalent radius of aerosol particles. Description of the particle radius r in terms of the water uptake per unit mass of the dry particle is based upon the assumption of volume additivity. This means the total volume of the particle is the sum of the volumes of the dry particle and of the pure water condensed on it.

$$(2.28) \qquad V = V_0 + V_w$$

V is the volume of the wet particle, V_0 the volume of the dry particle and V_w the volume of pure water being condensed on the particle.

Since $V_0 = m_0/\rho_0$ and $V_w = m_w/\rho_w$, where ρ_0 and ρ_w are the densities of the dry particle and of pure water, respectively, the volume ratio V/V_0 is

$$(2.29) \qquad V/V_0 = 1 + V_w/V_0 = 1 + \frac{\rho_0}{\rho_w} \cdot \frac{m_w}{m_0}$$

For generality the equivalent radius of the particle is introduced. This is the radius of a sphere with the same volume as the particle. The equivalent radius of the wet particle is given by $r = (3V/4\pi)^{1/3}$ and that of the dry particle $r_0 = (3V_0/4\pi)^{1/3}$. Then Eq. (2.29) becomes

$$(2.30) \qquad \frac{r}{r_0} = \left(1 + \frac{\rho_0}{\rho_w} \frac{m_w}{m_0}\right)^{1/3}$$

or

$$(2.31) \qquad r = r_0 \left(1 + \frac{\rho_0}{\rho_w} \frac{m_w}{m_0}\right)^{1/3}$$

Because the mean density ρ_0 of the dry particles can be determined on samples (cf. the following sections), Eqs. (2.30) and (2.31) contain only measurable properties.

2.1.2.3. Curvature correction and final equation. Combining Eqs. (2.27) and (2.31) the curvature correction (2.2) for an atmospheric aerosol particle reads

$$(2.32) \quad \delta = \exp\left(\frac{2v_w}{R_w T} \cdot \frac{\gamma_\sigma[\sigma_w(T_0) + a(T_0 - T) + b(m_s/m_0)/(m_w/m_0)]}{r_0[1 + (\rho_0/\rho_w) \cdot (m_w/m_0)]^{1/3}}\right)$$

Within this equation, the correction γ_σ due to the presence of surface-active substance is usually set equal to one. This neglect of surface-active material within atmospheric aerosol particles and cloud droplets cannot be justified. In the present stage of research it is based only on a lack of experimental data on the surface tension of atmospheric aerosol particles or cloud droplets. Therefore experiments are now being carried out by the author to measure whether surface-active materials play a role in the thermodynamic properties of aerosol particles and cloud droplets. The importance of this question will be demonstrated later with model calculations and the first experimental results (Section 6).

Together with Eqs. (2.23) and (2.32) the final equation is obtained from (2.4):

$$(2.33) \qquad f = \frac{\exp\left[\dfrac{2v_w}{R_w\,T} \cdot \dfrac{\sigma_w(T_0) + a(T_0 - T) + b(m_s/m_0)/(m_w/m_0)}{r_0[1 + (\rho_0/\rho_w) \cdot (m_w/m_0)]^{1/3}}\right]}{1 + \mu/(m_w/m_0)}$$

where $\gamma_\sigma = 1$.

Equation (2.33) will be used for model calculations of the physicochemical properties of atmospheric aerosol particles as functions of the relative humidity (cf. Section 6). It gives the equilibrium relative humidity f as a function of the water uptake per unit mass of the dry material m_w/m_0 for each value of the equivalent radius r_0 in dry state chosen. All the other parameters can be measured on samples of atmospheric aerosols; these are the linear mass increase coefficient $\mu = \mu(m_w/m_0)$, the mean density ρ_0 in dry state, the ratio m_s/m_0 of the dry water-soluble substance to the total dry substance, and the property b describing the dependence of the surface tension of the particle's liquid cover from the solute concentration.

Although the curvature correction δ, which is described by the exponential function in (2.33) has a complicated structure, the main features of it can be seen easily: Its value is always larger than or equal to unity, and its largest values are attained for the smallest equivalent radii r_0 and the smallest water uptakes m_w/m_0. On the other hand, the water activity a_w increases with the water uptake m_w/m_0. At least the relative humidity f, which is equal to the product $\delta \cdot a_w$, increases with the water uptake m_w/m_0 since it has reached a maximum value above unity. After that, f decreases with increasing m_w/m_0 until it reaches one at infinite dilution (i.e., $m_w/m_0 \to \infty$).

In cases where errors of the surface tension do not have a large influence on the mass of water taken up by the particle, i.e., for $f \leq 0.90$ to 0.95 and $r_0 \geq 0.1$ μm (Hänel, 1970a), Eq. (2.33) is valid even when a portion of the solute material has not been dissolved. A more detailed discussion of Eq. (2.33) will be given in Section 6.2.

2.2. Thermodynamic Equilibrium When No Complete Liquid Cover Exists

Up to now only the case of an aerosol particle with a liquid cover has been discussed. In nature this case is fulfilled only at large relative humidities. However the water uptake by atmospheric aerosol particles at small and moderate relative humidities sometimes is of interest for special applications. Therefore this case will be considered now.

2.2.1 The Threshold Water Activity. A particle at low relative humidity has no complete liquid cover. Here the water condensation is due to adsorption, capillary effects, hydration, fractional dissolution of the solute material, and interaction between the solution and the insoluble material (Glasstone, 1964). All these phenomena depend on chemical composition as well as on the structure of the dry particle. It is well known (cf. Orr *et al.*, 1958; Defay *et al.*, 1966) that the solubility of a solute particle increases when its spatial dimensions become smaller.

When the relative humidity is sufficiently high, a liquid cover develops and the water activity can be defined with respect to a plane surface. Thus the smallest water activity for which a complete liquid cover exists and the solute material regarded to be dissolved completely will be denoted by "threshold water activity" a_{wt}. At water activities $a_w \geqq a_{wt}$ Eq. (2.33) holds. The threshold water activity a_{wt} is a function of chemical composition and structure of the dry particle, and of the mass concentration of water-soluble material within the dry particle. Defining the volume ratio V/V_0 at the threshold water activity a_{wt} to be $V(a_{wt})/V_0 = 1 + \alpha$, Eq. (2.29) gives

$$(2.34) \qquad m_w(a_{wt})/m_0 = \alpha \rho_w/\rho_0$$

If the value of α is known, this equation contains only measurable properties. Estimates from the measurements of Orr *et al.* (1958) show that $\alpha = 1$ is a good value for practical purposes, which is in agreement with the author's own measurements on aerosol samples. From these measurements of water vapor condensation on samples of atmospheric aerosol particles one can derive an alternate a_{wt}-value for each sample. The threshold water activity is defined as the largest water activity for which the linear mass increase coefficient attains a relative maximum. This last maximum indicates that the water-soluble material is being totally dissolved. A threshold humidity has not been defined, because this would lead to very complicated considerations without deriving any benefit from it.

2.2.2. Conclusions From BET-Adsorption Isotherm. Phenomena like adsorption, capillary condensation, or condensation within pores can be described by an adsorption isotherm e.g., that by Brunauer, Emmett, and Teller (1938), the so-called BET-adsorption isotherm. This isotherm can easily be transformed into an equation like (2.22) for the water uptake per unit mass of dry substance: $m_w/m_0 = \mu \cdot f/(1 - f)$ where μ is the linear mass increase coefficient for the phenomena regarded above. This example shows that it is appropriate to define the linear mass increase coefficient in the whole range of water activity as has been done in the previous sections.

Thus it is defined at low relative humidity for a real aerosol particle

$$(2.35) \qquad m_w/m_0 = \mu \cdot a_w/(1 - a_w)$$

implying that

$$(2.36) \qquad f = a_w$$

Here the linear mass increase coefficient has to be interpreted as an empirical coefficient, correlating m_w/m_0 and $a_w = f$. The latter two can be obtained by measurement. The linear mass increase coefficient thus includes all effects leading to water vapor condensation on the particle.

2.2.3. Transition Regime of Water Activity. If it were defined that $f = a_w$ for water activities a_w smaller than the threshold activity a_{wt}, a jump in relative humidity would take place at $a_w = a_{wt}$ due to introduction of the curvature correction $\delta(a_{wt})$ being larger than unity. For larger particles, this jump is smaller than for smaller ones because of the decrease of the curvature correction with particle radius r_0 in dry state. This effect does not correspond with reality. Even the dissolution process of the solute material has not been found to occur abruptly at a specific relative humidity (cf. Orr *et al.*, 1958; Winkler, 1969, Winkler and Junge, 1971). To avoid the jump in relative humidity at the threshold water activity it is appropriate to define a transition regime Δa_{wt} within which the value of the curvature correction goes from unity to its value $\delta(a_{wt})$ at the threshold water activity. With this procedure the physical process occurring in nature shall be approximated.

For smaller particles the curvature correction is greater than for large ones. Thus, it is fitting to define

$$(2.37) \qquad \Delta a_{wt} = a_{wt} - a_{wt}/\delta(a_{wt})$$

In this equation Δa_{wt} increases with $\delta(a_{wt})$, i.e., increases when the particle becomes smaller. Moreover, Δa_{wt} cannot be larger than a_{wt} even for the smallest particles. The simplest way to increase the δ-value within the transition regime $a_{wt} - \Delta a_{wt} \leq a_w \leq a_{wt}$ is given by a linear rise

$$(2.38) \qquad \delta(a_w) = \delta(a_{wt}) \cdot a_w/a_{wt}$$

It must be emphasized that other procedures which define the transition regime of water activity and the raise of curvature correction are possible. Moreover it must be stated that this transition regime is of some importance only if the radius of the dry particle becomes smaller than 0.1 μm.

2.2.4. Summary. Over the whole range of water activity the linear mass increase coefficient describes the water uptake of aerosol particles per unit mass of dry matter, including all effects leading to water vapor condensation. This can be described by equations of a type like Eq. (2.35). The values of the curvature correction are:

$$\delta(a_w = f) = 1 \qquad \text{for} \quad 0 \leqq a_w \leqq a_{wt} - \Delta a_{wt}$$

$$(2.39) \qquad \delta(a_w) = \delta(a_{wt}) \cdot a_w/a_{wt} \qquad \text{for} \quad a_{wt} - \Delta a_{wt} \leqq a_w \leqq a_{wt}$$

$$\delta(a_w) \text{ from Eq. (2.32)} \qquad \text{for} \quad a_{wt} \leqq a_w \leqq 1$$

Thus the relative humidity is related to the water uptake per unit mass of dry substance in the whole range of values it can attain. The most important connecting link between those two parameters is the linear mass increase coefficient, which can be measured. Other properties occurring within the expression of the curvature correction can also be measured, or even estimated.

2.3. Thermodynamic Equilibrium between a Sample of Aerosol Particles and Its Surrounding Moist Air

Up to now the thermodynamic equilibrium between a single atmospheric aerosol particle and the surrounding moist air has been described. The most important parameter within this description is the linear mass increase coefficient. This coefficient can hardly be measured on single aerosol particles, but its measurement is possible on samples of a large number of atmospheric aerosol particles. The basic theoretic considerations for such measurements will be given now.

2.3.1. Analytic Characterization of a Sample. As it has been described, even a single aerosol particle can be composed of various substances, the same being true for a sample of a great number of aerosol particles. Thus, a deposit of atmospheric aerosol particles hanging together is considered to behave like a very large particle and the equations derived above are valid. Using a sample with spatial dimensions of the order of 1 mm in dry state, the curvature correction can be set equal to unity. The error introduced by this procedure is so small, that it cannot be measured.

Therefore it is appropriate to define for a deposit of aerosol particles

$$(2.40) \qquad f = a_w, \qquad 0 \leqq a_w \leqq 1$$

Hence the water uptake per unit mass of dry substance is given by

$$(2.41a) \qquad m_w/m_0 = \bar{\mu} f/(1 - f)$$

$\bar{\mu}$ denotes the mean linear mass increase coefficient of the sample of atmospheric aerosol particles. This coefficient can be evaluated in the same manner as has been done for a single particle (Eq. 2.21):

$$(2.41b) \qquad \bar{\mu} = \bar{\mu}_s \, \bar{\gamma} \, m_s/m_0$$

Here $\bar{\gamma}$ is the correction factor describing deviations from idealized conditions of water uptake ($\bar{\gamma} = 1$).

2.3.2. Application of the Results Obtained from Samples. A sample of aerosol particles and a single aerosol particle are in principle distinguished by their mass only, with regard to the chemical and structural complexity of both. Thus, the mean linear mass increase coefficient of a sample can be used for model calculations on the behavior of populations of single particles with the following presuppositions:

(1) In dry state, each single particle has the same chemical composition and structure as the sample.
(2) The threshold water activity and the transition regime are given in the same manner as has been derived above in Sections 2.2.1 and 2.2.3.

2.3.3. Extrapolation to Infinite Dilution. Due to instrumental limitations (cf. Section 6) the linear mass increase coefficient cannot be determined for the largest water activities near unity. Nevertheless, for model calculations based on measured mean mass increase coefficients, the latter must be known for infinite dilution, i.e., $a_w = 1$. Thus, the measured coefficients have to be extrapolated to $a_w = 1$ from the largest water activity for which a measurement has been taken. This largest water activity usually lies within the range 0.95 to 0.99995.

As the linear mass increase coefficient has a finite value at infinite dilution $a_w = 1$, such an extrapolation can be made without large errors involved. The maximum water activity, for which a measurement is available, will be denoted by a_{wm}. Thus for a water activity in the range $a_{wm} < a_w \leq 1$ it follows from Eq. (2.41b) that

$$\frac{\bar{\mu}(a_w)}{\bar{\mu}(a_{wm})} = \frac{\bar{\mu}_s(a_w)}{\bar{\mu}_s(a_{wm})} \cdot \frac{\bar{\gamma}(a_w)}{\bar{\gamma}(a_{wm})}$$

If at those large water uptakes per unit mass of dry material as are found near $a_w = 1$, the interface phenomena between the liquid and the solid parts within the sample are not changed severely, it can be assumed $\bar{\gamma}(a_w) = \bar{\gamma}(a_{wm})$ and therefore

$$(2.42) \qquad \bar{\mu}(a_w) \cong \bar{\mu}(a_{wm}) \cdot \bar{\mu}_s(a_w)/\bar{\mu}_s(a_{wm})$$

The ratio $\bar{\mu}_s(a_w)/\bar{\mu}_s(a_{wm})$ can be evaluated if the chemical composition of the solute material is known. Then either tabulated coefficients can be used (e.g., for sea salt) or the concept of the idealized mixed solute (cf. Section 2.1.1.2). The maximum water activity a_{wm} for the weighing system for the determination of the linear mass increase coefficient is 0.99. At present a new apparatus is under construction which allows the measurement of the linear mass increase coefficient up to $a_{wm} = 0.99995$.

In the previous sections the thermodynamic equilibrium between an aerosol particle or a sample of aerosol particles and the surrounding moist air has been described. Within the foregoing equations the relationship between the relative humidity, on the one hand, and the size or the mass of condensed water per unit mass of dry substance, on the other hand, are given for both single aerosol particles and samples of aerosol particles. Only those parameters were used which can be measured or determined by calculation.

The mean density and the mean refractive index will be discussed in the following sections. The mean density is important for the understanding of the transport of particles within the atmosphere, and the mean refractive index for problems concerning radiative transfer within the atmosphere.

2.4. The Mean Density as a Function of the Relative Humidity

The mean density of a sample of aerosol particles as well as of an airborne aerosol particle is defined as the ratio of the total mass to the total volume

$$(2.43) \qquad \rho = \frac{m}{V} = \rho_0 \frac{m/m_0}{V/V_0}$$

Here $m = m_0 + m_w$ is the mass of the moist particle or the moist sample and ρ_0 is the mean density of the dry particle or the dry sample. Assuming volume additivity for all the components the particle or the sample consists of one has

$$V = \sum_i V_i \qquad \text{or} \qquad \frac{m}{\rho} = \sum_i \frac{m_i}{\rho_i}$$

V_i, m_i, and ρ_i denoting the volumes, masses, and densities of the pure components, respectively. Thus the mean density is given by

$$(2.44) \qquad \frac{1}{\rho} = \sum_i \frac{1}{\rho_i} \frac{m_i}{m}$$

This is a useful equation for the calculation of mean density of the dry substance if its composition is known.

The humidity dependence of the mean density for a single particle or a sample is given by introducing Eq. (2.29) into Eq. (2.43)

$$(2.45a) \qquad \rho = \rho_0 \left(1 + \frac{m_w}{m_0}\right)\left(1 + \frac{\rho_0}{\rho_w} \cdot \frac{m_w}{m_0}\right)^{-1}$$

After adding ρ_w to and subtracting ρ_w from this equation, it follows after a simple rearrangement

$$(2.45b) \qquad \rho = \rho_w + (\rho_0 - \rho_w)\left(1 + \frac{\rho_0}{\rho_w} \cdot \frac{m_w}{m_0}\right)^{-1}$$

These are the most general equations that can be transformed for an aerosol sample with the help of the mean linear mass increase coefficient (Eq. 2.41a) into

$$\rho = \rho_0\left(1 + \bar{\mu}\,\frac{f}{1 - f}\right)\left(1 + \frac{\rho_0}{\rho_w}\,\bar{\mu}\,\frac{f}{1 - f}\right)^{-1}$$

$$(2.46) \qquad\qquad\qquad \text{and}$$

$$\rho = \rho_w + (\rho_0 - \rho_w)\left(1 + \frac{\rho_0}{\rho_w}\,\bar{\mu}\,\frac{f}{1 - f}\right)^{-1}$$

The mean density ρ_0 of the dry matter of a sample is computed following one of the Eqs. (2.45a or b) from a measurement of the density of the sample at a specific relative humidity and the related value of the water uptake per unit mass of dry matter.

2.5. The Mean Refractive Index as a Function of the Relative Humidity

For evaluating measurements of the extinction coefficient or of the scattering function of samples of atmospheric aerosol particles suspended in the air it is necessary to know about the functional relationship between their mean refractive index and the relative humidity (Bullrich, 1964; Hänel and Bullrich, 1970). Therefore, it has been tried in an early stage to estimate the change of the real part of the mean complex index of refraction with the relative humidity (Volz, 1954).

The individual aerosol particles as well as the samples of aerosol particles consist of various solid parts, solutions, and eventually also mixed liquids. Their mean refractive index, however, can be computed with the help of a "mixture rule" only if this rule applies to each of the systems involved. In addition, this rule must also be valid for arbitrary multiphase systems. It has been found (Eiden, 1971; Hänel and Bullrich, 1970) that the computation of the scattering and absorption of electromagnetic waves by atmospheric aerosol particles must be based upon values of the refractive index with three significant figures. This corresponds to the accuracy of the usual measurements of the radiative properties of atmospheric aerosol particles. Therefore, it is not necessary to compute the mean refractive index to four or five figures. On the other hand, the Mie theory for homogeneous spheres used for model calculations yields only an approximate computation of the radiative properties of atmospheric aerosol particles, so that it is again not necessary to obtain a very exact value of the mean refractive index.

The essential mixture rules for the real refractive index or, more exactly, the real part of the complex refractive index of materials with negligible absorption have been investigated by Dieterici (1922, 1923), Heller (1945, 1965), Heller and Pugh (1957), Nakagaki and Heller (1956), and Bodmann (1969a, b). They have proved that Dale-Gladstone's mixture rule (1858, 1864) is as qualified as that established by Lorentz (1880) and Lorenz (1880) but more advantageous because of its formal simplicity. Dale-Gladstone's mixture rule can be generalized (Bodmann, 1969a, b) as follows:

$$(2.47) \qquad (n^* - 1)v = \sum_i (n_i^* - 1)v_i' \frac{m_i}{m} \qquad \text{with} \quad m = \sum_i m_i$$

where n^* is the mean real refractive index, v is the specific volume, and m is the mass of the system under consideration; n_i^*, v_i' and m_i are the real refractive index, the partial specific volumes, and the masses of the components, respectively. Since the refractive indices depend on the wavelength of electromagnetic radiation, a calculation of the real part of the mean complex refractive index by Eq. (2.47) as well as by one of the following equations must be performed separately for each wavelength.

For volume additive systems it holds

$$(2.48) \qquad mv = \sum_i m_i v_i \qquad \text{because of} \quad v_i' = v_i$$

where v_i denotes the specific volumes of the pure components. With the help of Eq. (2.48), Eq. (2.47) can be written as follows

$$(2.49) \qquad n^* = \sum_i n_i^* \frac{V_i}{V}$$

with $V = mv$ representing the total volume of the system and $V_i = m_i v_i$ representing the individual volumes of the components before they are added. The last equation is the mixing rule established by Arago and Biot (1806).

It is possible to derive equations for the computation of both the real part and the imaginary part of the mean complex index of refraction, if the complex refractive index $n_c = n - ik$ is substituted in Dale-Gladstone's

Eq. (2.47). This procedure is appropriate because the description of absorbing media is formally the same as the description of nonabsorbing media if the complex refractive index is introduced instead of the real refractive index.

Thus Eq. (2.49) is also valid for the complex refractive index, giving

$$n_c = \sum_i n_{ci} \frac{V_i}{V}$$

and separately for the real and the imaginary parts

$$n = \sum_i n_i \frac{V_i}{V} \quad \text{and} \quad k = \sum_i k_i \frac{V_i}{V}$$

where $n_c = n - ik$ is the mean complex refractive index of the sample or a single particle and $n_{ci} = n_i - ik_i$ the complex refractive indices of the pure components. n and n_i are the relevant real parts, k and k_i the appropriate imaginary parts.

By splitting the sum in the equation above for the real part of the mean complex refractive index into the contribution of condensed water and the contribution of dry matter (Hänel, 1968, 1971), the application of Eq. (2.29) yields for a sample of aerosol particles or for a single aerosol particle

$$(2.50) \qquad n = n_w + (n_0 - n_w)\left(1 + \frac{\rho_0}{\rho_w} \cdot \frac{m_w}{m_0}\right)^{-1}$$

where n and n_0 are the real parts of the complex index of refraction at the relative humidities $f \neq 0$ and $f = 0$, respectively, and n_w is the real part of the refractive index of water. For an aerosol sample, Eq. (2.50) can be transformed into

$$(2.51) \qquad n = n_w + (n_0 - n_w)\left(1 + \frac{\rho_0}{\rho_w} \bar{\mu} \frac{f}{1 - f}\right)^{-1}$$

with the help of Eq. (2.41a).

Equation (2.51) enables one to compute the real part of the mean refractive index of the dry matter of the sample from a measurement of the coefficient of mass increase and a measurement of the real part of the mean refractive index at a specific relative humidity. For this calculation the mean density of the dry matter must be known. The assumption of volume additivity yields maximum errors of 0.01 to 0.02 in n_0, the numerical values of which range between 1.5 and 1.7. An analogous consideration gives for the imaginary part of the mean complex refractive index of a sample of aerosol particles or even a single particle

$$(2.52) \qquad k = k_w + (k_0 - k_w)\left(1 + \frac{\rho_0}{\rho_w} \cdot \frac{m_w}{m_0}\right)^{-1}$$

and for the specific case of a sample

$$(2.53) \qquad k = k_w + (k_0 - k_w)\left(1 + \frac{\rho_0}{\rho_w}\, \bar{\mu}\, \frac{f}{1-f}\right)^{-1}$$

(Hänel, 1968, 1970a) where k and k_0 represent the imaginary parts of the mean complex index of refraction at the relative humidities $f \neq 0$ and $f = 0$, and k_w represents the imaginary part of the complex refractive index of water. Equation (2.53) has been verified experimentally by Fischer (1971) for the wavelengths of light between 0.4 and 1.0 μm for samples of atmospheric aerosol particles.

2.6. Conclusions and Requirements for Measuring Accuracy

The previous section dealt with equations for relative humidity, equivalent radius, mean density, and mean complex refractive index as a function of water uptake per unit mass of dry substance of an atmospheric aerosol particle. The volume, mean density, and mean refractive index of an aerosol sample could be evaluated as functions of relative humidity directly. Presupposition for these equations was thermodynamic equilibrium between the particle or the sample and the surrounding moist air.

A sample of a large number of aerosol particles can be regarded as a special case of an aerosol particle, namely, that of a very large particle. Thus it becomes possible to determine from measurements on aerosol samples all properties that are needed to calculate the equivalent radius, mean density, or mean refractive index of single aerosol particles as functions of relative humidity.

From a consolidation of the foregoing equations these properties are:

(1) the linear or the exponential mass increase coefficient as a function of water uptake per unit mass of dry substance,

(2) the mean density of the dry substance, and

(3) the mean complex refractive index of the dry substance.

It must be emphasized that from the measurements on aerosol samples the behavior of single particles cannot be reconstructed but a "mean" behavior of all particles within the sample is considered.

The following sections will report on: (1) measurement techniques with which the required data on aerosol samples can be obtained, (2) the first data sets, and (3) model calculations based on these data sets.

In this way for the first time the influence of relative humidity on some important physical parameters at atmospheric aerosol particles can be investigated with optimum reliability.

Before we proceed to sampling and measuring methods, some measuring requirements will be discussed: The measurement of the water vapor condensation on aerosol particles must meet the requirement for being used in model computations describing the formation of fog and clouds. The mean linear mass increase coefficient of an aerosol sample (Eq. 2.41a)

$$\bar{\mu} = \frac{m_{\mathrm{w}}}{m_0} \cdot \frac{1 - f}{f}$$

at $f = 0.995$ must not have an error exceeding 10 to 20%—the detailed reasoning for this requirement will be given below. Since it is easy to collect samples containing about 1 mg aerosol particle matter, even commercial microscales enable one to have weighing errors of less than 0.5% at very high relative humidities. This implies that at $f = 0.995$, the error in f must not be greater than 0.0005 to 0.001. Then, the temperatures of the aerosol particle sample and the water vapor source within a laboratory weighing system must be kept constant with an accuracy of 0.005° to 0.01°C, because it is necessary to measure their difference with an accuracy of up to 0.01° to 0.02°C.

The volume ratio of an aerosol sample

$$\frac{V}{V_0} = 1 + \frac{\rho_0}{\rho_{\mathrm{w}}} \frac{m_{\mathrm{w}}}{m_0} = 1 + \frac{\rho_0}{\rho_{\mathrm{w}}} \bar{\mu} \frac{f}{1 - f}$$

at $f = 0.995$ must not yield an error exceeding 30% in order to be used in model computations of the physical properties of clouds as well as the standard visibility in fog to determine the droplet radii with an accuracy of up to 10%. If the coefficient of mass increase has an error of 20%, the mean density ρ_0 of the dry matter of the sample must be known with an accuracy of 10%.

According to Section 2.5 an accuracy of up to 0.02 is sufficient for the real part of the mean complex refractive index of the dry matter of the sample to compute the scattering and absorption processes of the electromagnetic radiation due to aerosol particles with sufficient accuracy by means of the Mie theory.

Equations (2.46) and (2.51) imply that the errors of the mean density and the real part of the mean complex refractive index at relative humidities $f > 0$ will be less than at $f = 0$. For example, if it is assumed $\rho_0 = 2.0$ gm/cm^3 $\pm 10\%$, at $\bar{\mu} \cdot f/(1 - f) = 5$, i.e., at $f > 0.9$, the mean density will be $\rho = 1.09$ gm/cm^3 $\pm 1\%$. And for example, in case of $n_0 = 1.55 \pm 0.02$ and $\rho_0 = 2.0$ gm/cm^3 $\pm 10\%$, $\bar{\mu} \cdot f/(1 - f) = 5$ results in a real part of the mean complex refractive index $n = 1.350 \pm 0.003$.

3. SAMPLING METHOD

3.1. Description of the Jet Impactor

When air that contains airborne particles is forced through a jet directed toward a surface normal to the direction of flow, the air is more likely to flow around the obstacle than the particles; therefore, particles can be collected on a plate (Fig. 4).

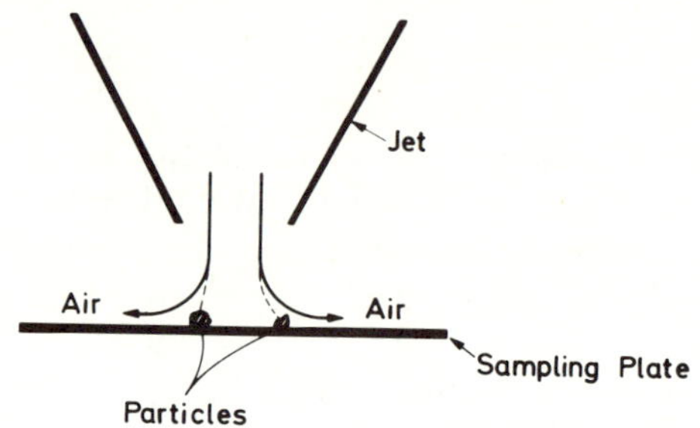

FIG. 4. Schematic diagram of a jet impactor.

Following the model of an automatic jet impactor (Hänel, 1966), two collecting devices have been built with a rate of air flow amounting to 25 m³/hr each. Both of these collecting devices contain three impactors, each of which has one glass plate, 7.5 cm in length and 2.5 cm in width;

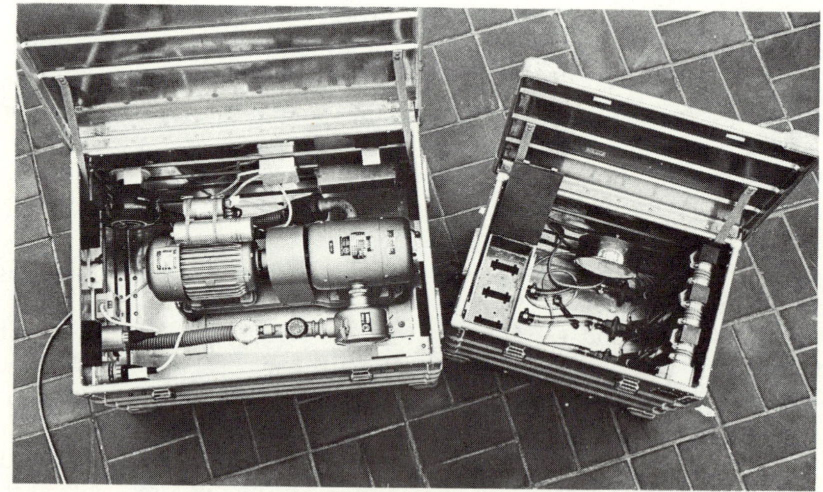

FIG. 5. Device for the sampling of atmospheric aerosol particles.

for a simultaneous uniform depositing of aerosol particles, the glass plates are reciprocated rhythmically underneath the jets, which are 7 cm in length and 0.3 mm in width.

The impactors together with the accessories are each built in two portable chests (Fig. 5). In order to guarantee an easy sampling at distance localities for covering regional differences in the physical properties of atmospheric aerosol particles, the equipment has been made so that it can be assembled and disassembled in about 2 hr and can be operated without maintenance for about half a year except for exchange of the glass plates and cleaning of the impactor nozzle.

3.2. Settling Rate and Size of the Particles under Investigation

The equations given by Hänel (1970b) have been used for computing the settling velocity and the equivalent radius of those particles within the samples under investigation which had been collected with a 50% probability; these computations have been carried out following the operational conditions of the impactor used for sampling. The computational results are listed in Table II together with the density values which the computation of the equivalent radius had been based upon. The density values which have been applied come from density measurements made with the relevant samples of aerosol particles. Particles with larger values of settling rate and equivalent radius than those compiled in Table II are collected with a probability larger than 50%.

TABLE II. Settling velocity v_{sett} and equivalent radius r of the particles collected with a 50% probability

Sampling time and location	v_{sett} (cm/sec)	r (μm)	Density (gm/cm^3)
Summer 1966, Mainz	0.0015	0.15	2.0
12–16 Jan. 1970, Mainz	0.0015	0.15	1.45
18 Jun.–27 Aug. 1970, Hohenpeissenberg	0.0011	0.2	1.75
16–25 Apr. 1969, "Meteor," Atlantic	0.0014	0.2	1.7
13–16 Apr. 1969, "Meteor," Atlantic	0.0013	0.25	1.7

4. MEASURING TECHNIQUES

4.1. Measuring the Mass as a Function of the Relative Humidity

Measurements of water vapor condensation on artificial and natural aerosol particles have already been carried out by Dalal (1947), Dessens (1947), Woodcock and Gifford (1948), and Junge (1952a, b). In particular, the latter conducted comprehensive experimental investigations into atmospheric aerosol particles with equivalent radii greater than about 1 μm.

With a microscope, Junge determined the changes of their radii with relative humidity. The particles were hung on spider threads. Because of the inaccuracy of the radius measurements and the necessity to include smaller particles in the measurements, Winkler (1969) built a mechanical microbalance, which he used for measuring the mass of water condensed on artificial and natural samples of aerosol particles as a function of the relative humidity. The reproducibility of the balance was about 0.5 to 1 μg, the relative humidity above the aerosol particle sample was known with an accuracy of about 0.01. The balance was merely an indicating system. However, a recording system is necessary for controlling thermodynamic equilibrium between the sample and the surrounding moist or dry air.

It has been found by the author that even the drying out of anhydrous salts in a vacuum has to be controlled by a recording system. The time for drying out sodium chloride and ammonium sulfate was 20 min and 1 hr, respectively, in a vacuum system. The drying out of ammonium sulfate occurred stepwise.

Winkler dried out his samples by ventilation with air that had been dried floating through phosphorus pentoxide. The time of dehumidifying was approximately 5 to 10 min. This procedure gives rise to objections: (1) The author's own experimental results were that even in a vacuum system where the drying process is more rapid by far, the drying process lasted up to 20 min or more. (2) Streaming dry air takes up water vapor from the walls of the weighing system before it reaches the sample, slowing down the drying process of the sample. (3) The water molecules from the interior of the sample must migrate to its surface by diffusion. This also prevents the drying process.

Because of the large error in relative humidity and the missing recording of the balance indication, Winkler's weighing system does not comply with the requirements that are indispensible for measurements at very high relative humidities. Moreover, his technique of measuring the mass of an aerosol sample in dry state is open to objections. Therefore, a new weighing system has been built.

4.1.1. Construction of the Measuring System with Accessories. The main part of the measuring system (Fig. 6) is an electronic microbalance

(Sartorius, Göttingen, Germany) the output of which can be recorded. The balance is connected to a water vapor source. The aerosol particle sample is kept in one of the two weighing cases of the balance at a temperature of 24°C. The relative humidity within the weighing case is given by the ratio of the water vapor pressure within the weighing case to the equilibrium water vapor pressure over a plane surface of water at the temperature of the weighing case. The water vapor pressure within the weighing case is the same as the equilibrium water vapor pressure at the water vapor source, which can be changed by alteration of the temperature of the source. Hence the relative humidity is changed by changing the temperature of the water vapor source. The temperature of the water vapor source is measured with a high precision thermometer; the temperature difference between the weighing case and the water vapor source is measured with a thermocouple. The reading of the balance and the voltage of the thermocouple are recorded simultaneously with a two-channel recording potentiometer.

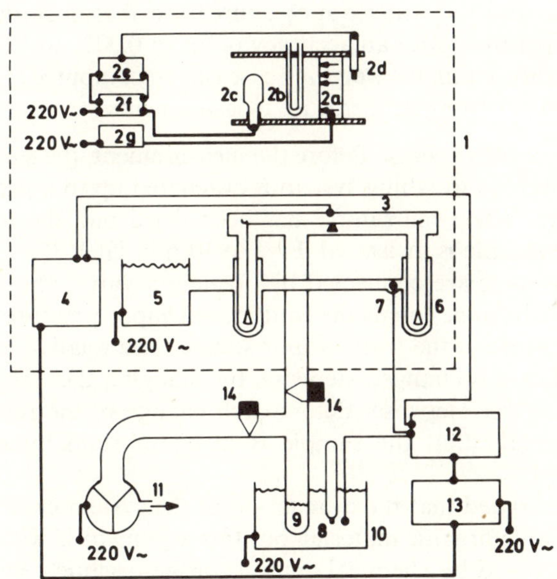

FIG. 6. Schematic diagram of the microbalance system. 1—case for the microbalance; 2—thermostat for the case; 2a—ventilator; 2b—cooling coil; 2c—bulb for heating; 2d—resistance thermometer; 2e—temperature controller; 2f—relais box for heating; 2g—speed-regulating device for ventilator; 3—electric microbalance; 4—control gear for microbalance; 5—thermostat for glass tubes; 6—glass tubes containing weighing dishes; 7—thermocouple; 8—standard thermometer; 9—water vapor source; 10—thermostat for water vapor source; 11—vacuum pump; 12—recorder; 13—voltage stabilizer; 14—valves. (———) instrument leads; (——————) 220V power leads.

The entire weighing system is fitted into a casing thermostatically kept at a temperature of 26°C. During the measurement, the temperature within the casing is monitored by means of an extra recorder; for a long run, it varies by no more than 1/10°C.

In this way, displacements of the zero mark of the balance due to temperature changes are eliminated. Furthermore, faulty adsorption or desorption of water on the walls of the weighing system due to short-term temperature variations can be avoided. The latter effects vapor pressure changes in the weighing system which at $f = 0.75$ already result in considerable errors of the measurements.

The total pressure within the weighing system is currently monitored by means of a diaphragmanometer. This manometer has an accuracy of 0.02 torr and thus enables one to estimate the relative humidity with an accuracy of up to 0.001. The weighing system is set in operation 12 to 24 hr before each measurement. Thus, the apparatus is in thermal equilibrium and the measurement is not affected by drifty voltages of the electronic component parts. The weighing is reproducible with an accuracy of 0.5 to 1 μg. The weighing case and the water vapor source are thermostatically kept at a constant temperature with an accuracy of up to 0.005° to 0.01°C. Thus, the relative humidity within the weighing case can be computed with an accuracy of 0.0005 to 0.001 from temperature measurements.

4.1.2. Measuring Process. Before the measurement, the sample is put into the balance and the weighing system is evacuated up to a pressure of about 0.1 to 0.3 torr. When the sample has been dried out, the mass of the dry matter of the sample is measured. Five to 10 min later, the balance and the water vapor source are connected by opening a valve, and then the vapor pressure in the balance equals the equilibrium vapor pressure corresponding to the temperature of the water vapor source. The relative humidity within the weighing case is changed stepwise, by changing the temperature of the water vapor source stepwise. Before each change of the relative humidity, it is made certain that the sample is in equilibrium by monitoring the balance record.

It could be argued that the evacuation of the system could lead to a loss of readily volatile organic material, but this argument is without substance. This can be proved by chemical analysis of the organic compounds within samples of aerosol particles. Ketseridis (1972) found within atmospheric aerosol particles organic substances with boiling points between 196° and 438°C. None of these materials are readily volatile and they have such low vapor pressures at 24°C that during the short time, when the mass of the dry substance is determined at the total pressure 0.1 to 0.3 torr, only a negligible amount of organic material can be lost—only 1 % of the total dry material, even in the worst case.

4.1.3. Temperature Measurement. The temperature of the water vapor source is measured with high precision thermometers, which have been aged and then tested by the German Physikalisch-Technische Bundesanstalt. These thermometers are graduated in increments of 1/50 °C and each has a zero mark. In order to account for eventual raises of the zero point, the zero marks of the thermometers are checked. Each temperature measurement includes the measurement of the mean temperature of the mercury column of the thermometer for the sake of thermometer correction (Kleinschmidt, 1935).

The temperature difference between the weighing case and the water vapor source is measured by means of a nickel chromium-constantan (NiCr-Co) thermocouple. The calibration of the thermocouple with the high precision thermometers yielded for temperature differences $0°C \leqq \Delta\vartheta \leqq 6°C$:

$$\Delta\vartheta = 15.896 \cdot \Delta U + 0.233 \cdot \Delta U^2 \pm 0.005°C, \qquad U \text{ in } mV;$$

and for $6°C \leqq \Delta\vartheta \leqq 32°C$:

$$\Delta\vartheta = 15.862 \cdot \Delta U + 0.246 \cdot \Delta U^2 \pm 0.012°C, \qquad U \text{ in } mV$$

The zero-point deviations of the system thermocouple–recorder are very small, indeed, for days. They always corresponded to a temperature difference less than 0.01°C.

4.1.4. Checking the Measuring System. In Fig. 7 the coefficients of mass increase obtained from two consecutive measurements of the same aerosol particle sample have been plotted for relative humidities increasing from $f = 0$. Differences in the measured results of 30 % occur only in the range of relative humidities less than 0.3 from inaccuracies of ± 1 μg resulting from measurements of the moisture dependence of the zero point and from inaccuracies of ± 0.5 to 1 μg resulting from the instrument error of the balance.

The measuring accuracy of the weighing system has been checked by measurements of the concentrations of hydrous solutions of NaCl and $(NH_4)_2SO_4$ at various relative humidities. The results compiled in Table III prove the reliability of the weighing system.

4.2. Determination of the Mean Density

Indirect measurements of the density of aerosol particle samples have already been carried out (Hänel, 1968). This method, however, provides density values with an accuracy of 15 to 30 % at best, however the errors in the measurements should not exceed 10 %. Therefore, it is necessary to apply another technique to the density measurements. There are already numerous pycnometric techniques that are based upon a displacement

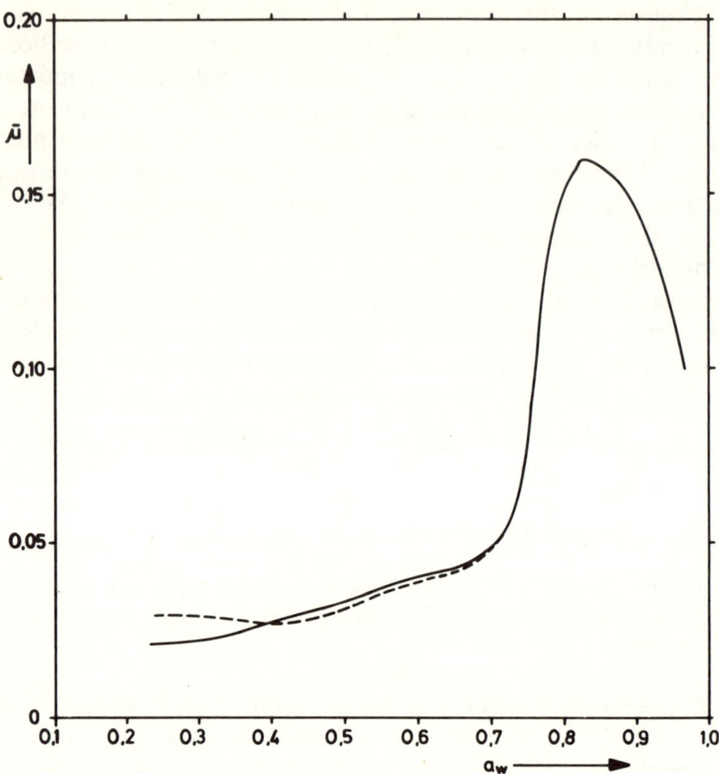

FIG. 7. Mean mass increase coefficients $\bar{\mu}$ of a sample of atmospheric aerosol particles as a function of water activity a_w for two consecutive measurements.

TABLE III. Comparison with the measurements of vapor pressure lowering over hydrous salt solutions taken by Bousfield (1923)[a] and Robinson and Stokes (1959)

		Measurements of m_w/m_0 taken by		
Substance	a_w	Author 24°C	Robinson and Stokes (1959) 25°C	Bousfield (1923) 18°C
NaCl	0.888	5.42	5.47	5.50
NaCl	0.794	3.23	3.26	3.26
$(NH_4)_2SO_4$	0.850	1.69	1.69	—

[a] *Proc. Roy. Soc., Ser. A* **103**, 429.

measurement of liquids or gases by solids. Pycnometers based on the displacement of liquids are not applicable here, because gas retained within capillaries would reduce the mean bulk density values; and since due to the mixing with the liquid the sample can be used only once, no further parameters can be measured on it.

The commercial gas pycnometers (see Müller, 1964, p. 258) yield exact measurements only if powder volumes of at least 20 to 200 cm^3 are applied. Therefore, another gas pycnometer has been built by the author which enables one to measure volumes of 20 to 30 mm^3 with an accuracy of 10%. The schematic structure of this apparatus is shown in the Fig. 8.

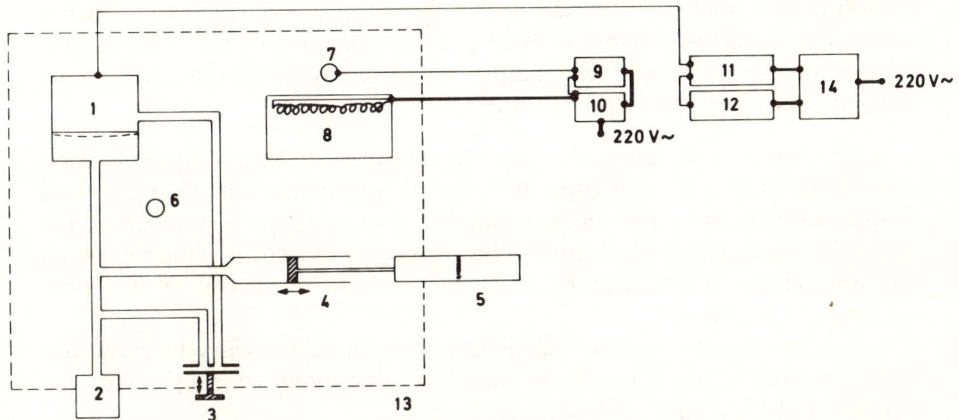

FIG. 8. Schematic diagram of the micropycnometer system. 1—sensing head of the capacitance manometer; 2—measuring chamber; 3—disc valve; 4—microliter syringe; 5—micrometer; 6—thermometer; 7—resistance thermometer; 8—ventilator with heating wire; 9—temperature controller; 10—relais box; 11—capacitance manometer; 12—recorder; 13—case for the pycnometer; 14—voltage stabilizer. (———) instrument leads; (———) 220V power leads.

4.2.1. Measuring Principle and Structure of the Microliter Gas Pycnometer. Basically, a gas pycnometer consists of a closed measuring chamber that allows well-defined changes of its volume. The pressure change that is caused by the volume change is also measured.

The sample to be measured is put into the measuring chamber after its mass has been determined. After the measuring chamber is shut, the volume is changed by moving a piston. Since the sample has displaced a specific volume of air in the measuring chamber, the partial volume of the air has been reduced. According to the Boyle-Mariotte law, the pressure change due to a specific shifting of the piston increases with an increase in the volume of the sample. Thus, the pressure change is a measure of the volume of the sample. A calibrated pycnometer enables one to compute the density from the pressure change if the mass of the sample is known.

The pressure in the gas pycnometer is measured with a capacitance manometer (Granville Philips, Boulder, Colorado). This manometer has a diaphragm that constitutes one of the two plates of a capacitor. If there is a difference in pressure on both sides of the diaphragm, it is bent toward the lower pressure side. This results in a change in capacity, which is calibrated in pressure differences. Before each measurement, equalization of pressure on both sides of the diaphragm is obtained by opening and closing a disc valve.

The volume change in the measuring chamber is made by means of a gas-proof microliter syringe, the piston of which is connected to a micrometer and is thus adjustable with an accuracy of up to 1 μm. The density measurements are made with samples of aerosol particles in equilibrium with the moist air within the laboratory.

4.2.2. Theory of the Microliter Gas Pycnometer. The following discussions are based upon the acceptance that the air represents a perfect gas and undergoes isothermal processes during the shifting of the piston. Then, the Boyle-Mariotte law is applicable. The constancy of the air temperature in the pycnometer is obtained by both the use of a thermostat and a slow shifting of the piston.

For determining a volume that is unknown, it is necessary to carry out three measurements: (1) with the empty measuring chamber, (2) with the sample, and (3) with a calibrated volume.

The three related processes are as follows: (1) Empty measuring chamber: $p_o V_o = pV$, where p_o is the pressure before the piston is shifted, V_o is the volume of the measuring chamber before the piston is shifted, with equal pressure at both sides of the diaphragm. $p = p_o + \Delta p_k + \Delta p_m$ is the pressure after the piston has been shifted, where Δp_k is the pressure change due to the movement of the piston, and Δp_m is the pressure change due to the movement of the diaphragm. $V = V_o + \Delta V_k + \Delta V_m$ is the volume after the piston has been shifted, where ΔV_k is the volume change due to the movement of the piston, and ΔV_m is the volume change due to the movement of the diaphragm. Hence

$$(4.1) \qquad V_o = -(\Delta V_k + \Delta V_m)\left(1 + \frac{p_o}{\Delta p_k + \Delta p_m}\right)$$

(2) Measuring chamber with the sample: following Eq. (4.1), with V_p representing the volume of the sample:

$$(4.2) \qquad V_o - V_p = -(\Delta V_k' + \Delta V_m')\left(1 + \frac{p_o'}{\Delta p_k' + \Delta p_m'}\right)$$

(3) Measuring chamber with the calibrated volume V_e:

$$(4.3) \qquad V_o - V_e = -(\Delta V_k'' + \Delta V_m'')\left(1 + \frac{p_o''}{\Delta p_k'' + \Delta p_m''}\right)$$

The three measurements are carried out with the same initial pressure, i.e.,

$$(4.4) \qquad p_o = p_o' = p_o''$$

and with the same pressure difference at both sides of the diaphragm after the piston has been shifted, i.e.,

$$(4.5) \qquad \Delta V_m = \Delta V_m' = \Delta V_m''$$

and

$$(4.6) \qquad \Delta p_k + \Delta p_m = \Delta p_k' + \Delta p_m' = \Delta p_k'' + \Delta p_m''$$

The cross-sectional area in the microliter syringe is F, hence:

$$(4.7) \qquad \Delta V_k = Fl, \qquad \Delta V_k' = Fl_p, \qquad \Delta V_k'' = Fl_e$$

l, l_p, and l_e are the distances the piston has been shifted during the three measurements.

Subtracting Eq. (4.1) from (4.2) and (4.3), respectively, and taking into account (4.5) and (4.6) yields the ratio

$$(4.8) \qquad \frac{V_p}{V_e} = \frac{l - l_p}{l - l_e}$$

Eq. (4.8) has the advantage that it does not involve the volume of the empty measuring chamber, the volume change due to the movement of the diaphragm, and the pressure change in the measuring chamber. Thus, these quantities need not be known, but they must be reproducible.

4.2.3. Performance of the Measurements and Calibrations. The temperature in the pycnometer is kept constant with an accuracy of $1/50\,°C$. The measurements are restricted to almost constant atmospheric pressure situations (± 0.2 to 0.5 mb). The capacitance manometer as well as a recorder for the pressure difference are connected to a voltage stabilizer. Twenty-four hours before each measurement, the measuring equipment and the thermostat are turned on to avoid temperature changes in the pycnometer and deviations of the zero marks of the measuring devices.

The calibrated volume is represented by a high-grade steel cylinder with a volume of 328.3 mm^3. An experimental check of Eq. (4.8) with spheres of known volumes yielded a mean measuring error of 2.1 mm^3.

Test measurements have been conducted with powdered quartz polluted with some iron originating from abrasion of grinding-mill material. This powder had a mean density of 2.7 gm/cm^3 determined by chemical analysis: From volumes between 24 and 29 mm^3, a density of 2.8 ± 0.3 gm/cm^3 and from a volume of 79 mm^3 a density of 2.7 ± 0.1 gm/cm^3 were measured.

From these calibrations it is clear that a microliter gas pycnometer is a useful tool in aerosol research. However there is a restriction to the applicability of the instrument described above. This restriction is the minimum volume of 20 to 30 mm^3 of a sample of aerosol particles from which a measurement can be taken to assure an error smaller than 10% in the mean density. Thus, long sampling periods are necessary to obtain enough material from the atmosphere. To overcome this deficiency, a new microliter gas pycnometer has been developed (Thudium, 1976) with a minimum volume smaller than 3 mm^3 for the aerosol sample.

4.3. Measuring the Mean Complex Index of Refraction

A method for measuring the real part of the mean complex index of refraction of aerosol particle samples at laboratory humidity has already been developed (Hänel, 1968). It is based upon measurement of the difference in the refractive index of a pure liquid and that of a mixture of this liquid with an aerosol sample. These measurements are performed with various liquids differing in the real part of the refractive index and with particles taken from one and the same sample. The real part of the refractive index of a liquid mixture that lowers the difference in the refractive index down to zero represents the real part of the mean complex index of refraction of the aerosol particle sample.

The mixtures are constituted in such a way that the ratio of the mass of the aerosol particles to the volume of the liquid is kept constant. Then, a simple formula (Hänel, 1968) can be used for interpolating the real part of the mean complex index of refraction from the measured differences in the refractive index as follows:

$$(4.9) \qquad n = n_1 + (n_1' - n_1) \frac{(n_{12} - n_1)}{(n_{12} - n_1) - (n_{12}' - n_1')}$$

where n is the real part of the mean complex index of refraction of the aerosol particle sample at the relative humidity existing in the laboratory, n_1 and n_1', respectively, are the real parts of the refractive indices of the pure liquid, and n_{12} and n_{12}', respectively, are the real parts of the refractive indices of the liquid mixtures. This technique eliminates the influence of particle size on the measured results obtained with an Abbe refractometer, which has been demonstrated by Heller and Pugh (1957) and by Nakagaki and Heller (1956).

The liquids used are kerosene and monochloride naphthalene as well as mixtures of both of them (Butler, 1933). The real parts of their refractive indices are 1.43, 1.47, 1.51, 1.59, and 1.63. These refractive indices cover the range of the real parts of the mean complex indices of refraction of aerosol particle samples.

For increasing the measuring accuracy, during the measurements the Abbe refractometer is thermostatically kept at a temperature of about 24°C with an accuracy of 1/20°C, and the measurements are performed only in monochromatic light of wavelength 0.589 μm. Test measurements with powdered quartz proved that the measuring accuracy of the real part of the mean complex index of refraction is 0.02. The wavelength dependence of the refractive index in the visible spectrum is almost of the same order of magnitude as the measuring error. Therefore, the above-described technique is not deemed appropriate for determining the wavelength dependence.

The ratio k/ρ of the imaginary part of the mean refractive index and the mean density has been measured by Fischer (1971, 1973) within an integrating sphere. These measurements have been taken at the relative humidity existing in the laboratory. Fischer's values are used for application of the author's results in computing the dependence of extinction, scattering, and absorption coefficients of atmospheric aerosol particles upon relative humidity.

5. Results of the Measurements

To determine the volume, mean density, and real part of the mean complex index of refraction of an aerosol particle sample as functions of the relative humidity, it is sufficient (according to Section 2) to measure their mass as a function of the relative humidity and to measure both their mean density as well as the real part of their mean complex index of refraction at a specific relative humidity.

The newly developed equipment, which has been described in Sections 3 and 4, has been used for carrying out the above-mentioned measurements with one and the same aerosol particle sample. Our purpose here is not the publication of large series of investigations into atmospheric aerosol particles, but the presentation of newly developed theory, sampling and measuring techniques, and application of the measured results. For this reason, only a limited number of samples have been investigated. This, however, has been an essential contribution to the advancement of knowledge of the physical properties of atmospheric aerosol particles.

Up to now, different aerosol particle samples had to be used for measuring the mass as a function of the relative humidity, on the one hand, and for measuring the mean density as well as the real part of the mean complex index of refraction at laboratory relative humidity, on the other; this has been done with samples that had been collected in Mainz in the summer of 1966 (Winkler, 1969, about 30 individual measurements after 1-hr sampling time each; Hänel, 1968, 8 individual measurements with a total sampling time of 40 days). Furthermore, the reduction in vapor pressure over hydrous sea-salt solutions has been measured (Keith and Arons, 1954) and the water vapor condensation upon seawater samples has been weighed (Winkler and Junge, 1971). The results of these measurements can be utilized, because the mean density and the real part of the mean complex index of refraction of dry sea salt can be computed from its mean chemical composition (Dietrich, 1954).

The most interesting recent values have been obtained from measurements of four aerosol particle samples collected during a winter fog situation from 12–16 January 1970 at Mainz, during the Atlantic expedition with the research vessel "Meteor" from 13–16 and 16–25 April 1969 over central Atlantic, and from 18 June–27 August 1970 on top of the Hohenpeissenberg in the Alpine foothills. The first three aerosol particle samples represent averages for specific weather situations. The last sample is to be regarded as an average for a season.

5.1. Discussion of the Results

Previous and recent measurements have been evaluated by means of the equations presented in Section 2. The results can be seen in Tables IV–VI. In Table IV the mean linear mass increase coefficient $\bar{\mu}$ is given as a function of water uptake per unit mass of dry matter m_w/m_0 and of water activity a_w for increasing and decreasing a_w. When studying these results it should be kept in mind that for samples of aerosol particles the water activity a_w equals relative humidity f according to Eq. (2.40). Table V lists the mean densities ρ_0 in dry state and Table VI, the real parts n_0 and the imaginary parts k_0 of the mean refractive indices. The results are numbered in the following way:

Model 1: Average aerosol in summer 1966 at Mainz

Model 2: Sea-spray aerosol

Model 3: Maritime aerosol over the Atlantic, 13–16 April 1969

Model 4: Maritime aerosol over the Atlantic containing Saharan dust, 16–25 April 1969

Model 5: Urban aerosol at Mainz in January 1970

Model 6: Aerosol on top of the Hohenpeissenberg, elevation 1000 m above MSL in the foothills of the Alps, in summer 1970.

The measurements of all samples have shown that the mass of the water condensed upon the aerosol particle samples at low water activities a_w is very small. Only after the values of a_w equal 0.5 to 0.7 will the samples take in greater amounts of water. This implies that the dependence of the mean density and the real part of the mean complex index of refraction

TABLE IV. Mean linear mass increase coefficient $\bar\mu$ versus water activity a_w and water uptake per unit mass of dry material m_w/m_0 for increasing and decreasing a_w

	Model 1[a]			Model 2[b]			Model 3[c]			Model 4[c]			Model 5[c]			Model 6[c]		
	a_w	$\bar\mu$	$\dfrac{m_w}{m_0}$	a_w	$\bar\mu$	$\dfrac{m_w}{m_0}$	a_w	$\bar\mu$	$\dfrac{m_w}{m_0}$	a_w	$\bar\mu$	$\dfrac{m_w}{m_0}$	a_w	$\bar\mu$	$\dfrac{m_w}{m_0}$	a_w	$\bar\mu$	$\dfrac{m_w}{m_0}$
increasing a_w	.1	.050	.006	.3	0	0	.204	.032	.008	.204	.032	.008	.195	.033	.008	.231	.025	.008
	.2	.065	.016	.35	.023	.012	.305	.026	.011	.349	.024	.013	.304	.030	.013	.301	.026	.011
	.3	.080	.034	.4	.090	.060	.399	.027	.018	.457	.021	.018	.398	.029	.018	.394	.027	.018
	.4	.095	.063	.5	.108	.108	.497	.030	.030	.590	.024	.034	.493	.030	.029	.485	.032	.030
	.5	.110	.110	.6	.136	.204	.557	.031	.039	.648	.033	.060	.556	.042	.053	.544	.037	.044
	.6	.127	.191	.65	.162	.301	.604	.042	.064	.7	.088	.206	.605	.068	.104	.590	.040	.058
	.65	.141	.262	.7	.216	.504	.653	.077	.145	.751	.130	.392	.655	.091	.174	.625	.042	.070
	.7	.160	.373	.725	.274	.722	.710	.155	.379	.789	.133	.497	.704	.109	.259	.677	.045	.094
	.75	.159	.477	.742	.854	2.450	.772	.369	1.249	.843	.120	.644	.772	.180	.607	.731	.056	.152
	.8	.156	.624	.790	.782	2.943	.834	.328	1.648	.896	.107	.922	.829	.187	.907	.768	.120	.397
	.85	.152	.861	.852	.699	4.018	.900	.280	2.520	.9	.105	.945	.886	.180	1.400	.827	.160	.765
	.9	.146	1.314	.9053	.617	5.899	.950	.243	4.617	.955	.075	1.604	.952	.171	3.392	.885	.150	1.154
	.95	.140	2.660	.9478	.548	9.950	.962	.234	5.924	.971	.072	2.411	.975	.177	6.903	.950	.105	1.995
	.975	.146	5.694	.9714	.512	17.38	.975	.225	8.775	.976	.071	2.887	.980	.180	8.820	.972	.103	3.576
	.986	.153	10.78	.9819	.490	26.56	.986	.211	14.86	.986	.067	4.719	.986	.186	13.10	.986	.115	8.099
	.99	.158	15.64	.99159	.468	54.01	.99	.206	20.39	.99	.065	6.435	.99	.192	19.01	.99	.117	11.58
	.9969	.181	58.21	.9958	.459	108.6	.9969	.207	66.57	.9969	.065	20.90	.9969	.213	68.50	.9969	.131	42.13
	1	.219	∞	1	.572	∞	1	.256	∞	1	.081	∞	1	.256	∞	1	.163	∞
decreasing a_w	.9	.146	1.314	.725	.889	2.344	.701	.428	1.003	.753	.158	.482	.860	.194	1.192	.827	.212	1.013
	.8	.161	.644	.7	.938	2.189	.599	.085	.127	.701	.175	.410	.820	.214	.975	.716	.096	.242
	.7	.178	.415	.65	.933	1.733	.516	.071	.076	.623	.072	.119	.789	.230	.860	.582	.067	.093
	.6	.199	.299	.6	.481	.722	.400	.057	.038	.506	.047	.048	.704	.174	.414	.517	.067	.072
	.5	.168	.168	.5	.301	.301	.317	.046	.021	.294	.045	.019	.597	.115	.170	.362	.069	.039
	.4	.146	.097	.4	.234	.156	.202	.056	.014	.224	.049	.014	.513	.101	.106	.264	.079	.028
	.3	.120	.051	.3	.197	.084							.401	.079	.053			
	.2	.095	.024	.2	.197	.049							.317	.068	.032			
													.200	.061	.015			

[a] Measurements from Winkler (1969).
[b] Measurements from Keith and Arons (1954) for $a_w \geqq 0.742$ and adjusted values from Winkler and Junge (1971) for lower a_w.
[c] Author's measurements.

TABLE V. Mean density ρ_0 in dry state

Model	ρ_0 (gm cm^{-3})	References, remarks
1	$3.0_3{}^a$	Hänel (1968, 1969)
2	2.25	Computed from the composition
3	2.4_5	Computed from the composition
4	2.6_4	Measurement with the microliter gas pycnometer
5	2.8_5	Measurement with the microliter gas pycnometer
6	1.8_5	Measurement with the microliter gas pycnometer

[a] The last digit is printed in a lower position because the uncertainty of ρ_0 is larger than 0.01.

upon relative humidities f at low relative humidities is very small; whereas at high relative humidities, their values approach those for water.

Furthermore, it is striking that differences occur in the values at increasing and decreasing relative humidity: During decreasing relative humidity the sample contains more water than during increasing relative humidity. This effect, called the hysteresis effect, has been described by Wall (1942), Junge (1952a), Orr *et al.* (1958), and Winkler (1969). Special experiments made by the author and considerations by Defay *et al.* (1966) imply that a preplanned antecedent change in humidity enables one to penetrate into the inner portion of the hysteresis loop. Thus, within the range of relative humidity which is subject to hysteresis effects, the physical parameters of the aerosol particles depend on the preceding humidity change, too.

The results given in Table IV prove that the hysteresis effect varies from one sample to the other. Its maximum occurs at different relative humidities. Because this hysteresis effect influences strongly the values of the physical parameters, it must not be neglected.

This implies the necessity of further measurements. A thorough investigation into the hysteresis effect is a challenge since the relative humidities to which it is related occur frequently in the atmosphere.

5.2. Discussion of the Coefficients of Mass Increase

The coefficients of mass increase at increasing values of a_w from $a_w = 0$, at a_w greater than 0.5 show a similar trend of the coefficients of mass increase for seawater aerosol particles (Model 2) and for the aerosol particle samples collected on board the research vessel "Meteor" (Models 3 and 4). However, lower relative humidities yield differences due to the presence of water-insoluble substances or additional salts as, e.g., $CaCl_2$ (Wall, 1942; Winkler, 1970).

TABLE VI. Mean complex refractive indices in dry state[a-e]

$\lambda(\mu m)$:	.3	.4	.5	.55	.6	.7	.8	1.0	References, remarks
Water									
n_w	1.358	1.343	1.336	1.334	1.332	1.330	1.328	1.324	—
k_w	0	0	0	0	0	0	0	0	—
Model 1									
n_0	1.68	1.64	1.63	1.62	1.62	1.61	1.61	1.61	Hänel (1968, 1969)
k_0	.040	.044	.049	.051	.054	.058	.063	.072	Fischer (1971)
Model 2									
n_0	1.61	1.57	1.56	1.55	1.55	1.54	1.54	1.54	Computed from chemical composition
k_0	0	0	0	0	0	0	0	0	
Model 3									
n_0	1.61	1.57	1.56	1.55	1.55	1.54	1.54	1.54	Computed from chemical composition
k_0	.089	.082	.066	.055	.047	.045	.050	.066	Fischer and Hänel (1972)
Model 4									
n_0	1.62	1.58	1.57	1.56	1.56	1.55	1.55	1.55	Fischer and Hänel (1972)
k_0	.015	.013	.012	.011	.011	.011	.012	.012	
Model 5									
n_0	1.61	1.57	1.55	1.55	1.55	1.54	1.54	1.53	—
k_0	.037	.042	.046	.048	.050	.055	.059	.068	Fischer (1973)
Model 6									
n_0	1.57	1.53	1.51	1.51	1.50	1.50	1.50	1.49	—
k_0	.012	.013	.014	.015	.015	.016	.018	.020	Fischer (1973)

[a] To make further model calculations more realistic, a mean dispersion of n_0 has been assumed for models 1, 3, 4, 5, and 6.
[b] n_w denotes the real part of the refractive index of water.
[c] k_w denotes the imaginary part of the refractive index of water.
[d] n_0 denotes the real part of the mean refractive index in dry state. The n_0 values have been measured at wavelength of light 0.589 μm.
[e] k_0 denotes the imaginary part of the mean refractive index in dry state.

The aerosol particle sample collected from 13–16 April 1969 on board the "Meteor," yielded a mass ratio m_s/m_0 of dry salt to total dry matter equal to 0.45. This value has been obtained from Eq. (2.41b) assuming idealized conditions of water uptake and that the water-soluble fraction consists of sea salt. Thus, about half of this sample consists of water-soluble matter.

The same way of estimation resulted in a ratio $m_s/m_0 = 0.14$ for a sample collected on board the "Meteor" from 16–25 April 1969. The absence of the well-marked maximum of the coefficient of mass increase expected around $a_w = 0.76$ renders this estimate debatable. This maximum marks that water activity for which a saturated solution of NaCl with plane surface is in equilibrium with the surrounding moist air. NaCl is the main constituent of sea salt. The threshold water activity is usually found near that maximum. The sample had been collected during a weather period when the influence of continental particles from Africa prevailed. It cannot be established whether the disappearance of the well-marked maximum is due to a small change in the chemical composition of the water-soluble substance or only due to the prevalence of the water-insoluble substance.

For the remaining three samples, the striking fact is the dissimilarity of the coefficient of mass increase of the samples as a function of water activity.

For application of the measured results to model calculations considering relative humidities around one, the linear mass increase coefficient must be known up to $a_w = 1$, i.e., $m_w/m_0 \to \infty$. The measurements, however, were carried out for $a_w \leqq 0.97$ only. Therefore an extrapolation to $a_w = 1$ has been undertaken with Eq. (2.42). For models 1, 5, and 6 this was carried out with the data for $(NH_4)_2SO_4$ suggesting that the water-soluble matter consists mainly of this salt. The data for sea salt were used for models 3 and 4. The errors due to the assumption of the chemical composition of the water-soluble matter is smaller than 10% for models 1, 5, and 6 and smaller than 5% for models 3 and 4. A further discussion rising from application of the tabulated data will follow in Section 6.2.2.

5.3. *Mean Densities and Real Parts of the Mean Complex Index of Refraction*

The equations given in Section 2 include the mean densities and the real parts of the mean complex index of refraction of the dry matter of aerosol particle samples. The measured values of both these parameters have been reduced to the dry state following Eqs. (2.46), (2.51), and (2.53); these reduced values are listed in Tables V and VI.

The range of mean density values in dry state is the same as that reported by Hänel (1968). Volz (1972) has reported on densities of water solubles from different rainfalls and locations ranging between 1.76 and 1.96 gm cm^{-3}. These results cannot be compared with the data presented here, because these measurements represent the water-soluble material only.

The sample collected from 13–16 April 1969 over the Atlantic was too small in mass for measuring density and refractive index. However, the value of the ratio $m_s/m_0 = 0.45$ and the knowledge of the refractive indices and densities of quartz, soot, and dry sea salt enabled the author to estimate the value of ρ_0 with a maximum error of 15% and n_0 with a maximum error of 0.03; the values of k_0/ρ_0 were measured by Fischer and Hänel (1972) directly.

Eiden (1966) has estimated the mean refractive index of airborne aerosol particles by means of model computations based upon measurements of the elliptical polarization of these particles in very hazy weather situations during winter. He obtained the mean values $n_c = 1.50 - 0.01i$ at f between 0.5 and 0.6 as well as $n_c = 1.44 - 0i$ at f between 0.8 and 0.9. The comparison of these values with the results from the sample collected from 12–16 January 1970 shows a good agreement at f between 0.5 and 0.6, whereas at f between 0.8 and 0.9 Eiden's mean real part of the refractive index is greater than the author's. The physical explanation might be the difference in the uptake of water by large aerosol particles or the presence of a great number of very small particles which at f between 0.8 and 0.9 grew only a little, thus increasing the mean refractive index that can be derived from the elliptical polarization.

5.4. *Applicability of the Results*

Within this section relevant aerosol data are compiled for six types of continental and maritime aerosol. Despite some minor assumptions, these data allow, for the first time, realistic computations on the influence of relative humidity on those physical properties of atmospheric aerosol particles, which are important to the problems mentioned in the beginning of this review.

Some results of such model calculations have been discussed earlier (Fischer and Hänel, 1972; Bullrich *et al.*, 1972; Hänel and Gravenhorst, 1974; Hänel 1969, 1970a, 1971, 1972a, b). They cover the humidity dependence of mean density, mean complex refractive index, aerosol size, aerosol extinction coefficient, and aerosol scattering function. Moreover, the sampling of aerosol particles at different relative humidities has been discussed. Grassl (1973b) has applied parts of the results for computation of the aerosol influence on radiative cooling.

In the following section the most important past model calculations are outlined and the new computational results are also given. In particular, the application of the results for cloud physics computations and new results on scattering and absorption coefficients of atmospheric aerosol particles versus relative humidity will be discussed.

6. Model Computations and Approximation Formulas Based upon Measured Properties

6.1. Basic Considerations

6.1.1. Discussion of the Model Assumptions. It was stated in the introduction that, in principle, each particle has its own characteristic of water uptake with relative humidity (cf. Meszaros, 1971) as well as its own mean density and mean complex index of refraction. This is due to the variety of chemical composition and structure of the different particles. Instead of data representing results from single particles we will use "mean" values obtained from measurements on samples of a large number of originally single particles. Consequently the first model assumption is that of the uniform chemical composition and structure of single particles in a dry state. Thus the "mean" values obtained from samples are regarded to be representative of the behavior of all these particles together, flowing separately from each other within moist air. This model assumption cannot be proved directly, since the present knowledge on atmospheric aerosol particles is still poor. The assumption can be supported by coagulation and coalescence processes occurring within the atmosphere (e.g., Walter, 1973; Whitby, 1974). Both processes have the tendency of smoothing out differences in chemical composition and structure.

Second, we are interested in the influence of relative humidity alone on the physical parameters of the particles. Thus we assume that during changes of relative humidity the particles do not coagulate and are neither added to nor removed from the sample considered. All these excluded effects occur in the atmosphere. Therefore, the results of the model calculations cannot be applied to those cases where these effects play an important role. An example is a fog situation, where the largest particles tend to be removed from the sample by gravitational settling.

A third assumption is the thermodynamic equilibrium between the particles and the surrounding moist air. There may occur rapid humidity changes due to turbulent mixing processes. Moreover, the growth of aerosol particles to fog or cloud droplets by the condensation process is initiated by thermodynamic nonequilibrium (Howell, 1949; Keith and Arons, 1954; Zebel, 1956; Mordy, 1959; Neiburger and Chien, 1960; Kornfeld, 1970; Paluch, 1971; Chien, 1971). Smaller particles can follow humidity changes more readily than larger ones. For all these reasons the following results are strictly applicable only in those situations where the humidity changes per unit time interval are so small that even the largest particles are not disadvantageously in nonequilibrium with the surrounding moist air.

In particular, the discussion of the dependence of the standard visual range upon relative humidity will show that despite objections, the foregoing assumptions can lead to reliable results.

6.1.2. The Use of the Equivalent Radius for Description of the Particle Size. It has been demonstrated in the introduction that a unique description of particle size is necessary for the solution of a variety of problems involving meteorology and aerosol measuring techniques. Therefore, the equivalent radius has been chosen. The equivalent radius is the radius of a sphere having the same volume as the particle. Since the volume is a conservative characteristic, i.e., independent of the particle's orientation in a streamline pattern or radiation field, the equivalent radius is unique.

It is evident that the equivalent radius describes the particle size very well, when the particle can be described to be spherical with a liquid shell. This is true for large relative humidities, when the importance of the atmospheric aerosol particles in meteorologic problems is most pronounced.

At the lowest relative humidities, a change of the total volume must not lead to a change in the apparent size of the particle, since condensation occurs in gaps or hidden cavities. However, the mean density, mean complex index of refraction, structure, and shape all change. The available experimental data on the humidity dependence of extinction and scattering coefficients of samples of airborne particles show that a theoretic description of these effects, based on the particle equivalent radius, gives reliable results even at low relative humidities. The same is true for a theoretic description of the settling velocity at low relative humidities, based on the equivalent radius and the dynamic shape factor. For all these reasons, the equivalent radius is the unique property for description of particle size over the whole range of relative humidity.

6.1.3. Remarks on the Applicability of the Mie Theory. For calculations of the optical properties of atmospheric aerosol particles, the theory of scattering and absorption of electromagnetic radiation by homogeneous spheres (Mie, 1908) is used. Thus, it must be regarded that the presupposition of homogeneous spheres does not match the present picture on atmospheric aerosol particles. However, measurements of the wavelength dependence of the extinction coefficient of atmospheric aerosol particles as well as measurements of its humidity dependence can be explained with Mie theory (Bullrich, 1964; Kasten, 1969; Kurtz, 1972; Hänel, 1971, 1972a, b). Moreover, Grassl (1971) could determine realistic size spectra of cloud and fog droplets from measurements of the wavelength dependence of the extinction coefficient using the Mie theory. Measurements from Quiney and Carswell (1972) of the humidity dependence of the scattering function of model aerosols give in principle the same results as Bullrich *et al.* (1972) predicted with the Mie theory. At moderate and high relative humidities Eiden (1966) has measured the elliptical polarization of light scattered by aerosol particles and could explain the results with the Mie theory. Contrary to these findings the measurements by Zerull (1973) on irregular particles indicate for dry atmospheric aerosol particles that only the extinction, the scattering, and the absorption coefficients should be calculable by the Mie theory without expecting large errors. These examples show that the Mie theory cannot be applied without critical judgment. From the foregoing discussion it seems most likely that the Mie theory should give reliable results when a liquid cover has been established on the particles.

6.2. Particle Size as a Function of Relative Humidity

The equivalent radius r as a function of relative humidity f has been computed on the basis of the values compiled in Tables IV and X combining Eqs. (2.31), (2.33), and (2.39). For these computations, the equivalent radius r_0 of the dry particle is a parameter, since all the other properties within these equations are known from experiment. In the following sections, the humidity dependence of the equivalent radius (r–f relationship) and that of the aerosol size distribution are discussed.

6.2.1. Particle Size at Relative Humidities Below 0.99. The dependence of the equivalent radius r of particles upon relative humidity f at different values of the equivalent radius r_0 in the dry state is demonstrated in Table VII for a maritime (Model 3) and a continental (Model 5) aerosol type at increasing and decreasing relative humidity. For a better understanding, in Table VII there are compiled values for the ratio r/r_0 as a function of f and r_0 indicating three main features:

(1) Below a certain relative humidity, the values of the ratio r/r_0 are independent of r_0. In this range no, or no complete, liquid cover has been established on the particle and the water uptake per unit mass of dry substance is the same for a single particle and for a sample of particles. It is stipulated that chemical composition and structure of both are identical in dry state (cf. Sections 2.2 and 2.3).

(2) For moderate and large relative humidities the ratio r/r_0 is smaller for those particles with smaller sizes in dry state, i.e., smaller values of r_0. This effect comes from the influence of the curvature correction which describes the increase of the equilibrium relative humidity over the curved

surface of a particle compared to that relative humidity over the corresponding plane surface.

This increase becomes larger when the particle size becomes smaller. Thus at a given water uptake per unit mass of dry substance the equilibrium relative humidity over a particle becomes larger when the equivalent radius r_0 of the dry particle becomes smaller. Conversely, the water uptake per unit mass of dry substance at a given relative humidity decreases with decreasing equivalent radius r_0 of the dry particle (cf. Eq. 2.33). From that the ratio r/r_0 decreases with r_0 at a given relative humidity f (cf. Eq. 2.30). This argument holds for relative humidities smaller than one; the situation for relative humidities larger than one will be discussed later.

(3) At moderate and low relative humidities the particles have taken up a smaller amount of water at increasing relative humidity compared to that at decreasing relative humidity. This hysteresis effect (cf. Section 5.1) cannot be neglected as it is demonstrated in Table VII, by the values of r/r_0 at $f = 0.7$.

Due to the complexity of the r–f relationship given by Eqs. (2.31), (2.33), and (2.39), there is a need for simple approximations. The simplest one is given by neglecting the curvature correction, as is essentially true for an infinitely large particle or a sample of particles. Thus, we assume the relative humidity f to be equal to the water activity a_w, $f = a_w$. Then, combining Eqs. (2.22) and (2.31) yields the approximation

$$(6.1) \qquad r \cong r_0\left(1 + \frac{\rho_0}{\rho_w}\mu(f)\frac{f}{1-f}\right)^{1/3}$$

This is, indeed, a formula that can be derived from the volume change with relative humidity of a sample (Hänel, 1968). From Table VII it is seen that the approximation is valid for relative humidities smaller than about 0.90 to 0.95 and for those particles with equivalent radii in dry state larger than about 0.1 μm.

For practical purposes, however, this range of applicability is too small. Therefore a better approximation is wanted, taking the curvature correction partially into account. Such a formula has been derived from Eqs. (2.31) and (2.33), which is valid for relative humidities larger than 0.70 but less than 0.99 and for equivalent radii in dry state larger than 0.04 μm:

$$r \cong r_0\left[\left(1 + \frac{\rho_0}{\rho_w}\mu(f)\frac{f}{1-f}\right)^{1/3} - \frac{2\sigma v_w}{R_w T r_0}\frac{f}{1-f}\left(1 + \frac{\rho_0}{\rho_w}\mu(f)\frac{f}{1-f}\right)^{-1/3}\right]$$

$$(6.2)$$

and assuming $f = a_w$ for the determination of $\mu(f)$ from the tables. When the surface tension is assumed to be that of pure water, i.e., $\sigma = \sigma_w$, we find

TABLE VII. Ratio r/r_0 of the particle equivalent radius at the relative humidity f to that in dry state $(f=0)$ as a function of f and r_0 for two aerosol models at the temperature 20° C and with $\gamma_\sigma=1$

Model 3: Maritime aerosol over the Atlantic 13–16 April 1969

r_0(cm):	2.69×10^{-6}	4.78×10^{-6}	8.47×10^{-6}	1.51×10^{-5}	4.78×10^{-5}	1.51×10^{-4}	∞^a [Eq. (6.1)]
inc. f^b				r/r_0			
0.2	1.007	1.007	1.007	1.007	1.007	1.007	1.007
0.4	1.014	1.014	1.014	1.014	1.014	1.014	1.014
0.6	1.047	1.047	1.047	1.047	1.047	1.047	1.047
0.7	1.131	1.162	1.180	1.192	1.204	1.207	1.208
0.8	1.603(4)^c	1.618(2)	1.627(1)	1.632(1)	1.637	1.639	1.639
0.9	1.824(4)	1.867(2)	1.893(1)	1.908(1)	1.922	1.926	1.928
0.95	2.093(5)	2.182(2)	2.236(1)	2.268(1)	2.296	2.305	2.309
0.975	2.437(7)	2.588(3)	2.687(1)	2.745(1)	2.798	2.816	2.824
0.99	2.874(8)	3.164(1)	3.366(−1)	3.494(−1)	3.636	3.685	3.708
dec. f^b							
0.2	1.011	1.011	1.011	1.011	1.011	1.011	1.011
0.4	1.030	1.030	1.030	1.030	1.030	1.030	1.030
0.6	1.056	1.079	1.092	1.100	1.100	1.100	1.100
0.7	1.449	1.479	1.494	1.501	1.507	1.509	1.510

Model 5: Urban aerosol at Mainz in January 1970

r_0(cm):	4.472×10^{-6}	8.735×10^{-6}	1.706×10^{-5}	3.332×10^{-5}	1.017×10^{-4}	3.103×10^{-4}	∞[a] [Eq. (6.1)]
inc. f				r/r_0			
0.2	1.008	1.008	1.008	1.008	1.008	1.008	1.008
0.4	1.017	1.017	1.017	1.017	1.017	1.017	1.017
0.6	1.085	1.085	1.085	1.085	1.085	1.085	1.085
0.7	1.194	1.195	1.195	1.195	1.195	1.195	1.195
0.8	1.422(2)[c]	1.439(1)	1.448(1)	1.452	1.455	1.456	1.456
0.9	1.706(3)	1.728(1)	1.748	1.756	1.767	1.770	1.772
0.95	1.963	2.059	2.109	2.134	2.152	2.158	2.161
0.975	2.415(1)	2.568	2.652	2.696	2.727	2.737	2.742
0.99	3.095(−3)	3.409(−3)	3.584(−2)	3.688(−1)	3.765	3.791	3.804
dec. f							
0.2	1.014	1.014	1.014	1.014	1.014	1.014	1.014
0.4	1.048	1.048	1.048	1.048	1.048	1.048	1.048
0.6	1.142	1.143	1.143	1.144	1.144	1.144	1.144
0.7	1.255	1.271	1.280	1.284	1.287	1.288	1.288
0.8	1.505	1.516	1.521	1.524	1.525	1.526	1.526

[a] The values for $r_0 = \infty$ are valid for samples of a large number of particles.
[b] inc. = increasing, dec. = decreasing.
[c] The percentage error (in parentheses) is given when Eq. (6.2) is used.

$2\sigma_w v_w R_w^{-1} T^{-1} = 1.056 \times 10^{-7}$ cm at 25°C. Values of the percentage errors of Eq. (6.2) are compiled in Table VII. They show that the r–f relationship can be approximated by (6.2) with $\sigma = \sigma_w$ when $f \leq 0.99$ and $r_0 \geq 0.04$ μm.

Another simple r–f relationship of importance for a quantitative visibility forecast (Kasten, 1969; Hänel, 1971, 1972a) has been given by Kasten (1969):

$$(6.3) \qquad r \cong r_0(1 - f)^{-\varepsilon}$$

Generally ε is a function of f and r_0, but to make this equation useful for some rough approximation ε can be regarded as a constant for each type of aerosol. These constant ε values are 0.25, 0.25, 0.255, 0.18, 0.285, and 0.20 for the models 1 to 6, respectively, (Hänel, 1972a). Using these values, Eq. (6.3) is restricted to relative humidities above 0.4, 0.75, 0.75, 0.75, 0.75, and 0.8, respectively, for models 1 to 6, and to those below 0.95. Moreover, only particles with equivalent radii in dry state larger than 0.1 μm can be treated.

In this section the dependence of the particle equivalent radius upon relative humidity has been treated for particles with equivalent radii in dry state larger than 0.04 μm and relative humidities smaller than 0.99. This range of applicability covers all particle sizes of immediate meteorologic interest and the relative humidities for which the assumption of thermodynamic equilibrium between the particles and the surrounding moist air is appropriate for particles flowing within the atmosphere.

6.2.2. Particle Size at Relative Humidities around One. At first some approximate formulas and, after that, the results of model calculations are discussed. In contrast to the formulas given above, the following only allow the computation of relative humidity from particle size.

In deriving approximate formulas, we use the exponential mass increase coefficient instead of the linear one. The following equations give the relative humidity f, the equivalent radius r, and an approximation for the exponential mass increase coefficient η as functions of the mass ratio m_0/m_w, i.e., the inverse water uptake per unit mass of dry substance. The first of these equations is given by introduction of the first of Eqs. (2.23) and (2.32) into Eq. (2.4):

$$(6.4) \quad f = \exp\left(\frac{2v_w}{R_w T r_0} \frac{\gamma_\sigma\left[\sigma_w(T_0) + a(T_0 - T) + b\dfrac{m_s}{m_0} \cdot \dfrac{m_0}{m_w}\right]}{[1 + (\rho_0/\rho_w)/(m_0/m_w)]^{1/3}} - \eta\frac{m_0}{m_w}\right)$$

Equation (2.31) shows that

$$r = r_0\left(1 + \frac{\rho_0}{\rho_w}\bigg/\frac{m_0}{m_w}\right)^{1/3}$$

An approximation for the exponential mass increase coefficient is

$$(6.5) \qquad \eta \cong \eta^0 - A\left(\frac{m_0}{m_w}\right)^{1/2} + B\left(\frac{m_0}{m_w}\right)^{C} + D\left(\frac{m_0}{m_w}\right)^{E}$$

In Eq. (6.5) η^0 is the limiting value for the exponential mass increase coefficient at infinite dilution ($m_0/m_w \to 0$) and A, B, C, D, and E are constants. The value of A can be computed from the theory of interionic interaction (cf., e.g., Harned and Owen, 1958). By experiment, this theory has been found to be reasonable (e.g., Kortüm, 1966, p. 167). In Tables VIII and IX approximations analogous to Eq. (6.5) are given: (a) for the mean exponential mass increase coefficient of atmospheric aerosol particles (models 1 to 6) and (b) for the exponential mass increase coefficient of pure salts [NaCl and $(NH_4)_2SO_4$] and for mixtures of pure salts with insoluble material, supposing idealized conditions of water uptake. Together with the formulas, the numerical values of the constants are given.

TABLE VIII. Constants for approximate calculation of the mean exponential mass increase coefficient for atmospheric aerosols after

$$\bar{\eta} \cong \bar{\eta}^0 - \bar{A}\left(\frac{m_0}{m_w}\right)^{1/2} + \bar{B}\left(\frac{m_0}{m_w}\right)^{\bar{C}} + \bar{D}\left(\frac{m_0}{m_w}\right)^{\bar{E}}$$

deduced from the experimental results given in Table IV

Model	$\bar{\eta}^0$	$\bar{A}^a$	$\bar{B}$	$\bar{C}$	$\bar{D}$	$\bar{E}$	Range of validity $m_0/m_w \leqq$	Maximum error (%)
1	0.219	2.520	2.433	0.519	0	0	2.1	5.6
2	0.572	5.996	6.604	0.552	0	0	0.4	4.6
3	0.256	1.810	2.067	0.589	−0.185	1.80	0.8	5.6
4	0.081	0.314	0.328	0.604	−0.006	1.45	0.4 (2.0)[b]	2.4 (7.3)[b]
5	0.265	3.654	3.569	0.524	0	0	0.7	2.5
6	0.163	1.173	1.110	0.539	0.064	3.50	0.9	3.3

[a] $\bar{A}$ is calculated from the theory of interionic interaction.

[b] In parentheses are given another range of validity and the corresponding maximum error.

TABLE IX. Constants for approximate calculation of the exponential mass increase coefficient of pure salts after

$$\eta_s \cong \eta_s^0 - A_s\left(\frac{m_s}{m_w}\right)^{1/2} + B_s\left(\frac{m_s}{m_w}\right)^{C_s} + D_s\left(\frac{m_s}{m_w}\right)^{E_s}$$

deduced from the values given by Robinson and Stokes (1959). For the conditions of idealized water uptake the exponential mass increase coefficients for mixtures of pure salts with insoluble material is

$$\eta^{id} \cong \eta^{id,\,0} - A^{id}\left(\frac{m_0}{m_w}\right)^{1/2} + B^{id}\left(\frac{m_0}{m_w}\right)^{C_s} + D^{id}\left(\frac{m_0}{m_w}\right)^{E_s}$$

$$\eta^{id,\,0} = \eta_s^0\frac{m_s}{m_0} \qquad A^{id} = A_s\left(\frac{m_s}{m_0}\right)^{3/2} \qquad B^{id} = B_s\left(\frac{m_s}{m_0}\right)^{C_s+1} \qquad D^{id} = D_s\left(\frac{m_s}{m_0}\right)^{D_s+1}$$

Substance	η_s^0	$A_s{}^a$	B_s	C_s	D_s	E_s	Range of validity $m_s/m_w \leqq$	Maximum error (%)
NaCl	0.616	5.40	5.88	0.539	0	0	$0.35\ (0.32)^b$	$2.1\ (2.8)^b$
$(NH_4)_2SO_4$	0.409	7.10	7.50	0.540	−0.590	1.00	0.40	2.1

^{a} A_s is calculated from the theory of interionic interaction.
^{b} In parentheses are given another range of validity and the corresponding maximum error.

In the literature (e.g., Junge, 1963; Dufour and Defay, 1963; Defay et al., 1966; Fletcher, 1962; Mason, 1963) approximate equations of the type

$$(6.6) \qquad f \cong 1 + \frac{\alpha}{r} - \frac{\beta}{(r/r_0)^3 - 1}$$

are given. Often $(r/r_0)^3 - 1$ is approximated by $(r/r_0)^3$. α and β are regarded as constants. Indeed, we come to Eq. (6.6), combining Eqs. (6.4), (2.31), and (6.5) and linearizing the exponential function. This procedure yields

$$\alpha = \frac{2v_w}{R_w T}\gamma_\sigma\left[\sigma_w(T_0) + a(T_0 - T) + b\frac{m_s}{m_0}\frac{\rho_0}{\rho_w}\left\{\left(\frac{r}{r_0}\right)^3 - 1\right\}^{-1}\right]$$

and

$$\beta = \eta\frac{\rho_0}{\rho_w} \cong \frac{\rho_0}{\rho_w}\Bigg\{\eta^0 - A\left(\frac{\rho_0}{\rho_w}\right)^{-1/2}\left[\left(\frac{r}{r_0}\right)^3 - 1\right]^{-1/2}$$

$$(6.7)$$

$$+ B\left(\frac{\rho_0}{\rho_w}\right)^C\left[\left(\frac{r}{r_0}\right)^3 - 1\right]^{-C} + D\left(\frac{\rho_0}{\rho_w}\right)^E\left[\left(\frac{r}{r_0}\right)^3 - 1\right]^{-E}\Bigg\}$$

Thus rigorous derivation indicates that α and β should not be regarded as constants. This is shown in Fig. 9c where the assumption of a constant mass increase coefficient leads to remarkably higher equivalent radii than can be calculated without this assumption, especially in the case of ammonium sulfate. For interpolation purposes however, Eq. (6.6) with constant α and β can be used leading to reliable results.

Due to the linearization of the exponential function, Eq. (6.6) together with Eq. (6.7) gives reliable results only when $|f - 1| \leq 0.05$, since the error is of the order of $\frac{1}{2}(f - 1)^2$. With these approximations the particle size at relative humidities near one can be calculated with reasonable accuracy even for small particles. The treatment of the latter is important for the study of the liquid water content in fog and in clouds (cf. Mordy, 1959). This discussion of errors is important because an error of 25% of the particle radius causes an error of 95% of its liquid water content.

Computations of the r–f relationship have been performed for the six types of atmospheric aerosol particles as well as for some hypothetical aerosols consisting of pure NaCl, pure $(NH_4)_2SO_4$, and mixtures of each of these salts with SiO_2. For these latter mixtures the concept of idealized water uptake [Eqs. (2.18) and (2.20)] has been used. The computations were done for the temperature 20°C. For pure NaCl only a second temperature, 0°C, was chosen. In each case the correction γ_σ due to the presence of surface-active substance was set equal to one.

In Fig. 9a–d, the equilibrium relative humidity f is given as a function of the particle equivalent radius r. The value of the equivalent radius in dry state is $r_0 = 0.1$ μm. The curves show that the relative humidity increases with particle equivalent radius until it reaches a maximum value. A further increase of the equivalent radius causes a decrease of relative humidity. At infinite particle size the relative humidity approaches the value one.

The maximum relative humidity of the f–r curve is called the critical relative humidity f_c, and the expression $f_c - 1$ is the critical supersaturation. At the critical relative humidity, the particle attains the so-called critical radius r_c. Values for the critical supersaturation $f_c - 1$ and the ratio r_c/r_0 of the critical radius to the equivalent radius in dry state are given in Table X.

The critical relative humidity f_c or the critical supersaturation $f_c - 1$ is of importance for cloud physics considerations. This will be demonstrated now.

A particle will grow due to water vapor condensation if the water vapor density ρ_a in the ambient air is larger than the equilibrium water vapor density ρ_r over the particle's curved surface. It will shrink due to evaporation of water when ρ_a is smaller than ρ_r. The water vapor density ρ is related to the water vapor pressure p by $\rho = pR_w^{-1}T^{-1}$. Thus, if there is no temperature difference between the particle and the surrounding moist air,

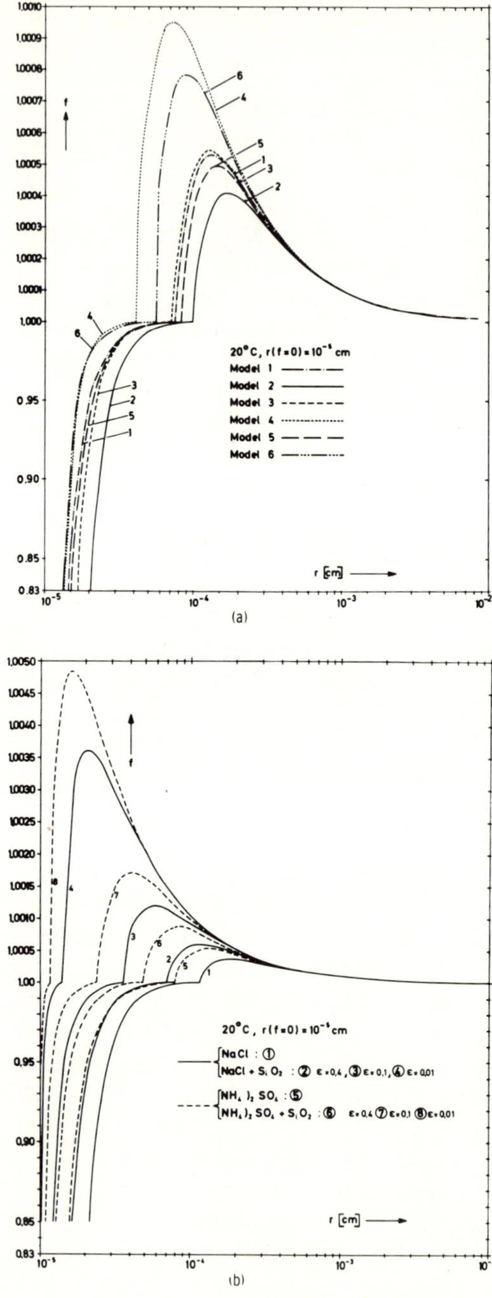

FIG. 9a and b. See facing page for legend.

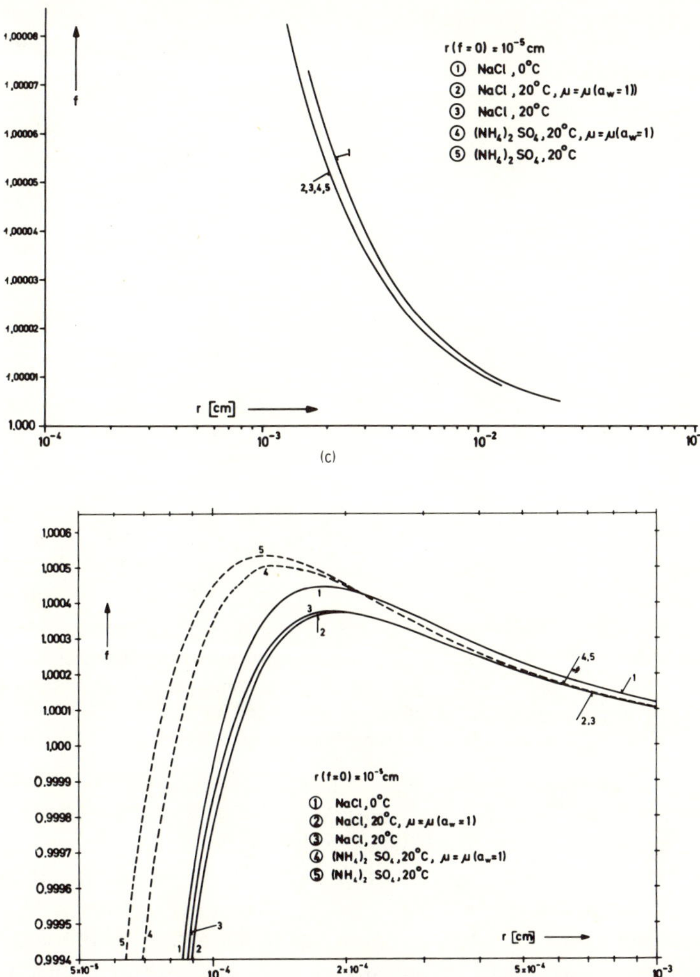

FIG. 9. Particle equivalent radius r versus relative humidity f for atmospheric aerosol particles models 1 to 6 (a). For hypothetical aerosols and pure NaCl and $(NH_4)_2SO_4$ aerosols at 0°C and 20°C (b–d).

The curves with the supplement $\mu = \mu(a_w = 1)$ indicate a constant value of the linear mass increase coefficient, namely, that at infinite dilution. No correction for the surface tension due to surface-active materials or films has been made ($\gamma_\sigma = 1$).

TABLE X. Critical supersaturation $f_c - 1$ and critical equivalent radius r_c over equivalent radius r_0
equivalent radius r_0 in dry state together

Model	1	2	3	4	5	6	NaCl	NaCl	NaCl $\mu_s = \mu_s^0$
t (°C):	20	20	20	20	20	20	0	20	20
b (dynes/cm):	16.4	27.6	27.6	27.6	16.4	16.4	27.6	27.6	27.6
ρ_0 (gm/cm³):	3.03	2.25	2.45	2.64	2.85	1.85	2.153	2.153	2.153
m_s/m_0:	.5	1	.45	.14	.64	.3	1	1	1
r_0 (cm)					$f_c - 1^b$				
10^{-6}	1.99	1.34	1.94	3.21	1.86	2.95	1.46	1.24	1.19
10^{-5}	5.31	4.09	5.48	9.53	4.96	7.87	4.46	3.77	3.72
10^{-4}	1.67	1.20	1.71	2.94	1.56	2.47	1.39	1.18	1.18
10^{-3}	5.27	3.79	5.41	9.27	4.93	7.81	4.40	3.72	3.72
					r_c/r_0				
10^{-6}	3.47	5.43	3.79	2.42	3.70	2.38	5.49	5.81	6.05
10^{-5}	13.3	16.6	13.0	7.26	14.4	9.04	17.8	18.9	19.2
10^{-4}	43.0	59.4	41.8	24.0	45.9	29.0	57.4	60.7	60.9
10^{-3}	136	190	132	77.3	145	91.7	182	192.4	192.5

[a] The quantity ε is the ratio of the volumes of pure dry salt to that of total dry matter, μ_s^0 is the linear mass increase coefficient at infinite dilution. The f_c and r_c values have been obtained with a double precision computer program based on a method of interval-partition with a mathematical accuracy of one part in a thousand. For the computations, the correction γ_σ has been assumed to be 1.

a particle will grow when the water vapor pressure p_a of the ambient air is larger than the equilibrium water vapor pressure p_r over the particle's curved surface. The particle will shrink when p_a is smaller than p_r.

From the definition of relative humidity (2.3), we obtain the critical equilibrium water vapor pressure $p_c = f_c p_w$, when the particle is in thermodynamic equilibrium with the surrounding moist air. p_w is the equilibrium water vapor pressure over a plane surface of water saturated with air. Thus a particle will grow infinitely, when $p_a > p_c$. If $p_a \leq p_w < p_c$ the particle will grow or shrink, until p_r is equal to p_a. Then it has reached a stable size. When $p_w < p_a < p_c$ the isobar $p_a = $ const intersects the curve for the equilibrium water vapor pressure p_r of the particle at two points (Fig. 10) marking two different equivalent radii r_e and r_a. There are now two possibilities. The particle will grow infinitely when its equivalent radius is larger than r_a, since p_a remains larger than p_r. If the particle equivalent radius is smaller than r_a, it will grow or shrink until it reaches the stable size r_e.

It is evident from the curves in Fig. 9 and the values given in Table X that the critical relative humidity f_c and therefore the critical supersaturation in dry state for the aerosol models 1 to 6 and some hypothetical model aerosols as a function of with the constants necessary for Eq. (2.33)[a]

(NH$_4$)$_2$SO$_4$	(NH$_4$)$_2$SO$_4$ $\mu_s = \mu_s^0$	NaCl + SiO$_2$ $\varepsilon = 0.4$	$\varepsilon = 0.1$	$\varepsilon = 0.01$	(NH$_4$)$_2$SO$_4$ + SiO$_2$ $\varepsilon = 0.4$	$\varepsilon = 0.1$	$\varepsilon = 0.01$
20	20	20	20	20	20	20	20
16.4	16.4	27.6	27.6	27.6	16.4	16.4	16.4
1.769	1.769	2.451	2.600	2.645	2.298	2.562	2.641
1	1	.351	.0828	.00814	.308	.0691	.00670
			$f_{\mathrm{c}} - 1$				
1.91	1.61	1.96	3.77	7.68	3.03	5.38	8.81×10^{-2}
5.32	5.04	5.98	12.0	36.2	8.49	17.2	48.5×10^{-4}
1.61	1.59	1.86	3.73	11.8	2.56	5.14	16.4×10^{-5}
5.07	5.04	5.89	11.8	37.2	8.06	15.9	50.6×10^{-7}
			r_{c}/r_0				
3.67	4.49	3.71	2.04	1.23	2.36	1.52	1.13
13.2	14.2	11.9	5.93	2.08	8.30	4.09	1.62
44.2	44.9	38.4	19.2	6.06	27.8	13.8	4.32
142	142	122	60.9	19.2	89.9	44.9	14.1

[b] The values of $f_{\mathrm{c}} - 1$ for each line are given in the order of magnitude indicated in the last column.

$f_{\mathrm{c}} - 1$, too, are highly dependent on the chemical composition and structure of the particles considered. The critical supersaturation is larger for those particles with smaller values of exponential mass increase coefficient. Thus, it increases with an increasing amount of insoluble material within the particle. The effect of the chemical composition of the water-soluble material on the critical supersaturation can be of the same order as a decrease of the volume fraction of insoluble material from 1 to 0.4. Junge and McLaren (1970) have performed calculations of critical supersaturation and critical radius, but they did not use measured aerosol data and their study is mainly related to the influence of the volume fraction of insoluble material.

At large equivalent radii compared to the critical radius, the differences between the curves become negligible. In this region the equilibrium relative humidity over the curved surface of a particle is governed by the curvature correction term. This is demonstrated in Fig. 9c. Thus the assumption of independence of the equilibrium relative humidity from chemical composition and structure of the dry particle can be justified only for large equivalent radii compared to the critical radius. It is not true to

state this as a generality for relative humidities around one, as is often found
in the literature.

From the computational results it can be seen that the chemical
composition and structure of the particles have an influence on the relation between equilibrium relative humidity and particle size, unless the
particle equivalent radius does not considerably exceed the critical radius.
Thus chemical composition and structure influence the condensational
growth to cloud droplets, especially when the deviation from thermodynamic equilibrium is small.

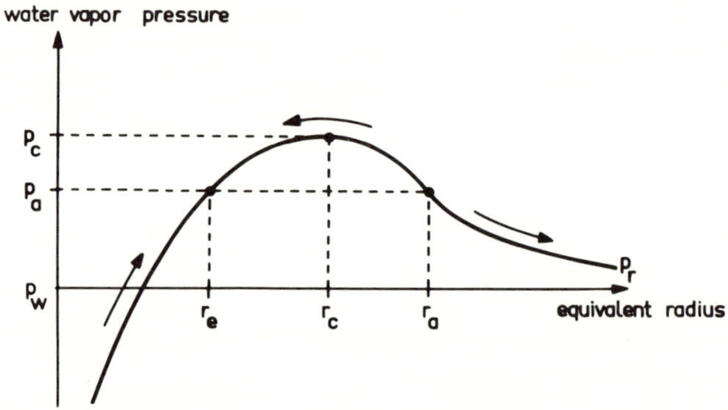

FIG. 10. Schematic representation of particle growth and shrinking due to differences between
the equilibrium water vapor pressure p_r over the particle's surface and the water vapor pressure
p_a of the ambient air.

6.2.3. The Influence of Surface Tension on the Equilibrium Particle Size.
When the surface tension is larger than zero, the equilibrium water vapor
pressure over the liquid cover of a particle is larger than that over a plane
surface with the same physicochemical properties. Thus the surface tension
of the particle's liquid cover at the liquid-air interphase is an important
property in thermodynamic considerations on atmospheric aerosol particles
when relative humidities around unity are considered. It must be known to
what amount the surface tension deviates from that of pure water.

In the foregoing discussions (Section 2) we mentioned briefly the influence of dissolved electrolytes, surface-active materials, and films on the
surface tension. Some additional details need to be discussed now. At
large concentrations the surface tension of an electrolyte solution is larger
than that of water, and at very small concentrations it goes a little below
that of pure water [to a maximum of 0.02%, after the measurements by
Jones and Ray (1941)]. The surface tension of the solution approaches

that of pure water at infinite dilution. However, for all practical and theoretic purposes involving atmospheric aerosol particles, it can be assumed that an electrolyte solution has a larger surface tension than pure water. This effect has been treated in a simple manner by Eq. (2.24). Surface-active materials and films have a large influence on the value of the surface tension even if they are present in small amounts (Osipov, 1962; Kortüm, 1963; Gaines, 1966; Defay *et al.*, 1966). These materials have been found on the sea surface by Jarvis *et al.* (1967) and Garrett (1967a, b). Because the sea surface is a source of aerosol particles, surface-active materials should be found on marine aerosols too. This is, indeed, true as has been shown by Goetz and Pueschel (1965), Blanchard (1968), and Barger and Garrett (1970). From the comprehensive measurements of organic compounds within atmospheric aerosol particles from different sites by Ketseridis (1972), it must be concluded that surface-active material and films should be expected for each aerosol type all over the world.

The presence of organic materials generally should cause a more or less pronounced decrease of surface tension to a value below that of pure water. Recent unpublished measurements of the author on the surface tension of rainwater samples and on highly diluted aerosol samples in water, both taken at Mainz, Germany, showed such a decrease, usually amounting from 10 to 30%.

For these reasons, the effect of a deviation of the value of the surface tension from that of pure water on the equilibrium equivalent radius of a particle has to be discussed. This can be done with the help of the Figs. 9c and 9d, where the relative humidity f is plotted against the particle equivalent radius r. These f–r curves at thermodynamic equilibrium are given for NaCl solution droplets 0°C and 20°C. The curve for 20°C can be obtained also from that of 0°C, when the surface tension at 0°C is lowered by a factor 0.9. Such a lowering of the surface tension leads to the following effects at relative humidities near unity:

(1) A decrease in critical supersaturation and an increase in critical radius (cf. Table X).

(2) When the particle equivalent radius is smaller than the critical radius, an increase of the particle radius between about 5 to 25%.

(3) When the particle radius is larger than the critical radius, a decrease of the particle radius of about 25%.

Thus the mass of water taken up by the particle is increased nearly twofold or reduced about one-half due to a decrease of surface tension by 10%. This indicates that the results presented up to this point, as far as they are concerned with relative humidities larger than about 0.995, have to be corrected in the future. In particular, the calculations of critical supersaturations and critical radii are open to objections.

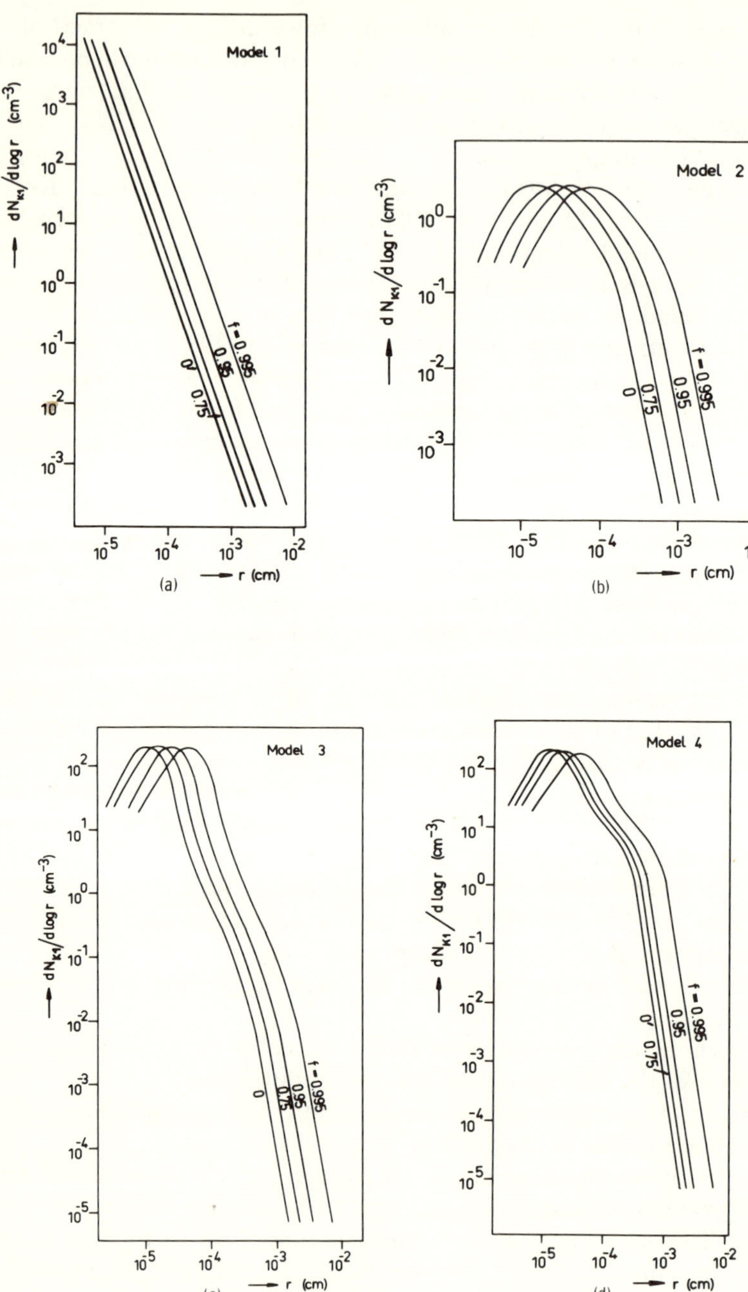

FIG. 11a–d. See facing page for legend.

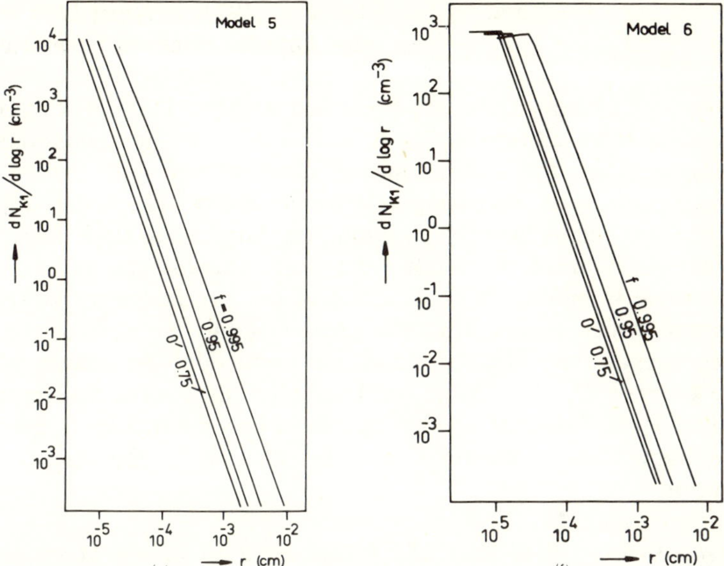

FIG. 11. Aerosol size distribution $dN_{K1}/d \log r$ versus relative humidity f for the aerosol models 1 to 6. N_{K1} is the particle number per cubic centimeter, r the equivalent radius of the particles. The curves are valid for increasing relative humidity. Originally published by Hänel (1972b).

Realistic cloud physics calculations cannot be performed until the surface tension of atmospheric aerosol particles has been investigated thoroughly. In this context, it must be determined how the surface-active materials are scattered on the particles of different sizes. In addition the effect of water-soluble, surface-active materials and that of films on the growth and shrinking processes of atmospheric aerosol particles under realistic conditions must be considered.

6.3. Particle Size Frequency Distribution as a Function of Relative Humidity

The following computations of the aerosol size frequency distribution versus relative humidity are based on measured size distributions. These have been combined with the results from the measurements of the mean density and those of water uptake versus relative humidity.

For models 1, 5, and 6 the usual power-law size frequency distributions (Bullrich, 1964) have been used. These distributions are means for continental aerosols in central Europe. From Junge (1963) the size distribution for model 2 has been derived. The size frequency distributions for models 3 and 4 have been measured by Jaenicke *et al.* (1971) in April 1969 over the Atlantic at the same time the pertinent samples of aerosol particles were taken.

The computational results are presented in Figs. 11a–f. They are valid for increasing relative humidity. As it is expected from the discussions in the foregoing sections, the size frequency distributions shift toward larger equivalent radii with increasing relative humidity. The most important radius increase occurs for relative humidities larger than 0.95. They do not vary their shape until the relative humidity exceeds the value 0.95. At larger relative humidities a broadening of the size frequency distribution is observed, which comes from the greater radius increase of the larger particles due to their smaller curvature correction. To ensure constant particle number N_{K1} per cubic centimeter (cf. the model assumptions in Section 6.1.1) the values of $dN_{K1}/d \log r$ must become smaller at the highest relative humidities, when there is a broadening of the size frequency distribution.

6.4. Mean Density, Real Part and Imaginary Part of the Mean Complex Refractive Index versus Relative Humidity

The relationship between the relative humidity and the mean density as well as the real part and imaginary part n and k of the mean complex refractive index has been computed for single particles and samples of aerosol particles. For this formulas (2.45b), (2.50), and (2.52) have been used. They read

$$\rho = \rho_{\mathrm{w}} + (\rho_0 - \rho_{\mathrm{w}})\left(1 + \frac{\rho_0}{\rho_{\mathrm{w}}}\frac{m_{\mathrm{w}}}{m_0}\right)^{-1}$$

$$n = n_{\mathrm{w}} + (n_0 - n_{\mathrm{w}})\left(1 + \frac{\rho_0}{\rho_{\mathrm{w}}}\frac{m_{\mathrm{w}}}{m_0}\right)^{-1} \tag{6.8}$$

$$k = k_{\mathrm{w}} + (k_0 - k_{\mathrm{w}})\left(1 + \frac{\rho_0}{\rho_{\mathrm{w}}}\frac{m_{\mathrm{w}}}{m_0}\right)^{-1}$$

The subscript "w" denotes pure water and "0" the dry substance. m_{w}/m_0 is the water uptake per unit mass of dry substance. From Eqs. (6.8) it is seen that the amounts of the differences $\rho - \rho_{\mathrm{w}}$, $n - n_{\mathrm{w}}$, and $k - k_{\mathrm{w}}$ decrease with increasing water uptake m_{w}/m_0, i.e., increasing relative humidity. This means that the values of ρ, n, and k tend toward those of pure water. Since all three formulas are mathematically similar, the main features of the humidity dependence of the three properties considered are the same.

In Table XI the mean density ρ, the real part n, and the imaginary part k of the mean refractive index are given as functions of the relative humidity f. The calculations are performed for single particles as well as for samples of particles. Two examples, a maritime (model 3) and an urban (model 5) aerosol, have been chosen. From the values given in Table XI, the following conclusions can be drawn:

(1) There are only small changes of ρ, n, and k when f is smaller than 0.6. The largest changes occur within the humidity range $0.6 \leqq f \leqq 0.95$.

(2) The differences between increasing and decreasing relative humidity generally cannot be neglected.

(3) In the region of relative humidity where the curvature correction has an influence, the values of ρ, n, and k are larger, the smaller the particles are in dry state. However, the values for the single particles do not differ remarkably from those for the sample. Thus, the formulas for ρ, n, and k as functions of relative humidity f for samples (Eqs. 2.46, 2.51, and 2.53) can be used when the errors, which can be derived from Table XI, are tolerable for the specific problem considered. The relevant equations read

$$\rho = \rho_{\rm w} + (\rho_0 - \rho_{\rm w})\left(1 + \frac{\rho_0}{\rho_{\rm w}}\,\bar{\mu}\,\frac{f}{1-f}\right)^{-1}$$

(6.9)
$$n = n_{\rm w} + (n_0 + n_{\rm w})\left(1 + \frac{\rho_0}{\rho_{\rm w}}\,\bar{\mu}\,\frac{f}{1-f}\right)^{-1}$$

$$k = k_{\rm w} + (k_0 - k_{\rm w})\left(1 + \frac{\rho_0}{\rho_{\rm w}}\,\bar{\mu}\,\frac{f}{1-f}\right)^{-1}$$

When $f > 1$ it can be assumed $\rho = \rho_{\rm w}$, $n = n_{\rm w}$, and $k = k_{\rm w}$.

(4) ρ, n, and k decrease with relative humidity f. This is not the case generally. At large wavelengths of light (cf. Table XV) $k_0 - k_{\rm w}$ can be negative. Then k increases with relative humidity f.

The values of the mean density, the real part and the imaginary part of the mean complex refractive index as functions of relative humidity given in Table XI are quoted as an illustration. They show the main features of the trend of these properties in relation to relative humidity, which is not different for samples and for single particles.

6.5. Extinction, Scattering, and Absorption Coefficients of Atmospheric Aerosol Particles as Functions of Relative Humidity

The aerosol particles within the atmosphere scatter and absorb the incident radiation. Thus they influence visibility (Middleton, 1952) and the heat budget of the atmosphere (e.g., Grassl, 1973b, 1974). For a quantitative description of these phenomena, the extinction, the scattering, and the absorption coefficients of the aerosol particles within the atmosphere must be known. They will be calculated as functions of relative humidity for

TABLE XI. Particle mean density ρ, real part n, and imaginary part k of the mean complex refractive index at the wavelength of light 0.55 μm as functions of the relative humidity f and of the equivalent radius r_0 of the dry particle. The values for $r_0 = \infty$ are valid for samples of a large number of aerosol particles[a]

Model 3: Maritime aerosol over the Atlantic 13–16 April 1969

r_0 (cm)	2.69×10^{-6}	4.78×10^{-6}	8.47×10^{-6}	1.51×10^{-5}	4.78×10^{-5}	1.51×10^{-4}	∞
inc. f				ρ (gm/cm^3)			
0.2	2.420	2.420	2.420	2.420	2.420	2.420	2.420
0.4	2.391	2.391	2.391	2.391	2.391	2.391	2.391
0.6	2.263	2.263	2.263	2.263	2.263	2.263	2.263
0.7	2.002	1.924	1.883	1.856	1.831	1.825	1.823
0.8	1.352	1.342	1.337	1.334	1.331	1.329	1.329
0.9	1.239	1.223	1.214	1.209	1.204	1.203	1.202
0.95	1.158	1.140	1.130	1.124	1.120	1.118	1.118
0.975	1.100	1.084	1.075	1.070	1.066	1.065	1.064
0.99	1.077	1.048	1.038	1.033	1.030	1.030	1.028
dec. f							
0.2	2.403	2.403	2.403	2.403	2.403	2.403	2.403
0.4	2.327	2.327	2.327	2.327	2.327	2.327	2.327
0.6	2.231	2.154	2.114	2.089	2.089	2.089	2.089
0.7	1.477	1.448	1.444	1.429	1.424	1.422	1.421

Model 5: Urban aerosol at Mainz in January 1970

r_0 (cm)	4.472×10^{-6}	8.735×10^{-6}	1.706×10^{-5}	3.332×10^{-5}	1.017×10^{-4}	3.103×10^4	∞
inc. f				ρ (gm/cm^3)			
0.2	2.806	2.806	2.806	2.806	2.806	2.806	2.806
0.4	2.759	2.759	2.759	2.759	2.759	2.759	2.759
0.6	2.448	2.448	2.448	2.448	2.448	2.448	2.448
0.7	2.087	2.084	2.084	2.084	2.084	2.084	2.084
0.8	1.643	1.621	1.609	1.604	1.601	1.599	1.599
0.9	1.418	1.401	1.392	1.391	1.391	1.390	1.390
0.95	1.245	1.212	1.197	1.190	1.186	1.184	1.183
0.975	1.131	1.109	1.099	1.094	1.091	1.090	1.089
0.99	1.057	1.043	1.038	1.036	1.034	1.034	1.034
dec. f							
0.2	2.774	2.774	2.774	2.774	2.774	2.774	2.774
0.4	2.607	2.607	2.607	2.607	2.607	2.607	2.607
0.6	2.142	2.239	2.239	2.236	2.236	2.236	2.236
0.7	1.936	1.901	1.882	1.874	1.868	1.866	1.866
0.8	1.543	1.531	1.525	1.523	1.522	1.521	1.521

continued

Table XI—*Continued*

Model 3: Maritime aerosol over the Atlantic 13–16 April 1969

r_0(cm):	2.69×10^{-6}	4.78×10^{-6}	8.47×10^{-6}	1.51×10^{-5}	4.78×10^{-5}	1.51×10^{-4}	∞
inc. f				n			
0.2	1.546	1.546	1.546	1.546	1.546	1.546	1.546
0.4	1.541	1.541	1.541	1.541	1.541	1.541	1.541
0.6	1.522	1.522	1.522	1.522	1.522	1.522	1.522
0.7	1.483	1.472	1.465	1.462	1.458	1.457	1.457
0.8	1.386	1.385	1.384	1.384	1.383	1.383	1.383
0.9	1.370	1.367	1.366	1.365	1.364	1.364	1.364
0.95	1.358	1.355	1.353	1.353	1.352	1.352	1.352
0.975	1.349	1.346	1.345	1.344	1.344	1.344	1.344
0.99	1.345	1.341	1.340	1.339	1.339	1.339	1.338
dec. f							
0.2	1.543	1.543	1.543	1.543	1.543	1.543	1.543
0.4	1.531	1.531	1.531	1.531	1.531	1.531	1.531
0.6	1.517	1.506	1.500	1.496	1.496	1.496	1.496
0.7	1.405	1.401	1.400	1.398	1.397	1.397	1.397

Model 5: Urban aerosol at Mainz in January 1970

r_0(cm):	4.472×10^{-6}	8.735×10^{-6}	1.706×10^{-5}	3.332×10^{-5}	1.017×10^{-4}	3.103×10^{-4}	∞
inc. f				n			
0.2	1.545	1.545	1.545	1.545	1.545	1.545	1.545
0.4	1.539	1.539	1.539	1.539	1.539	1.539	1.539
0.6	1.503	1.503	1.503	1.503	1.503	1.503	1.503
0.7	1.461	1.461	1.461	1.461	1.461	1.461	1.461
0.8	1.409	1.406	1.405	1.405	1.404	1.404	1.404
0.9	1.383	1.381	1.380	1.380	1.380	1.380	1.380
0.95	1.363	1.359	1.357	1.356	1.356	1.355	1.355
0.975	1.349	1.347	1.346	1.345	1.345	1.345	1.344
0.99	1.341	1.339	1.338	1.338	1.338	1.338	1.338
dec. f							
0.2	1.541	1.541	1.541	1.541	1.541	1.541	1.541
0.4	1.522	1.522	1.522	1.522	1.522	1.522	1.522
0.6	1.479	1.479	1.479	1.478	1.478	1.478	1.478
0.7	1.443	1.439	1.437	1.436	1.435	1.435	1.435
0.8	1.397	1.396	1.395	1.395	1.395	1.395	1.395

continued

Table XI—*Continued*

Model 3: Maritime aerosol over the Atlantic 13–16 April 1969

r_0(cm)	2.69×10^{-6}	4.78×10^{-6}	8.47×10^{-6}	1.51×10^{-5}	4.78×10^{-5}	1.51×10^{-4}	∞
inc. f				k			
0.2	0.054	0.054	0.054	0.054	0.054	0.054	0.054
0.4	0.053	0.053	0.053	0.053	0.053	0.053	0.053
0.6	0.048	0.048	0.048	0.048	0.048	0.048	0.048
0.7	0.038	0.035	0.033	0.032	0.032	0.031	0.031
0.8	0.013	0.013	0.013	0.013	0.013	0.012	0.012
0.9	0.009	0.008	0.008	0.008	0.008	0.008	0.008
0.95	0.006	0.005	0.005	0.005	0.005	0.004	0.004
0.975	0.004	0.003	0.003	0.003	0.003	0.002	0.002
0.99	0.003	0.002	0.001	0.001	0.001	0.001	0.001
dec. f							
0.2	0.053	0.053	0.053	0.053	0.053	0.053	0.053
0.4	0.050	0.050	0.050	0.050	0.050	0.050	0.050
0.6	0.047	0.044	0.042	0.041	0.041	0.041	0.041
0.7	0.018	0.017	0.017	0.016	0.016	0.016	0.016

Model 5: Urban aerosol at Mainz in January 1970

r_0 (cm):	4.472×10^{-6}	8.735×10^{-6}	1.706×10^{-5}	3.332×10^{-5}	1.017×10^{-4}	3.103×10^{-4}	∞
inc. f				k			
0.2	0.047	0.047	0.047	0.047	0.047	0.047	0.047
0.4	0.046	0.046	0.046	0.046	0.046	0.046	0.046
0.6	0.038	0.038	0.038	0.038	0.038	0.038	0.038
0.7	0.028	0.028	0.028	0.028	0.028	0.028	0.028
0.8	0.017	0.016	0.016	0.016	0.016	0.016	0.016
0.9	0.011	0.010	0.010	0.010	0.010	0.010	0.010
0.95	0.006	0.005	0.005	0.005	0.005	0.005	0.005
0.975	0.003	0.003	0.003	0.002	0.002	0.002	0.002
0.99	0.001	0.001	0.001	0.001	0.001	0.001	0.001
dec. f							
0.2	0.046	0.046	0.046	0.046	0.046	0.046	0.046
0.4	0.042	0.042	0.042	0.042	0.042	0.042	0.042
0.6	0.032	0.032	0.032	0.032	0.032	0.032	0.032
0.7	0.024	0.023	0.023	0.023	0.023	0.022	0.022
0.8	0.014	0.014	0.014	0.014	0.014	0.014	0.014

[a] inc. = increasing, dec. = decreasing.

several wavelengths of electromagnetic radiation based on the experimental data for the aerosol models 1 to 6 presented previously (Section 5.1).

6.5.1. Basic Theory. The change that the irradiance of a plane parallel monochromatic beam of electromagnetic radiation undergoes along its path x is expressed by Lambert's law

$$(6.10) \qquad I = I_{o} \exp[-(\sigma_E + \sigma_{EM})x]$$

I_{o} denotes the irradiance of the incident radiation, I the irradiance of the radiation after its passage along the path x. σ_E denotes the extinction coefficient of the aerosol particles and σ_{EM} the extinction coefficient of the air molecules. They have the dimension one over unit length.

The extinction coefficient σ_E of a population of N_K particles within a volume V_K is the sum of the scattering and the absorption coefficients σ_S and σ_A of the particles, i.e.,

$$(6.11) \qquad \sigma_E = \sigma_S + \sigma_A$$

An analogous formula is true for the extinction coefficient of the air molecules. Thus Lambert's law (6.10) in a differential form states that the extincted radiation is the sum of the scattered and the absorbed radiations.

The extinction, the scattering, and the absorption coefficients of the clean air will not be considered in this context. It should only be mentioned that from the results of Owens (1967) it can be derived that the influence of relative humidity on the extinction coefficient σ_{EM} of the air molecules is smaller than 1 to 2 % in the wavelength region 0.3 to 1 μm of electromagnetic radiation. Thus, this influence is negligible for all practical visibility considerations, as will be demonstrated later (Section 6.6.1).

The extinction, the scattering, and the absorption coefficients of aerosol particles are properties related to a unit volume. They each can be represented in the same manner, namely, as a sum over the products of the geometric cross sections πr_j^2 of the equivalent spheres with the pertinent dimensionless efficiency factors κ_j of all N_K particles in the volume V_K of the air considered, divided by this volume:

$$(6.12) \qquad \sigma = \frac{1}{V_K} \sum_{j=1}^{j=N_K} \pi r_j^2 \kappa_j$$

Thus σ is a property related to the unit volume. In definition (6.12) multiple scattering is not considered. If σ and κ_j have the subscripts "E," "S," and "A," Eq. (6.12) is the formulation for the extinction, the scattering, and the absorption coefficients, respectively.

The efficiency factors of extinction κ_{Ej}, of scattering κ_{Sj}, and of absorption κ_{Aj} are functions of the particle equivalent radius r_j, the wavelength

λ of the electromagnetic radiation, and the particle mean complex refractive index $n_j(\lambda) - ik_j(\lambda)$.

The shape of the particle as well as its chemical composition and its structure and, moreover, its orientation in the radiation field influence the values of the efficiency factors. The relation between the efficiency factors is the same as the one between the related coefficients

$$(6.13) \qquad \kappa_{Ej} = \kappa_{Sj} + \kappa_{Aj}$$

The following computations of the extinction, the scattering, and the absorption coefficients versus wavelength of electromagnetic radiation and relative humidity are based upon Eq. (6.12).

For the computation of the equivalent radius and mean complex refractive index versus relative humidity, the equations derived in Section 2, including the curvature correction, have been used. In the computer program each particle was allowed its own refractive index as a function of wavelength of electromagnetic radiation and relative humidity. The efficiency factors are calculated from Mie's theory (1908) which is valid for homogeneous spheres. This latter assumption involved in the Mie theory has been discussed earlier in Section 6.1.3.

Together with the coefficients σ_E, σ_S, and σ_A we have calculated, for comparative purposes, the total geometric cross section, Q_K, of the equivalent spheres of the particles per unit volume. This property is defined as the sum of the geometric cross sections of the equivalent spheres of the N_K particles within the volume V_K of the air divided by V_K:

$$(6.14) \qquad Q_K = \frac{1}{V_K} \sum_{j=1}^{j=N_K} \pi r_j^2$$

Thus Q_K is the total geometric cross section per unit volume of air, and its dimension is the same as that of σ_E, σ_S, and σ_A. The pertinent number of particles per unit volume is

$$(6.15) \qquad N_{K1} = \frac{N_K}{V_K}$$

6.5.2. Computational Results for the Region 0.3 to 1 µm of the Wavelength of Electromagnetic Radiation. The extinction coefficient σ_E, the absorption coefficient σ_A, and the total geometric cross section Q_K per unit volume (1 cm^3) of collectives of atmospheric aerosol particles have been computed for models 1 to 6 for relative humidities between 0 and 0.995 and the wavelengths of light 0.3 µm, 0.55 µm, and 1 µm. The results are compiled in Tables XII and XIII, and in Fig. 12.

TABLE XII. Standardized total geometric cross section Q_{K0}/N_{K1}, standardized extinction coefficient σ_{E0}/N_{K1}, and standardized absorption coefficient σ_{A0}/N_{K1} versus wavelength λ of electromagnetic radiation for models 1 to 6[a]

Model	Q_{K0}/N_{K1}	$\lambda = 0.3\ \mu m$		$\lambda = 0.55\ \mu m$		$\lambda = 1.0\ \mu m$	
		σ_{E0}/N_{K1}	σ_{A0}/N_{K1}	σ_{E0}/N_{K1}	σ_{A0}/N_{K1}	σ_{E0}/N_{K1}	σ_{A0}/N_{K1}
1	1.53×10^{-10}	2.95×10^{-10}	6.23×10^{-11}	1.60×10^{-10}	4.38×10^{-11}	9.24×10^{-11}	3.27×10^{-11}
2	4.10×10^{-9}	9.59×10^{-9}	0	1.05×10^{-8}	0	1.03×10^{-8}	0
3	7.54×10^{-11}	2.09×10^{-9}	7.78×10^{-10}	1.71×10^{-9}	4.44×10^{-10}	9.79×10^{-10}	3.20×10^{-10}
4	4.03×10^{-9}	9.42×10^{-9}	2.91×10^{-9}	9.71×10^{-9}	1.94×10^{-9}	9.06×10^{-9}	1.51×10^{-9}
5	1.53×10^{-10}	2.72×10^{-10}	5.70×10^{-11}	1.46×10^{-10}	4.02×10^{-11}	8.44×10^{-11}	3.09×10^{-11}
6	3.99×10^{-10}	1.04×10^{-9}	1.09×10^{-10}	6.21×10^{-10}	7.63×10^{-11}	3.49×10^{-10}	5.67×10^{-11}

[a] All values in cm^2.

TABLE XIII. Ratios of the total geometric cross sections, the extinction coefficients, and the absorption coefficients of aerosol particles at the relative humidity f and in dry state as a function of relative humidity f for models 1 to 6[a,b]

Model 1 | Model 2

		$\lambda = 0.3\ \mu m$		$\lambda = 0.55\ \mu m$		$\lambda = 1.0\ \mu m$				$\lambda = 0.3\ \mu m$	$\lambda = 0.55\ \mu m$	$\lambda = 1.0\ \mu m$
	$\dfrac{Q_K(f)}{Q_{K0}}$	$\dfrac{\sigma_E(f)}{\sigma_{E0}}$	$\dfrac{\sigma_A(f)}{\sigma_{A0}}$	$\dfrac{\sigma_E(f)}{\sigma_{E0}}$	$\dfrac{\sigma_A(f)}{\sigma_{A0}}$	$\dfrac{\sigma_E(f)}{\sigma_{E0}}$	$\dfrac{\sigma_A(f)}{\sigma_{A0}}$		$\dfrac{Q_K(f)}{Q_{K0}}$	$\dfrac{\sigma_E(f)}{\sigma_{E0}}$	$\dfrac{\sigma_E(f)}{\sigma_{E0}}$	$\dfrac{\sigma_E(f)}{\sigma_{E0}}$
inc. f								inc. f				
0.2	1.03	1.03	1.01	1.03	1.01	1.03	1.01	0.35	1.02	1.06	1.00	1.03
0.4	1.13	1.12	1.04	1.12	1.04	1.11	1.03	0.5	1.16	1.17	1.14	1.13
0.6	1.36	1.37	1.08	1.34	1.07	1.31	1.09	0.6	1.29	1.31	1.26	1.32
0.7	1.63	1.64	1.12	1.64	1.13	1.61	1.16	0.7	1.64	1.67	1.60	1.61
0.75	1.79	1.80	1.14	1.81	1.16	1.78	1.19	0.75	3.54	3.45	3.42	3.54
0.8	1.99	2.02	1.17	2.07	1.20	2.00	1.22	0.8	3.97	3.62	3.75	3.99
0.85	2.29	2.42	1.23	2.44	1.23	2.37	1.26	0.85	4.61	4.38	4.23	4.69
0.9	2.79	3.04	1.27	3.13	1.29	3.06	1.34	0.9	5.70	5.52	5.29	5.65
0.95	3.95	4.61	1.36	5.04	1.41	4.96	1.44	0.95	8.27	7.71	7.04	7.81
0.975	6.00	7.64	1.49	8.89	1.55	9.03	1.58	0.975	12.3	11.0	11.2	12.1
0.99	10.5	14.5	1.62	18.9	1.68	20.6	1.75	0.99	20.9	19.2	18.1	19.5
0.995	15.4	21.7	1.74	31.3	1.81	36.5	1.85	0.995	32.9	25.9	28.2	28.4
dec. f								dec. f				
0.2	1.05	1.05	1.02	1.05	1.02	1.04	1.01	0.2	1.07	1.07	1.10	1.06
0.4	1.19	1.19	1.06	1.19	1.05	1.16	1.05	0.4	1.22	1.25	1.20	1.25
0.6	1.54	1.54	1.10	1.53	1.11	1.49	1.13	0.5	1.41	1.42	1.45	1.43
0.65	1.60	1.60	1.11	1.59	1.12	1.56	1.15	0.6	1.88	1.89	1.88	1.83
0.7	1.70	1.72	1.14	1.71	1.14	1.68	1.18	0.65	2.87	2.85	2.80	2.74
0.8	2.08	2.14	1.19	2.18	1.21	2.10	1.23	0.7	3.27	3.32	3.14	3.14

continued

Model 3 | | | | | | | Model 4

| | $\dfrac{Q_K(f)}{Q_{K0}}$ | $\lambda = 0.3$ μm | | $\lambda = 0.55$ μm | | $\lambda = 1.0$ μm | | | $\dfrac{Q_K(f)}{Q_{K0}}$ | $\lambda = 0.3$ μm | | $\lambda = 0.55$ μm | | $\lambda = 1.0$ μm | |
		$\dfrac{\sigma_E(f)}{\sigma_{E0}}$	$\dfrac{\sigma_A(f)}{\sigma_{A0}}$	$\dfrac{\sigma_E(f)}{\sigma_{E0}}$	$\dfrac{\sigma_A(f)}{\sigma_{A0}}$	$\dfrac{\sigma_E(f)}{\sigma_{E0}}$	$\dfrac{\sigma_A(f)}{\sigma_{A0}}$			$\dfrac{\sigma_E(f)}{\sigma_{E0}}$	$\dfrac{\sigma_A(f)}{\sigma_{A0}}$	$\dfrac{\sigma_E(f)}{\sigma_{E0}}$	$\dfrac{\sigma_A(f)}{\sigma_{A0}}$	$\dfrac{\sigma_E(f)}{\sigma_{E0}}$	$\dfrac{\sigma_A(f)}{\sigma_{A0}}$
inc. f								inc. f							
0.2	1.01	1.01	1.01	1.01	1.01	1.01	1.00	0.2	1.01	1.01	1.01	1.01	1.01	1.02	1.01
0.4	1.03	1.03	1.01	1.03	1.01	1.03	1.01	0.4	1.03	1.02	1.01	1.02	1.02	1.03	1.03
0.6	1.10	1.09	1.04	1.10	1.04	1.10	1.04	0.6	1.05	1.05	1.03	1.05	1.03	1.06	1.04
0.65	1.22	1.21	1.09	1.24	1.08	1.23	1.08	0.65	1.11	1.11	1.06	1.11	1.06	1.09	1.01
0.7	1.43	1.40	1.17	1.47	1.12	1.47	1.12	0.7	1.33	1.31	1.18	1.35	1.13	1.34	1.10
0.75	2.19	2.12	1.42	2.31	1.26	2.37	1.26	0.75	1.59	1.56	1.28	1.59	1.16	1.57	1.08
0.8	2.67	2.57	1.54	2.88	1.32	3.00	1.33	0.8	1.77	1.77	1.35	1.72	1.17	1.73	1.07
0.85	2.99	2.88	1.58	3.32	1.39	3.47	1.36	0.85	1.98	1.95	1.41	1.93	1.20	1.94	1.12
0.9	3.66	3.50	1.70	4.15	1.43	4.53	1.42	0.9	2.29	2.28	1.48	2.20	1.24	2.20	1.12
0.95	5.18	5.23	1.95	6.23	1.54	7.42	1.53	0.95	2.79	2.82	1.60	2.76	1.27	2.66	1.09
0.975	7.63	6.88	2.15	9.24	1.64	12.9	1.60	0.975	4.03	3.94	1.77	4.15	1.32	4.46	1.16
0.99	12.6	9.55	2.42	14.2	1.77	23.9	1.69	0.99	6.57	6.10	1.96	6.19	1.44	6.90	1.20
0.995	18.6	15.3	2.52	22.4	1.84	38.9	1.74	0.995	9.95	9.10	2.15	8.81	1.41	9.91	1.14
dec. f								dec. f							
0.2	1.02	1.02	1.01	1.02	1.01	1.02	1.01	0.2	1.02	1.02	1.01	1.02	1.01	1.03	1.03
0.4	1.06	1.06	1.03	1.06	1.03	1.06	1.02	0.4	1.05	1.05	1.03	1.05	1.03	1.06	1.04
0.5	1.11	1.11	1.04	1.12	1.05	1.11	1.04	0.5	1.08	1.08	1.05	1.07	1.03	1.07	1.00
0.6	1.21	1.20	1.08	1.23	1.08	1.22	1.08	0.6	1.13	1.13	1.07	1.14	1.08	1.12	1.04
0.65	1.80	1.76	1.32	1.83	1.17	1.88	1.18	0.65	1.38	1.35	1.19	1.38	1.11	1.37	1.05
0.7	2.26	2.19	1.43	2.39	1.26	2.45	1.27	0.7	1.62	1.60	1.30	1.62	1.18	1.62	1.11
0.75	2.43	2.32	1.45	2.60	1.29	2.66	1.28	0.75	1.72	1.71	1.34	1.69	1.19	1.71	1.11

Model 5		$\lambda = 0.3\ \mu m$		$\lambda = 0.55\ \mu m$		$\lambda = 1.0\ \mu m$		Model 6		$\lambda = 0.3\ \mu m$		$\lambda = 0.55\ \mu m$		$\lambda = 1.0\ \mu m$	
	$\dfrac{Q_K(f)}{Q_{K0}}$	$\dfrac{\sigma_E(f)}{\sigma_{E0}}$	$\dfrac{\sigma_A(f)}{\sigma_{A0}}$	$\dfrac{\sigma_E(f)}{\sigma_{E0}}$	$\dfrac{\sigma_A(f)}{\sigma_{A0}}$	$\dfrac{\sigma_E(f)}{\sigma_{E0}}$	$\dfrac{\sigma_A(f)}{\sigma_{A0}}$		$\dfrac{Q_K(f)}{Q_{K0}}$	$\dfrac{\sigma_E(f)}{\sigma_{E0}}$	$\dfrac{\sigma_A(f)}{\sigma_{A0}}$	$\dfrac{\sigma_E(f)}{\sigma_{E0}}$	$\dfrac{\sigma_A(f)}{\sigma_{A0}}$	$\dfrac{\sigma_E(f)}{\sigma_{E0}}$	$\dfrac{\sigma_A(f)}{\sigma_{A0}}$
inc. f								inc. f							
0.2	1.02	1.01	1.00	1.01	1.01	1.01	1.00	0.2	1.01	1.01	1.00	1.01	1.00	1.01	1.00
0.4	1.04	1.03	1.01	1.03	1.01	1.03	1.01	0.4	1.02	1.02	1.01	1.02	1.00	1.03	1.01
0.5	1.06	1.05	1.01	1.06	1.02	1.05	1.02	0.5	1.04	1.04	1.02	1.04	1.01	1.05	1.01
0.6	1.18	1.18	1.05	1.19	1.06	1.17	1.05	0.6	1.07	1.07	1.01	1.07	1.01	1.09	1.02
0.65	1.29	1.31	1.08	1.31	1.09	1.28	1.08	0.65	1.09	1.08	1.00	1.09	1.02	1.11	1.03
0.7	1.43	1.46	1.11	1.45	1.10	1.41	1.11	0.7	1.10	1.09	1.00	1.10	1.02	1.12	1.03
0.75	1.75	1.83	1.16	1.84	1.17	1.80	1.19	0.75	1.30	1.29	1.05	1.34	1.07	1.33	1.05
0.8	2.07	2.20	1.20	2.27	1.23	2.19	1.23	0.8	1.62	1.64	1.07	1.72	1.10	1.72	1.09
0.85	2.40	2.67	1.25	2.72	1.27	2.61	1.28	0.85	1.88	1.91	1.10	2.05	1.13	2.10	1.13
0.9	2.85	3.24	1.29	3.30	1.31	3.14	1.33	0.9	2.08	2.10	1.12	2.30	1.15	2.34	1.13
0.95	4.21	5.34	1.42	5.90	1.45	5.84	1.48	0.95	2.66	2.78	1.17	3.21	1.19	3.33	1.19
0.975	6.54	9.13	1.54	10.8	1.58	11.0	1.62	0.975	3.80	4.03	1.25	5.04	1.26	5.54	1.28
0.99	11.5	17.7	1.74	23.3	1.74	25.5	1.78	0.99	6.72	6.94	1.32	10.3	1.35	12.3	1.34
0.995	17.0	25.2	1.80	38.6	1.84	44.8	1.85	0.995	9.94	10.0	1.39	16.0	1.38	21.1	1.43
dec. f								dec. f							
0.2	1.03	1.03	1.01	1.03	1.01	1.03	1.01	0.2	1.03	1.03	1.02	1.03	1.01	1.04	1.01
0.4	1.10	1.09	1.02	1.10	1.03	1.09	1.03	0.4	1.05	1.05	1.02	1.05	1.01	1.06	1.02
0.6	1.31	1.32	1.09	1.33	1.09	1.29	1.08	0.6	1.10	1.08	1.00	1.10	1.02	1.11	1.03
0.7	1.61	1.67	1.13	1.69	1.15	1.64	1.15	0.65	1.11	1.10	1.01	1.12	1.02	1.13	1.03
0.75	1.99	2.11	1.19	2.17	1.22	2.10	1 22	0.7	1.21	1.20	1.03	1.24	1.05	1.25	1.03
0.8	2.29	2.51	1.24	2.56	1.25	2.46	1.26	0.8	1.87	1.90	1.11	2.03	1.12	2.08	1.13
0.85	2.56	2.86	1.22	2.95	1.29	2.82	1.30	0.85	2.02	2.04	1.12	2.21	1.13	2.26	1.13

[a] λ = wavelength of electromagnetic radiation.
[b] inc. = increasing, dec. = decreasing.

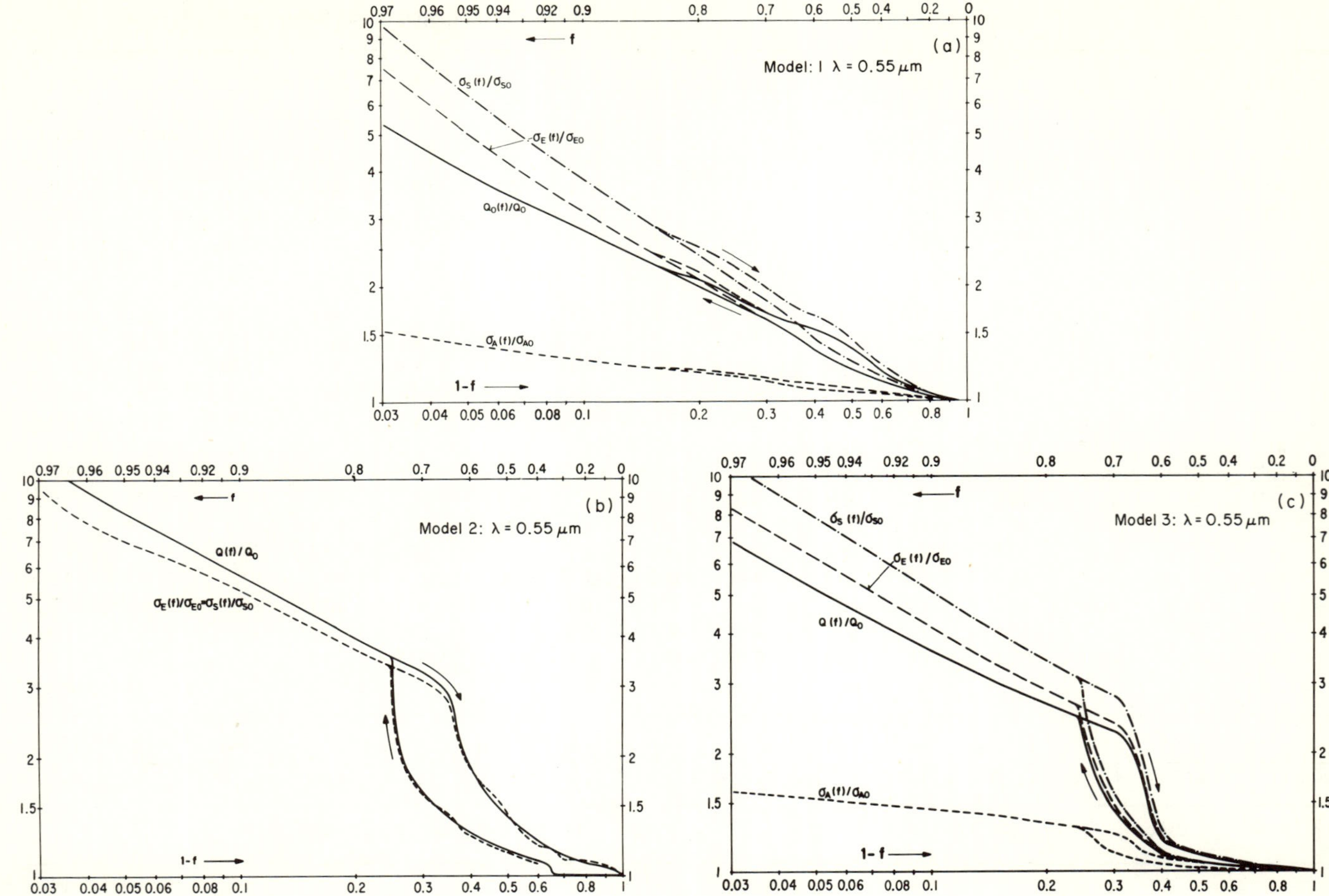

f
(a)
Model: 1 λ = 0.55 μm
σ_S(t)/σ_{SO}
σ_E(t)/σ_{EO}
a_0(t)/a_0
σ_A(t)/σ_{AO}
1-f
f
(b)
Model 2: λ = 0.55 μm
a(t)/a_0
σ_E(t)/σ_{EO}=σ_S(t)/σ_{SO}
1-f
f
(c)
Model 3: λ = 0.55 μm
σ_S(t)/σ_{SO}
σ_E(t)/σ_{EO}
a(t)/a_0
σ_A(t)/σ_{AO}
1-f

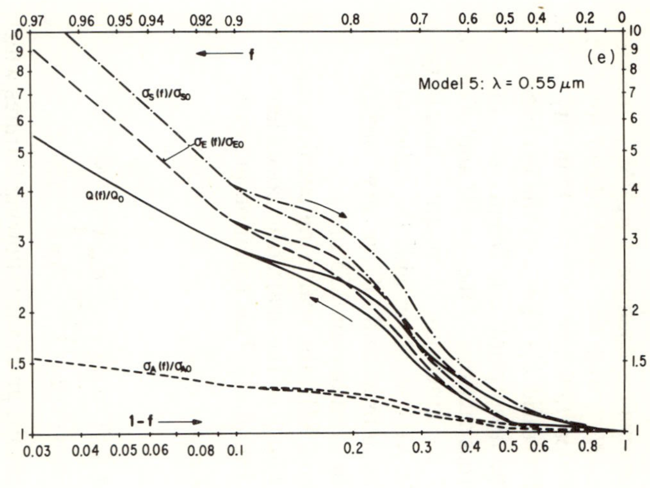
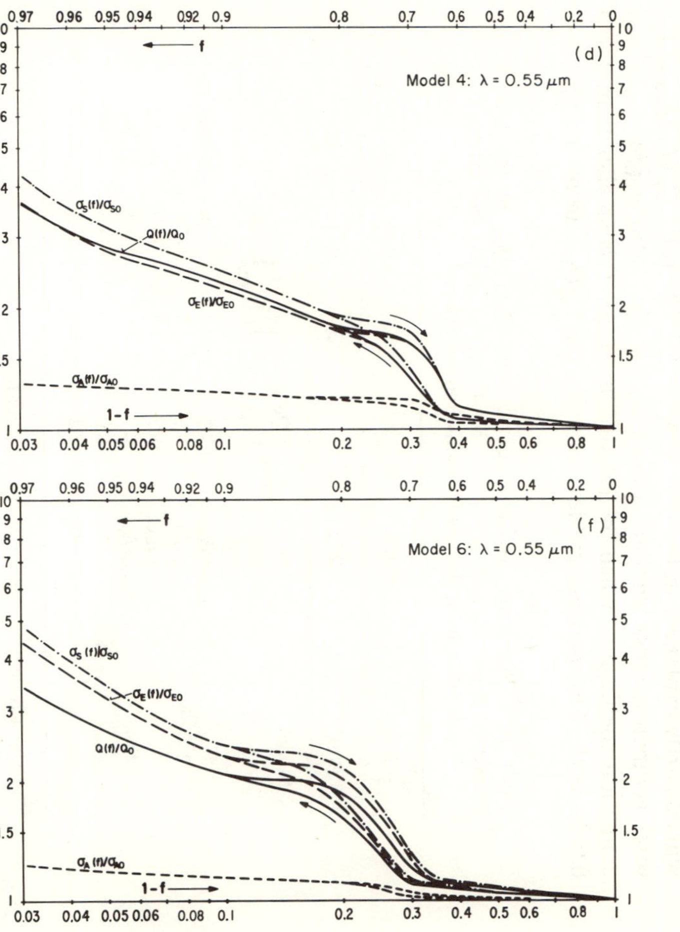

FIG. 12. Ratios $Q_K(f)/Q_{K0}$, $\sigma_E(f)/\sigma_{E0}$, $\sigma_S(f)/\sigma_{S0}$, and $\sigma_A(f)/\sigma_{A0}$ of the total geometric cross sections, the extinction coefficients, the scattering coefficients, and the absorption coefficients at the relative humidity f and in dry state at the wavelength of light 0.55 μm for the aerosol models 1 to 6.

In Table XII the standardized extinction coefficient σ_{E0}/N_{K1}, the standardized absorption coefficient σ_{A0}/N_{K1}, and the standardized total geometric cross section Q_{K0}/N_{K1} for dry particles, i.e., relative humidity $f = 0$, are given for all aerosol models. The ratio $\sigma_E(f)/\sigma_{E0}$ of the extinction coefficients, the ratio $\sigma_A(f)/\sigma_{A0}$ of the absorption coefficients and the ratio $Q_K(f)/Q_{K0}$ of the total geometric cross sections of the humid and of the dry particles are listed in Table XIII as functions of relative humidity f. Additionally the results from Table XIII together with the ratio $\sigma_S(f)/\sigma_{S0}$ of the scattering coefficients of the humid and of the dry particles are shown in Fig. 12 for the wavelength of radiation 0.55 μm.

From the results of Tables XII and XIII, the scattering coefficient σ_{S0} of the dry particles and the ratio $\sigma_S(f)/\sigma_{S0}$ of the scattering coefficients both can be computed with the help of Eq. (6.11). This equation indicates

$$(6.16a) \qquad \sigma_{S0} = \sigma_{E0} - \sigma_{A0}$$

and after simple transformations

$$(6.16b) \qquad \frac{\sigma_S(f)}{\sigma_{S0}} = \left(\frac{\sigma_E(f)}{\sigma_{E0}} - \frac{\sigma_A(f)}{\sigma_{A0}} \cdot \frac{\sigma_{A0}}{\sigma_{E0}} \right) \left(1 - \frac{\sigma_{A0}}{\sigma_{E0}} \right)^{-1}$$

A first rough examination of the results gives the following features which will be partly explained in the next section: (1) The ratio $Q_K(f)/Q_{K0}$ of the total geometric cross sections almost equals the ratio $\sigma_E(f)/\sigma_{E0}$ of the extinction coefficients, when the relative humidity f is smaller than about 0.9 to 0.95. The percentage deviation of $Q_K(f)/Q_{K0}$ from $\sigma_E(f)/\sigma_{E0}$ is

TABLE XIV. Percentage deviation of the ratio $Q_K(f)/Q_{K0}$ of the total geometric cross sections from the ratio $\sigma_E(f)/\sigma_{E0}$ of the extinction coefficients at the relative humidities $f = 0.9$ and $f = 0.95$

f:	0.9			0.95		
λ^a (μm):	0.3	0.55	1.0	0.3	0.55	1.0
Model	$100 \times \left(\dfrac{Q_K}{Q_{K0}} - \dfrac{\sigma_E}{\sigma_{E0}} \right) \Big/ \left(\dfrac{\sigma_E}{\sigma_{E0}} \right)$					
1	-8.2	-10.9	-8.8	-14.3	-21.6	-20.0
2	3.3	7.8	0.9	7.3	17.5	-5.9
3	4.6	-11.8	-19.2	-1.0	-16.9	-30.2
4	0.4	4.1	4.1	-1.1	1.1	4.9
5	-12.0	-13.6	-9.2	-21.2	-28.6	-27.9
6	-1.0	-9.6	-11.1	-4.3	-17.1	20.1

^{a} The wavelength of electromagnetic radiation.

listed in Table XIV for the relative humidities 0.9 and 0.95 and the wavelengths of radiation 0.3, 0.55, and 1.0 μm. At wavelength 0.55 μm, being of importance for visibility considerations, this deviation ranges from -13.6 to 7.8% for $f = 0.9$ and from -28.6 to 17.5% for $f = 0.95$. Thus

$$(6.17) \qquad \sigma_E(f)/\sigma_{E0} \cong Q_K(f)/Q_{K0}$$

can be used for practical purposes, especially when the visibility must be known as a function of relative humidity. It is an advantage that the use of Eq. (6.17) does not require the knowledge of the mean complex refractive index of the particles.

(2) The ratio $\sigma_A(f)/\sigma_{A0}$ of the absorption coefficients is smaller and the ratio $\sigma_S(f)/\sigma_{S0}$ of the scattering coefficients is larger than the ratio $\sigma_E(f)/\sigma_{E0}$ of the extinction coefficients. This has been demonstrated in Fig. 12 for the wavelength of radiation 0.55 μm and is also valid for the two remaining wavelengths 0.3 μm and 1.0 μm, as can be derived from the values given in Table XIII. From

$$\frac{\sigma_A(f)}{\sigma_{A0}} < \frac{\sigma_E(f)}{\sigma_{E0}} < \frac{\sigma_S(f)}{\sigma_{S0}}$$

it follows

$$(6.18) \qquad \frac{\sigma_A(f) - \sigma_{A0}}{\sigma_{A0}} < \frac{\sigma_E(f) - \sigma_{E0}}{\sigma_{E0}} < \frac{\sigma_S(f) - \sigma_{S0}}{\sigma_{S0}}.$$

This latter relation indicates that the percentage change with relative humidity attains the greatest values for the scattering coefficient and the smallest values for the absorption coefficient. At the relative humidity $f = 0.995$ the total geometric cross section is 9.9 to 32.9 times larger than in dry state. The corresponding factors range from 8.8 to 44.8 for the extinction coefficient, from 10.7 to 69.8 for the scattering coefficient, and from 1.0 to 2.5 for the absorption coefficient. Thus, the total geometric cross section, the extinction coefficient, and the scattering coefficient increase with relative humidity far more than the absorption coefficient does. As will be discussed later, this result is strictly limited to that wavelength of radiation with no or negligible absorption of radiation by pure water.

(3) The hysteresis effect can lead to significant differences of extinction and scattering coefficients between increasing and decreasing relative humidity. These differences are most significant for the maritime aerosols (models 2, 3, and 4) in the humidity range $0.6 \leq f \leq 0.75$. For these aerosol types the extinction and scattering coefficients may have twofold values at decreasing relative humidity compared to those at increasing relative humidity.

6.5.3. Results for the Wavelength Region 1.3 to 12 μm. Estimates of the aerosol influence on the divergence of the electromagnetic radiation in the atmosphere—determining radiative heating or cooling rates of the air— require supplementary data of the extinction and absorption coefficients of atmospheric aerosol particles at wavelengths ranging from 1 μm to 12 μm (cf., e.g., Grassl, 1973a, b, 1974). Thus these coefficients have been computed for the most important aerosol models 3, 5, and 6, i.e., marine aerosol, aerosol from a polluted area, and clean air aerosol. The calculations have been performed with the Mie theory in the same manner as described in Section 6.5.1.

Unfortunately, no measurements had been made of the mean complex refractive index in the wavelength region 1 to 12 μm on those specific aerosol samples which the preceding model calculations are based upon. Nevertheless, there exist numerous experimental data by Volz (1972) and Fischer (1975) that allow suitable extrapolations of the known mean complex refractive index data (Table VI) from 1 μm to 12 μm wavelength of electromagnetic radiation. The complex refractive indices of pure water in this wavelength region have been taken from Irvine and Pollack (1968) and Zolotarev *et al.* (1970). The extrapolated mean complex refractive indices of the dry particles and the complex refractive indices of water are compiled in Table XV together with the pertinent standardized extinction and absorption coefficients in the dry state.

In Table XVI the ratio $\sigma_E(f)/\sigma_{E0}$ of the extinction coefficients at the relative humidity f and in the dry state and the corresponding ratio $\sigma_A(f)/\sigma_{A0}$ for the absorption coefficients are compiled as functions of relative humidity f at the wavelengths of radiation 1.3, 1.65, 2.0, 2.5, 3.0, 9.25, 10.0, and 12.0 μm. The wavelength 9.25 μm has been chosen, because here a maximum of the imaginary part of the mean complex refractive index of atmospheric aerosol particles usually occurs.

The results can be subdivided into two groups. Within the first group, the results for the wavelengths of radiation 1.3, 1.65, 2.0, and 2.5 μm can be combined; within the second, the results for the remaining wavelengths.

The ratios $\sigma_E(f)/\sigma_{E0}$ and $\sigma_A(f)/\sigma_{A0}$ for the wavelengths of radiation 1.3 to 2.5 μm do not differ very much from those for wavelengths between 0.3 and 1.0 μm (cf. Table XIII). The reason is that the complex refractive indices of pure water and of dry particles show the same features in both wavelength regions. These features are a smaller real part and a considerably smaller imaginary part of refractive index of pure water than of the dry particles. This means a decrease of the real part and a considerable decrease of the imaginary part of mean complex refractive index with relative humidity.

For the second group of the results, i.e., the wavelengths of radiation

TABLE XV. Real part n_w and imaginary part k_w of the complex refractive index of water, real part n_0 and imaginary part k_0 of the mean complex refractive index of dry aerosols, standardized extinction coefficient σ_{E0}/N_{K1} and standardized absorption coefficient σ_{A0}/N_{K1} of the dry particles versus wavelength λ of electromagnetic radiation

$\lambda(\mu m)$:	1.3	1.65	2.0	2.5	3.0	9.25	10.0	12.0
n_w:	1.321	1.316	1.304	1.246	1.351	1.257	1.214	1.11
k_w:	1.12×10^{-6}	6.7×10^{-5}	0.00108	0.00165	0.259	0.0422	0.0532	0.244
Model 3								
n_0	1.52	1.55	1.47	1.47	1.34	1.65	1.75	1.75
k_0	0.086	0.055	0.132	0.132	0.264	0.528	0.198	0.198
$\sigma_{E0}/N_{K1}(cm^2)$	7.56×10^{-10}	5.87×10^{-10}	5.26×10^{-10}	4.39×10^{-10}	4.44×10^{-10}	2.75×10^{-10}	1.73×10^{-10}	1.39×10^{-10}
$\sigma_{A0}/N_{K1}(cm^2)$	3.00×10^{-10}	1.92×10^{-10}	2.79×10^{-10}	2.34×10^{-10}	3.20×10^{-10}	2.13×10^{-10}	9.71×10^{-11}	8.15×10^{-11}
Model 5								
n_0	1.51	1.49	1.47	1.47	1.33	1.55	1.70	1.70
k_0	0.088	0.112	0.136	0.136	0.272	0.544	0.340	0.204
$\sigma_{E0}/N_{K1}(cm^2)$	6.72×10^{-11}	5.53×10^{-11}	4.82×10^{-11}	3.87×10^{-11}	4.29×10^{-11}	2.57×10^{-11}	1.79×10^{-11}	1.19×10^{-11}
$\sigma_{A0}/N_{K1}(cm^2)$	2.94×10^{-11}	2.83×10^{-11}	2.76×10^{-11}	2.30×10^{-11}	3.33×10^{-11}	2.15×10^{-11}	1.33×10^{-11}	7.74×10^{-12}
Model 6								
n_0	1.49	1.48	1.47	1.47	1.43	1.60	1.70	1.70
k_0	0.026	0.033	0.040	0.040	0.080	0.160	0.060	0.060
$\sigma_{E0}/N_{K1}(cm^2)$	2.70×10^{-10}	2.15×10^{-10}	1.79×10^{-10}	1.44×10^{-10}	1.27×10^{-10}	6.17×10^{-11}	6.99×10^{-11}	4.12×10^{-11}
$\sigma_{A0}/N_{K1}(cm^2)$	5.43×10^{-11}	5.23×10^{-11}	5.10×10^{-11}	4.81×10^{-11}	5.88×10^{-11}	3.71×10^{-11}	1.67×10^{-11}	1.38×10^{-11}

TABLE XVI. Ratio $\sigma_E(f)/\sigma_{E0}$ of the extinction coefficients and ratio $\sigma_A(f)/\sigma_{A0}$ of the absorption coefficients at the relative humidity f and in the dry state at different wavelengths λ of electromagnetic radiation[a]

Model 3

$\lambda(\mu m)$:	1.3	1.65	2.0	2.5	3.0	9.25	10.0	12.0	1.3	1.65	2.0	2.5	3.0	9.25	10.0	12.0
inc. f				$\sigma_E(f)/\sigma_{E0}$								$\sigma_A(f)/\sigma_{A0}$				
0.2	1.01	1.01	1.01	1.01	1.02	1.01	1.01	1.02	1.01	1.00	1.01	1.01	1.02	1.01	1.01	1.02
0.4	1.02	1.03	1.02	1.02	1.04	1.02	1.02	1.03	1.01	1.00	1.01	1.01	1.04	1.02	1.03	1.06
0.6	1.08	1.10	1.08	1.07	1.14	1.06	1.07	1.11	1.04	1.02	1.04	1.04	1.13	1.05	1.07	1.18
0.65	1.18	1.23	1.19	1.16	1.31	1.13	1.14	1.25	1.08	1.04	1.08	1.08	1.30	1.12	1.15	1.42
0.7	1.40	1.47	1.41	1.36	1.66	1.26	1.30	1.54	1.15	1.09	1.16	1.15	1.62	1.22	1.29	1.91
0.75	2.28	2.32	2.21	2.06	2.96	1.67	1.85	2.71	1.32	1.20	1.35	1.31	2.85	1.53	1.84	3.74
0.8	2.91	2.91	2.75	2.50	3.87	1.93	2.20	3.51	1.39	1.26	1.42	1.39	3.69	1.69	2.18	4.95
0.85	3.41	3.36	3.16	2.84	4.57	2.13	2.48	4.11	1.42	1.28	1.48	1.44	4.29	1.81	2.44	5.84
0.9	4.57	4.37	4.09	3.62	6.09	2.60	3.13	5.45	1.49	1.33	1.58	1.53	5.58	2.07	3.03	7.83
0.95	7.32	7.36	6.44	5.64	9.77	3.83	4.83	8.92	1.59	1.40	1.77	1.71	8.80	2.71	4.56	12.8
0.975	12.7	13.1	11.3	9.70	16.4	6.29	8.21	15.4	1.75	1.45	2.04	2.01	14.5	3.92	7.46	22.1
0.99	26.4	26.4	24.3	19.4	32.6	12.8	17.1	31.4	1.80	1.54	2.57	2.66	27.2	7.06	15.0	44.1
0.995	43.6	48.1	42.6	35.2	54.3	23.1	31.4	55.2	1.92	1.72	3.27	3.61	43.7	11.8	26.5	76.0
dec. f																
0.2	1.02	1.02	1.02	1.02	1.03	1.02	1.01	1.02	1.01	1.00	1.01	1.01	1.03	1.02	1.02	1.04
0.4	1.05	1.06	1.05	1.05	1.09	1.04	1.04	1.07	1.03	1.01	1.03	1.03	1.08	1.04	1.03	1.11
0.5	1.09	1.11	1.09	1.08	1.15	1.07	1.07	1.12	1.04	1.02	1.05	1.04	1.15	1.06	1.08	1.20
0.6	1.18	1.22	1.18	1.16	1.30	1.12	1.14	1.24	1.08	1.04	1.08	1.08	1.29	1.12	1.15	1.41
0.65	1.78	1.85	1.76	1.66	2.26	1.44	1.54	2.06	1.24	1.15	1.26	1.23	2.19	1.37	1.54	2.73
0.7	2.35	2.39	2.26	2.10	3.06	1.70	1.88	2.80	1.34	1.21	1.35	1.32	2.96	1.54	1.87	3.87
0.75	2.57	2.59	2.45	2.25	3.40	1.79	2.01	3.08	1.36	1.22	1.38	1.35	3.25	1.60	2.00	4.30

Model 5:

$\lambda(\mu m)$:	1.3	1.65	2.0	2.5	3.0	9.25	10.0	12.0	1.3	1.65	2.0	2.5	3.0	9.25	10.0	12.0
inc. f				$\sigma_E(f)/\sigma_{E0}$								$\sigma_A(f)/\sigma_{A0}$				
0.2	1.01	1.02	1.01	1.01	1.02	1.01	1.01	1.02	1.00	1.01	1.00	1.01	1.02	1.01	1.01	1.03
0.4	1.03	1.03	1.02	1.02	1.05	1.02	1.02	1.04	1.01	1.01	1.01	1.01	1.05	1.02	1.03	1.07
0.5	1.05	1.05	1.04	1.04	1.08	1.03	1.04	1.07	1.02	1.02	1.02	1.02	1.07	1.03	1.04	1.11
0.6	1.16	1.15	1.13	1.12	1.26	1.09	1.13	1.24	1.06	1.06	1.06	1.06	1.24	1.09	1.12	1.35
0.65	1.26	1.25	1.22	1.20	1.43	1.15	1.21	1.41	1.09	1.09	1.09	1.09	1.41	1.14	1.20	1.60
0.7	1.39	1.37	1.33	1.30	1.64	1.22	1.29	1.62	1.12	1.13	1.13	1.12	1.61	1.19	1.28	1.90
0.75	1.76	1.71	1.64	1.57	2.22	1.38	1.52	2.19	1.16	1.20	1.21	1.20	2.15	1.32	1.47	2.70
0.8	2.13	2.06	1.96	1.84	2.81	1.54	1.74	2.77	1.26	1.27	1.28	1.27	2.70	1.43	1.65	3.52
0.85	2.54	2.45	2.31	2.13	3.45	1.71	1.96	3.42	1.31	1.32	1.34	1.32	3.31	1.54	1.84	4.42
0.9	3.03	2.90	2.72	2.47	4.27	1.91	2.23	4.20	1.35	1.37	1.40	1.38	4.06	1.68	2.06	5.54
0.95	5.66	5.43	5.07	4.45	8.03	2.97	3.64	8.10	1.52	1.55	1.63	1.60	7.37	2.30	3.12	10.7
0.975	10.7	10.3	9.63	8.30	15.2	5.09	6.36	15.5	1.65	1.72	1.90	1.91	13.5	3.43	5.12	20.5
0.99	25.3	24.8	23.4	20.2	34.3	11.6	14.7	36.5	1.85	1.94	2.51	2.67	29.6	6.74	11.0	47.6
0.995	45.9	45.8	43.7	38.5	60.3	22.2	28.0	66.9	1.94	2.11	3.34	3.75	50.5	11.8	19.9	85.7
dec. f																
0.2	1.03	1.02	1.02	1.02	1.04	1.02	1.02	1.04	1.01	1.01	1.01	1.01	1.04	1.02	1.02	1.06
0 4	1.09	1.08	1.07	1.07	1.13	1.05	1.07	1.13	1.03	1.03	1.03	1.03	1.13	1.05	1.07	1.19
0.6	1.28	1.26	1.23	1.21	1.45	1.16	1.22	1.43	1.09	1.10	1.10	1.09	1.43	1.14	1.21	1.63
0.7	1.62	1.58	1.52	1.46	1.99	1.32	1.44	1.96	1.17	1.18	1.18	1.17	1.93	1.27	1.40	2.38
0.75	2.04	1.98	1.89	1.78	2.68	1.51	1.69	2.64	1.25	1.26	1.27	1.25	2.58	1.41	1.61	3.34
0.8	2.38	2.30	2.17	2.01	3.22	1.64	1.88	3.18	1.28	1.30	1.32	1.30	3.08	1.50	1.77	4.09
0.85	2.73	2.62	2.48	2.26	3.76	1.79	2.07	3.72	1.32	1.34	1.36	1.34	3.59	1.60	1.92	4.84

continued

TABLE XVI—*Continued*

Model 6:

$\lambda(\mu m)$:	1.3	1.65	2.0	2.5	3.0	9.25	10.0	12.0	1.3	1.65	2.0	2.5	3.0	9.25	10.0	12.0
inc. f					$\sigma_E(f)/\sigma_{E0}$								$\sigma_A(f)/\sigma_{A0}$			
0.2	1.01	1.01	1.01	1.01	1.02	1.01	1.00	1.01	1.00	1.00	1.00	1.01	1.03	1.00	1.01	1.04
0.4	1.02	1.02	1.02	1.02	1.04	1.02	1.02	1.03	1.00	1.01	1.00	1.01	1.08	1.01	1.03	1.11
0.5	1.04	1.04	1.04	1.03	1.08	1.03	1.03	1.05	1.01	1.01	1.01	1.02	1.15	1.02	1.05	1.20
0.6	1.07	1.08	1.07	1.06	1.15	1.06	1.06	1.10	1.02	1.02	1.02	1.02	1.27	1.05	1.09	1.37
0.65	1.10	1.10	1.09	1.08	1.19	1.07	1.08	1.13	1.02	1.03	1.03	1.03	1.35	1.06	1.12	1.47
0.7	1.10	1.11	1.10	1.09	1.21	1.08	1.08	1.14	1.03	1.03	1.03	1.03	1.38	1.06	1.13	1.50
0.75	1.32	1.32	1.30	1.27	1.65	1.23	1.25	1.46	1.06	1.08	1.08	1.08	2.11	1.19	1.40	2.51
0.8	1.73	1.71	1.68	1.58	2.42	1.51	1.54	2.06	1.11	1.13	1.15	1.14	3.33	1.42	1.88	4.30
0.85	2.07	2.06	2.01	1.85	3.12	1.75	1.80	2.61	1.14	1.16	1.20	1.19	4.37	1.60	2.33	5.86
0.9	2.31	2.28	2.23	2.05	3.58	1.90	1.97	2.99	1.17	1.19	1.24	1.23	5.11	1.72	2.63	6.92
0.95	3.34	3.31	3.22	2.88	5.45	2.60	2.74	4.60	1.22	1.25	1.35	1.34	7.85	2.23	3.87	11.2
0.975	5.58	5.58	5.80	4.76	9.47	4.17	4.45	8.20	1.30	1.34	1.55	1.59	13.5	3.34	6.64	20.4
0.99	12.7	12.9	12.7	11.1	21.6	9.51	10.2	19.6	1.39	1.47	2.12	2.31	30.0	7.0	15.6	48.8
0.995	22.4	23.3	23.4	20.6	37.9	18.0	19.3	36.0	1.44	1.59	2.92	3.41	51.1	12.6	29.1	88.6
dec. f																
0.2	1.03	1.03	1.03	1.02	1.06	1.02	1.03	1.04	1.01	1.01	1.01	1.01	1.12	1.02	1.04	1.16
0.4	1.06	1.05	1.05	1.05	1.11	1.04	1.05	1.07	1.02	1.02	1.02	1.02	1.20	1.04	1.07	1.27
0.6	1.10	1.10	1.10	1.08	1.20	1.07	1.08	1.13	1.03	1.03	1.03	1.03	1.36	1.06	1.12	1.48
0.65	1.12	1.12	1.11	1.10	1.24	1.09	1.10	1.16	1.03	1.03	1.04	1.04	1.42	1.07	1.15	1.57
0.7	1.23	1.23	1.23	1.20	1.47	1.17	1.18	1.32	1.06	1.06	1.07	1.06	1.81	1.14	1.29	2.10
0.75	1.63	1.61	1.59	1.51	2.23	1.45	1.47	1.90	1.10	1.11	1.14	1.13	3.03	1.36	1.77	3.85
0.8	2.06	2.04	2.00	1.84	3.07	1.74	1.79	2.58	1.14	1.16	1.20	1.19	4.33	1.59	2.30	5.78
0.85	2.23	2.21	2.17	1.99	3.43	1.86	1.92	2.87	1.17	1.18	1.23	1.22	4.89	1.68	2.53	6.59

[a] inc. = increasing, dec. = decreasing.

3.0, 9.25, 10.0, and 12.0 μm, these relationships between the complex refractive indices of pure water and of dry particles do not exist. Thus the ratios $\sigma_E(f)/\sigma_{E0}$ and $\sigma_A(f)/\sigma_{A0}$ of the second group of the results are both considerably larger or smaller than those of the first group. In the next section, interpretations of the computational results will be given.

6.5.4. Interpretation of the Extinction Coefficients. In this section an explanation of the relationship between the change of the extinction co-efficient σ_E with relative humidity f and the corresponding change of the total geometric cross section Q_K of the equivalent spheres of the particles per unit volume of air will be given. From Eqs. (6.12) and (6.14) the ratios $\sigma_E(f)/\sigma_{E0}$ and $Q_K(f)/Q_{K0}$ are

$$\frac{\sigma_E(f)}{\sigma_{E0}} = \left(\sum_{j=1}^{j=N_K} \pi r_j^2(f)\kappa_{Ej}(f) \right) \Big/ \left(\sum_{j=1}^{j=N_K} \pi r_{0j}^2 \kappa_{E0j} \right)$$

and

$$\frac{Q_K(f)}{Q_{K0}} = \left(\sum_{j=1}^{j=N_K} \pi r_j^2(f) \right) \Big/ \left(\sum_{j=1}^{j=N_K} \pi r_{0j}^2 \right)$$

where r_{0j} and κ_{E0j} are the equivalent radius and the efficiency factor of extinction of the jth particle in dry state, respectively. Combination of both equations gives the following twofold ratio between the extinction coefficients and the total geometric cross sections at the relative humidity f and in the dry state

$$(6.19) \qquad R_{EQ} = \frac{\dfrac{\sigma_E(f)}{Q_K(f)}}{\dfrac{\sigma_{E0}}{Q_{K0}}} = \frac{\left(\sum_{j=1}^{j=N_K} \pi r_j^2(f)\kappa_{Ej}(f) \right) \Big/ \left(\sum_{j=1}^{j=N_K} \pi r_j^2(f) \right)}{\left(\sum_{j=1}^{j=N_K} \pi r_{0j}^2\kappa_{E0j} \right) \Big/ \left(\sum_{j=1}^{j=N_K} \pi r_{0j}^2 \right)}$$

The right side of this equation is the ratio of two weighted arithmetic means. The numerator is the weighted arithmetic mean of the efficiency factors of all particles at the relative humidity f with the pertinent geometric cross sections of the equivalent spheres as weighting factors. The denominator is the analogous weighted arithmetic mean in the dry state. Thus the twofold ratio R_{EQ} describes the change of the weighted mean efficiency factor of extinction with relative humidity.

In Table XVII values of the twofold ratio R_{EQ} are given for the most important aerosol models 3, 5, and 6 at selected relative humidities and wavelengths of electromagnetic radiation. They show that there exists a close relationship between the change of the extinction coefficient with relative humidity and that of the total geometric cross section for relative

TABLE XVII. Twofold ratio $R_{EQ} = [\sigma_E(f)/Q_K(f)]/(\sigma_{E0}/Q_{K0})$ where $\sigma_E(f)$ and $Q_K(f)$ are the extinction coefficient and the total geometric cross section of the particle equivalent spheres per unit volume of air at the relative humidity f, and σ_{E0} and Q_{K0} are the same properties in the dry state

$\lambda^a(\mu m)$:	0.55	1.3	2.5	3.0	9.25	10.0	12.0
f				R_{EQ}			
Model 3							
0.7^b	1.03	0.98	0.95	1.16	0.88	0.91	1.08
0.9	1.13	1.25	0.99	1.66	0.71	0.86	1.49
0.95	1.20	1.41	1.09	1.89	0.74	0.93	1.72
0.99	1.13	2.10	1.54	2.59	1.02	1.36	2.49
Model 5							
0.7^b	1.01	0.97	0.91	1.15	0.85	0.90	1.13
0.9	1.16	1.06	0.87	1.50	0.67	0.78	1.47
0.95	1.40	1.34	1.06	1.91	0.71	0.86	1.92
0.99	2.03	2.20	1.76	2.98	1.01	1.28	3.17
Model 6							
0.7^b	1.00	1.00	0.99	1.10	0.98	0.98	1.04
0.9	1.11	1.11	0.99	1.72	0.91	0.95	1.44
0.95	1.21	1.26	1.08	2.05	0.98	1.03	1.73
0.99	1.53	1.89	1.65	3.21	1.42	1.52	2.92

^{a} The wavelength of light.
^{b} Values at $f = 0.7$ are for increasing relative humidity.

humidities smaller than 0.9 to 0.95 at all wavelengths of radiation except 3.0 and 12.0 μm.

Indeed, for relative humidities smaller than 0.9 to 0.95 Eq. (6.19) can be simplified remarkably. From Section 6.2.1 it is known that in this range of relative humidities the influence of the curvature correction on the ratio $r_j(f)/r_{0j}$ can be neglected for particles with equivalent radii in dry state larger than 0.1 μm (cf. Table VII). In this context, we are able to apply this approximation also to particles with equivalent radii in dry state larger than 0.04 μm, since the particles smaller than 0.1 μm do not contribute to the extinction coefficient to a large extent. Therefore $r_j^2(f)$ in the numerator of Eq. (6.19) becomes with the help of Eq. (6.1)

$$(6.20) \qquad r_j^2(f) = \left(\frac{r_j(f)}{r_{0j}}\right)^2 r_{0j}^2 \simeq \left(1 + \frac{\rho_0}{\rho_w}\bar{\mu}\frac{f}{1-f}\right)^{2/3} r_{0j}^2$$

and $[1 + (\rho_0/\rho_w)\bar{\mu}(f/1 - f)]^{2/3}$ can be emitted from both summations. Now Eq. (6.19) can be written

$$(6.21) \qquad R_{EQ} \cong \frac{\left(\sum\limits_{j=1}^{j=N_K} \pi r_{0j}^2 \kappa_{Ej}(f)\right)\bigg/\left(\sum\limits_{j=1}^{j=N_K} \pi r_{0j}^2\right)}{\left(\sum\limits_{j=1}^{j=N_K} \pi r_{0j}^2 \kappa_{E0j}\right)\bigg/\left(\sum\limits_{j=1}^{j=N_K} \pi r_{0j}^2\right)} = \frac{\sum\limits_{j=1}^{j=N_K} r_{0j}^2 \kappa_{Ej}(f)}{\sum\limits_{j=1}^{j=N_K} r_{0j}^2 \kappa_{E0j}}$$

The right side of Eq. (6.21) again is a direct measure of the change of the weighted mean efficiency factor of extinction with relative humidity.

However, the weighting factors at the relative humidity f and in the dry state are the same, namely, the geometric cross sections of the dry particles. Thus, at relative humidities smaller than 0.9 to 0.95 only the dependences of the efficiency factors of extinction on relative humidity have to be considered.

In doing this we have to look upon the dependence of the efficiency factor κ_E of extinction on the complex refractive index $n - ik$ and on the particle equivalent radius r for each wavelength λ of electromagnetic radiation separately. From the Mie theory it is known that the equivalent radius r is not the relevant property for size but rather size parameter $\alpha = 2\pi r/\lambda$, describing the relation between the equivalent radius r and the wavelength λ of radiation. However, in this context, the size parameter α is not a useful property, but for interpretation of extinction coefficients, it is appropriate to plot the efficiency factor κ_E of extinction versus the property

$$(6.22) \qquad \alpha_E = 2\alpha[(n - 1)^2 + k^2]^{1/2}$$

α_E will be called here the generalized size parameter of extinction (cf. van de Hulst, 1957; Penndrof, 1958; Hänel, 1971).

Examples of the κ_E–α_E relationship are given in Fig. 13. Chosen were those complex refractive indices that are close to those the aerosol particles attain in the whole range of relative humidity at the wavelengths of electromagnetic radiation considered here. From these examples it can be seen that the maxima and minima of the κ_E–α_E curves are located at the same α_E values independent of the complex refractive indices. The curves lie close together when α_E is larger than two and approach the value $\kappa_E = 2$ at large α_E. The κ_E–α_E curves differ remarkably when α_E is much smaller than two. In this latter region the κ_E values increase strongly with increasing α_E and remain smaller than two. Therefore, the generalized size parameter of extinction is a useful tool for the classification of κ_E values.

Moreover, the generalized size parameter of extinction is only slightly affected by the relative humidity except at the wavelength of radiation 3.0 μm and at relative humidities larger than 0.95, as demonstrated in Table XVIII.

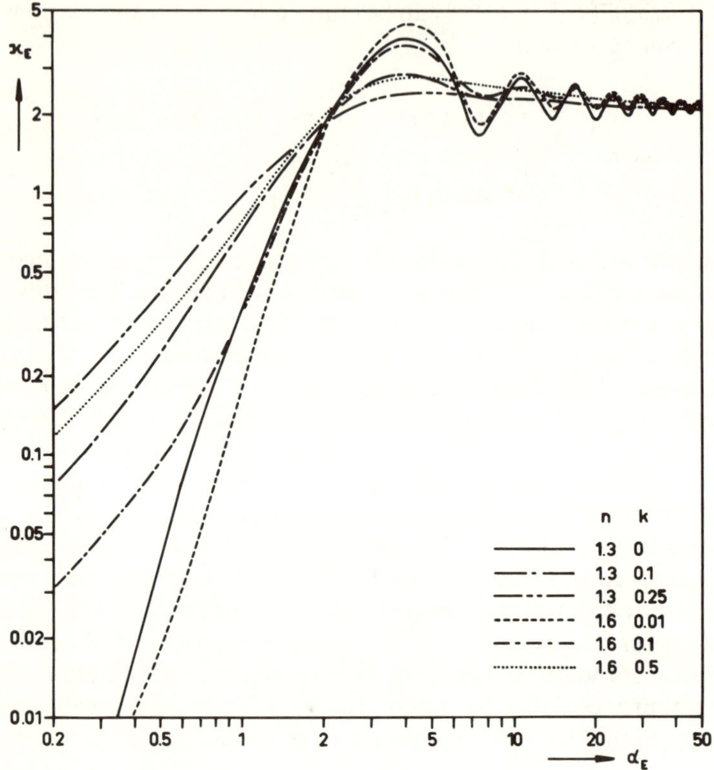

FIG. 13. Efficiency factor of extinction κ_E versus generalized size parameter α_E of extinction as computed from Mie's theory for several complex refractive indices $n - ik$.

The reason is a counterbalance between the change of the equivalent radius r with relative humidity and the pertinent opposite change of the absolute value of the refractive index minus 1 $[(n - 1)^2 + k^2]^{1/2}$, since in most cases the value of $[(n - 1)^2 + k^2]^{1/2}$ is smaller for pure water than for the dry particles. This compensation is deficient when the square root $[(n - 1)^2 + k^2]^{1/2}$ remains nearly constant. In these cases the percentage change of the generalized size parameter of extinction almost equals the percentage change of the equivalent radius.

After these preliminaries, the values of the twofold ratio R_{EQ} can be explained for relative humidities smaller than 0.9 to 0.95: (1) 0.3 μm $\leq \lambda \leq$ 2.5 μm: In this wavelength region the particles with equivalent radii in dry state around 0.5 μm are the most efficiently screening particles. In each case their generalized size parameters of extinction are larger than one (cf. Table XVIII). For these particles the κ_E–α_E relationship does not depend notably on relative humidity (cf. the curves for the complex refractive indices $1.3 - 0i$, $1.6 - 0.01i$, and $1.6 - 0.1i$ in Fig. 13). Moreover, the aerosol size distributions are broad enough that there exist regions of particle size, for which the efficiency factor of extinction slightly increases with relative humidity, and at the same time there are others for which it slightly decreases. These increases and decreases balance nearly perfectly, thus leading to the approximation (6.17), which can be written as

$$(6.23) \qquad\qquad R_{EQ} \cong 1$$

TABLE XVIII. Ratio $\alpha_E(f)/\alpha_{E0}$ of the generalized size parameters of extinction at the relative humidity f and in the dry state, as well as values for α_{E0} of a particle with an equivalent radius in the dry state $r_0 = 0.5\ \mu m^a$

$\lambda(\mu m)$:	0.55	1.3	2.5	3.0	9.25	10.0	12.0
f				$\alpha_E(f)/\alpha_{E0}$			
Model 3							
0.7	1.00	1.00	0.94	1.22	0.83	0.83	0.81
0.9	1.27	1.27	1.09	1.94	0.76	0.74	0.77
0.95	1.46	1.47	1.24	2.32	0.82	0.79	0.86
0.99	2.24	2.25	1.87	3.70	1.18	1.09	1.28
Model 5							
0.7	1.00	1.00	0.94	1.20	0.84	0.84	0.83
0.9	1.19	1.21	1.04	1.79	0.77	0.73	0.78
0.95	1.39	1.42	1.18	2.19	0.84	0.76	0.87
0.99	2.30	2.35	1.92	3.83	1.30	1.11	1.40
Model 6							
0.7	1.00	1.00	0.98	1.05	0.96	0.94	0.92
0.9	1.13	1.13	0.99	1.45	0.88	0.78	0.73
0.95	1.21	1.21	1.04	1.62	0.90	0.77	0.74
0.99	1.84	1.84	1.50	2.73	1.23	0.95	1.07
Model				α_{E0}			
3	6.32	2.54	1.22	0.90	0.56	0.48	0.40
5	6.30	2.50	1.22	0.90	0.52	0.48	0.38
6	5.82	2.38	1.18	0.92	0.42	0.44	0.36

^{a} This r_0 value is approximately the median value of the particle sizes in dry state for all size distributions considered in this review. The ratio $\alpha_E(f)/\alpha_{E0}$ is almost independent of particle size in dry state as long as the relative humidity remains smaller than 0.9 to 0.95, since in this region the influence of the curvature correction can be neglected.

(2) $\lambda = 3.0$ μm: The mean complex refractive index of dry particles and the complex refractive index of water are almost the same for aerosol models 3 and 5. Therefore, the κ_E–α_E relationship can be regarded as independent of relative humidity. Moreover the α_E values of the most important screening particles are smaller than three (Table XVIII). For these particles the κ_E–α_E relationship is almost linear (note in Fig. 13 the curve for the complex refractive index $1.3 - 0.25i$). Thus, we have on the right side of Eq. (6.21)

$$\kappa_{Ej}(f) \cong \frac{\alpha_{Ej}(f)}{\alpha_{E0j}} \kappa_{E0j} \cong \frac{r_j(f)}{r_{0j}} \kappa_{E0j} \cong \left(1 + \frac{\rho_0}{\rho_w} \bar{\mu} \frac{f}{1-f}\right)^{1/3} \kappa_{E0j}$$

when we use Eq. (6.1), and the twofold ratio becomes

$$(6.24) \qquad R_{EQ} \cong \left(1 + \frac{\rho_0}{\rho_w} \bar{\mu} \frac{f}{1-f}\right)^{1/3}$$

For aerosol model 6 one obtains the same result as for the other two models. Accidentally, the increases of the κ_E values for the small particles balance the decreases of the κ_E values of the large particles in such a way that Eq. (6.24) can be used as a good approximation.

(3) $\lambda = 9.25$ and 10.0 μm: Here the generalized size parameters of the majority of the screening particles are smaller than two. There is a slight decrease of α_E with relative humidity at first, followed by an increase when the relative humidity is larger than 0.9 to 0.95. At the same time, both the real part and the imaginary part of the mean complex refractive index decrease considerably, making, for α_E values smaller than two, the κ_E–α_E relationship nearly independent of relative humidity [this can be shown easily with the approximation formula for κ_E by Penndorf (1962a, b)]. For these reasons Eq. (6.23) is valid.

(4) $\lambda = 12.0$ μm: For models 3, 5, and 6 the approximation (6.24) is accidentally valid.

The foregoing considerations show that only two simple approximations are necessary to describe the humidity dependence of the twofold ratio R_{EQ}, provided that the relative humidity remains smaller than 0.9 to 0.95. Inasmuch as Eq. (6.20) is equivalent to

$$(6.25) \qquad \frac{Q_K(f)}{Q_{K0}} \cong \left(1 + \frac{\rho_0}{\rho_w} \bar{\mu} \frac{f}{1-f}\right)^{2/3}$$

the ratio of the extinction coefficients at the relative humidity f and in dry state

$$\frac{\sigma_E(f)}{\sigma_{E0}} = \frac{Q_K(f)}{Q_{K0}} R_{EQ}$$

is given as

$$(6.26a) \qquad \frac{\sigma_E(f)}{\sigma_{E0}} \simeq \frac{Q_K(f)}{Q_{K0}} \simeq \left(1 + \frac{\rho_0}{\rho_w}\,\bar{\mu}\,\frac{f}{1-f}\right)^{2/3}$$

for the wavelengths of radiation 0.3 to 2.5 μm, 9.25 μm, and 10.0 μm. It is

$$(6.26b) \qquad \frac{\sigma_E(f)}{\sigma_{E0}} \simeq \left(\frac{Q_K(f)}{Q_{K0}}\right)^{3/2} \simeq \left(1 + \frac{\rho_0}{\rho_w}\,\bar{\mu}\,\frac{f}{1-f}\right)$$

for the wavelengths of radiation 3.0 μm and 12.0 μm. These approximations contain only properties that can be directly measured on samples of atmospheric aerosol particles. It is an outstanding feature that no information on the complex refractive index of water and the mean complex refractive index of the dry particles is necessary. The maximum errors of the last two formulas at the relative humidity 0.95 can amount to $\pm 40\%$ in specific cases.

6.5.5. *Interpretation of the Absorption Coefficients.* The relationship between the change of the absorption coefficient with relative humidity f and the simultaneous changes of the total geometric cross section Q_K, the real part n and the imaginary part k of the mean complex refractive index with relative humidity shall be considered now. As the definition of the absorption coefficient is analogous to that of the extinction coefficient, formula (6.19) defines the twofold ratio between the absorption coefficients and the total geometric cross sections at the relative humidity f and in dry state as

$$(6.27) \qquad R_{AQ} = \frac{[\sigma_A(f)/Q_K(f)]}{(\sigma_{A0}/Q_{K0})}$$

too, only the subscript "E" standing for "extinction" has to be replaced by the subscript "A" standing for "absorption." The interpretation of R_{AQ} in analogous to that for R_{EQ} in the preceding section. Thus R_{AQ} describes the change of the weighted mean efficiency factor of absorption with relative humidity.

Values for the twofold ratio R_{AQ} are compiled in Table XIX at selected relative humidities and wavelengths of electromagnetic radiation for aerosol models 3, 5, and 6. The results can be divided into three groups. At the wavelengths of radiation 0.3 to 2.5 μm R_{AQ} decreases, and at 3.0 and 12.0 μm R_{AQ} increases strongly with relative humidity. The latter is also observed for model 6 at the wavelength 10.0 μm. In the remaining cases R_{AQ} first decreases and then, after the relative humidity has become larger than 0.9 to 0.95, increases with relative humidity.

TABLE XIX. Twofold ratio $R_{AQ} = (\sigma_A(f)/Q_K(f))/(\sigma_{A0}/Q_{K0})$ where $\sigma_A(f)$ and $Q_K(f)$ are the absorption coefficient and the total geometric cross section of the equivalent spheres per unit volume of air at the relative humidity f, and σ_{A0} and Q_{K0} are the same properties in the dry state

$\lambda^a(\mu m)$:	0.55	1.3	2.5	3.0	9.25	10.0	12.0
f				R_{AQ}			
Model 3							
0.7^b	0.783	0.804	0.804	1.13	0.853	0.902	1.34
0.9	0.391	0.407	0.418	1.52	0.566	0.828	2.14
0.95	0.297	0.307	0.330	1.70	0.523	0.880	2.47
0.99	0.140	0.143	0.211	2.16	0.560	1.19	3.50
Model 5							
0.7^b	0.769	0.783	0.783	1.13	0.832	0.895	1.33
0.9	0.460	0.474	0.484	1.42	0.589	0.723	1.94
0.95	0.344	0.361	0.380	1.75	0.546	0.741	2.54
0.99	0.151	0.161	0.232	2.57	0.586	0.957	4.14
Model 6							
0.7^b	0.927	0.936	0.936	1.25	0.964	1.03	1.36
0.9	0.553	0.563	0.591	2.46	0.827	1.26	3.33
0.95	0.376	0.459	0.504	2.95	0.838	1.45	4.21
0.99	0.201	0.207	0.344	4.46	1.04	2.32	7.26

[a] Wavelength of electromagnetic radiation.

[b] Values at $f = 0.7$ are valid for increasing relative humidity.

To explain these different results, the efficiency factor of absorption κ_A must be known as a function of the equivalent radius r, the real part n, and the imaginary part k of the mean complex refractive index of the particle as well as of the wavelength λ of radiation. Likewise, as in the foregoing section, a generalized size parameter of absorption α_A will be defined as a tool for the following interpretations of absorption coefficients.

For this purpose an approximation for the efficiency factor of absorption is used, which is derived from the difference of two approximations for the efficiency factors of extinction and of scattering given by Penndorf (1962a, b). This approximation for κ_A reads

$$\kappa_A \cong \frac{24nk\alpha}{(n^2 + 2)^2 + k^2(2n^2 + k^2 - 4)} \quad \text{with} \quad \alpha = 2\pi\frac{r}{\lambda}$$

when the terms of higher order of the size parameter α are omitted.

This approximation is valid for $1.25 \leq n \leq 1.75, 0 < k \leq 1$, and very small α values. This range of application covers the range of the mean complex refractive indices of atmospheric aerosol particles at the wavelengths of radiation 0.3 to 12.0 μm. Moreover, it is a good approximation for the efficiency factor of extinction of very small absorbing particles (Deirmendjian, 1969; Bergstrom, 1973).

Within the range of the mean complex refractive indices mentioned above, it is

$$n/\{(n^2 + 2)^2 + k^2(2n^2 + k^2 - 4)\} \cong 1/\{3(n^2 + 2)\}$$

Thus, κ_A can be simplified further into

(6.28a)
$$\kappa_A \cong \frac{8k\alpha}{n^2 + 2}$$

To facilitate the further considerations the generalized size parameter α_A of absorption is defined by

(6.28b)
$$\alpha_A = \frac{8k\alpha}{n^2 + 2}$$

The justification of this definition will be given in Fig. 14, which shows the efficiency factor of absorption κ_A for homogeneous spheres (Mie theory)

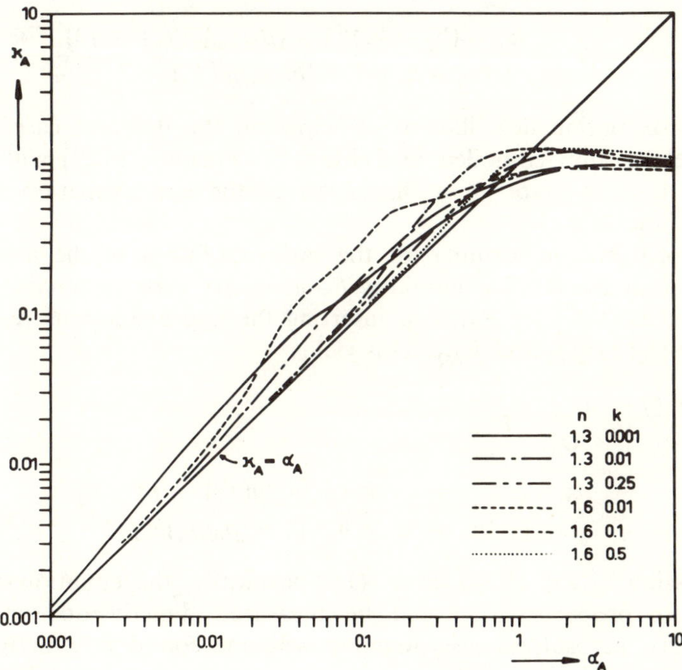

FIG. 14. Efficiency factor of absorption κ_A versus generalized size parameter α_A of absorption as computed from Mie's theory for several complex refractive indices $n - ik$.

plotted against the generalized size parameter of absorption α_A. Chosen were realistic complex refractive indices regarding atmospheric aerosol particles. It is seen from this figure that the approximations $\kappa_A \cong \alpha_A$ when $\alpha_A \lesssim 1$ and $\kappa_A \cong 1$ when $\alpha_A > 1$ can be used for rough estimates. At all wavelengths of electromagnetic radiation considered here the α_A values of the most important absorbing particles are smaller than one. Therefore, the approximation $\kappa_A \cong \alpha_A$ can be used for the explanation of the change of the twofold ratio R_{AQ} with relative humidity. This gives

$$\kappa_{Aj}(f) \cong \frac{\alpha_{Aj}(f)}{\alpha_{A0j}} \kappa_{A0j}$$

$$= \frac{r_j(f)}{r_{0j}} \cdot \frac{n_0^2 + 2}{k_0} \cdot \frac{k_w + (k_0 - k_w)(r_j(f)/r_{0j})^{-3}}{\{n_w + (n_0 - n_w)(r_j(f)/r_{0j})^{-3}\}^2 + 2} \kappa_{A0j}$$

Using $r_j(f)/r_{0j} \cong [1 + (\rho_0/\rho_w)\bar{\mu}(f/1 - f)]^{1/3}$, i.e., the approximation (6.1), we have

$$(6.29) \quad R_{AQ} \cong \left(1 + \frac{\rho_0}{\rho_w} \bar{\mu} \frac{f}{1 - f}\right)^{1/3} \frac{n_0^2 + 2}{k_0}$$

$$\times \frac{k_w + (k_0 - k_w)/[1 + (\rho_0/\rho_w)\bar{\mu}(f/1 - f)]}{\{n_w + (n_0 - n_w)/[1 + (\rho_0/\rho_w)\bar{\mu}(f/1 - f)]\}^2 + 2}$$

This approximation describes in principle all the different cases of the R_{AQ}-f relationship compiled in Table XIX. Formula (6.29) will not be analyzed here in favor of a discussion of the approximation formula following immediately.

An approximation formula for the ratio $\sigma_A(f)/\sigma_{A0}$ of the absorption coefficients at the relative humidity f and in dry state is obtained using the definition (6.27) for R_{AQ} and inserting the approximations (6.25) and (6.29) for $Q_K(f)/Q_{K0}$ and R_{AQ}. This yields

$$(6.30) \quad \frac{\sigma_A(f)}{\sigma_{A0}} = \frac{Q_K(f)}{Q_{K0}} R_{AQ}$$

$$\cong \frac{n_0^2 + 2}{k_0} \cdot \frac{k_w(\rho_0/\rho_w)\bar{\mu}(f/1 - f) + k_0}{\{n_w + (n_0 - n_w)/[1 + (\rho_0/\rho_w)\bar{\mu}(f/1 - f)]\}^2 + 2}$$

Discussion of Eq. (6.30) gives: (1) Specifically, the equation contains only those properties that can be measured directly on samples of atmospheric aerosol particles together with the complex refractive index of pure water. No information on the aerosol size frequency distribution is necessary. However, it must be kept in mind that the errors can reach $\pm 40\%$ at relative humidity 0.95.

(2) The absorption coefficient σ_A increases with relative humidity f when (a) there is absorption of electromagnetic radiation by pure water, i.e., $k_w > 0$, or (b) the real part of the mean complex refractive index in dry state is larger than the real part of the complex refractive index of pure water, i.e., $n_0 - n_w > 0$. σ_A increases with f in all the cases considered here.

(3) The absorption coefficient σ_A will decrease with relative humidity f if $n_0 - n_w < 0$ and, at the same time, $k_w = 0$ or $k_w[(\rho_0/\rho_w)\bar{\mu}(f/1 - f)] \ll k_0$. It is $n_0 - n_w < 0$ at the wavelength 3.0 μm for models 3 and 5 but, at the same time, there is strong absorption by pure water so that σ_A finally increases with f.

6.6. Applications of the Results

Not all applications of the results for practical use can be considered. For this reason only three of these will be discussed.

(1) The estimate of the standard visual range as a function of relative humidity, which is important for quantitative visibility forecasts.

(2) The humidity dependence of the ratio between the extinction coefficient and the corresponding mass of the aerosol particles. From this ratio the air pollution by particles can be estimated roughly from visibility observations.

(3) The humidity dependence of the single scattering albedo. The latter is used for calculations of heating and cooling rates in the atmosphere due to interaction between electromagnetic radiation and aerosol particles.

6.6.1. The Standard Visual Range as a Function of Relative Humidity. Provided that the aerosol particles are distributed homogeneously in a uniformly illuminated atmosphere and the threshold contrast for a blackbody, which still can be observed, is set equal to 0.02, the distance between the observer and the blackbody is given by

$$(6.31) \qquad V_S = \frac{\ln(1/0.02)}{(\sigma_E + \sigma_{EM})_{\lambda=0.55\,\mu m}} = \frac{3.91}{(\sigma_E + \sigma_{EM})_{\lambda=0.55\,\mu m}}$$

where V_S is the standard visual range (cf., e.g., Koschmieder, 1926; Middleton, 1952). The standard visual range thus is a direct measure of the extinction coefficient $(\sigma_E + \sigma_{EM})_{\lambda=0.55\,\mu m}$ of the turbid air at the wavelength of light $\lambda = 0.55$ μm. Its dependence on relative humidity (Kasten, 1969; Hänel, 1972a) has to be considered among other effects when a quantitative visibility forecast (Hänel, 1971) is needed.

Because for practical purposes, e.g., traffic matters, low visibilities are exclusively of interest, the preceding equation can be simplified, neglecting the extinction coefficient σ_{EM} of the air molecules compared to that σ_E of the aerosol particles. Then it is $V_S \cong 3.91/(\sigma_E)_{\lambda=0.55\,\mu m}$ and the ratio of the standard visual ranges $V_S(f_1)$ and $V_S(f_2)$ at the relative humidities f_1 and f_2 becomes

(6.32)
$$\frac{V_S(f_1)}{V_S(f_2)} \cong \left(\frac{\sigma_E(f_2)}{\sigma_E(f_1)}\right)_{\lambda=0.55\,\mu m}$$

Into this formula Eq. (6.26a) is introduced, yielding

$$\frac{V_S(f_1)}{V_S(f_2)} \cong \left(\frac{1 + (\rho_0/\rho_w)\bar{\mu}(f_2)(f_2/1 - f_2)}{1 + (\rho_0/\rho_w)\bar{\mu}(f_1)(f_1/1 - f_1)}\right)^{2/3}$$

This latter equation is too complicated for practical use, therefore we take the approximation formula (6.3) by Kasten (1969), instead of the approximation (6.1) for describing $Q_K(f)/Q_{K0}$ and obtain

$$\frac{V_S(f_1)}{V_S(f_2)} \cong \left(\frac{1 - f_1}{1 - f_2}\right)^{2\varepsilon}$$

The last equation however is valid only in the range of relative humidity 0.4 to $0.8 \leqq f_1, f_2 \leqq 0.9$ to 0.95. Therefore the range of applicability of this formula has been extended by definition of the so-called fitted exponent $2\varepsilon^*(\lambda)$ (Hänel, 1972a). Using this exponent it is

(6.33)
$$\frac{V_S(f_1)}{V_S(f_2)} \cong \left(\frac{1 - f_1}{1 - f_2}\right)^{2\varepsilon^*(\lambda=0.55\,\mu m)}$$

in the range of relative humidity 0.4 to $0.8 \leqq f_1, f_2 \leqq 0.999$. Thus the approximation is valid in the whole range of relative humidities, which is important for practical use. The values of $\varepsilon^*(\lambda = 0.55\,\mu m)$ are 0.325, 0.28, 0.26, 0.23, 0.345, and 0.28 for aerosol models 1 to 6, respectively. Equation (6.33) has been verified experimentally, e.g., by Pueschel *et al.* (1969), Kurtz (1972), and Covert *et al.* (1972).

It must be kept in mind that, as a result of the assumptions at the beginning of this section, the latter formula only contains the influence of relative humidity on the standard visual range. Not included are the influences due to coagulation of the particles or due to addition or removal of particles from the sample considered. Because of these preconditions an experimental study on the humidity dependence of the visibility, the standard visual range, or the extinction coefficient of the aerosol particles requires some precautions. When no further information on the aerosol size distribution and the physicochemical properties of the particles is available, only short-period variations of any of these properties can be combined with the corresponding changes of relative humidity. Thus the results from visibility statistics, which combine absolute values of visibility with absolute values of relative humidity (Neiburger and Wurtele, 1949; Buma, 1960; Fett, 1967), can be applied only with precautions. On the other hand, reliable results were obtained by Frankenberger (1967), Kasten (1968, 1969), Garland (1969), Pueschel *et al.* (1969), Eggleton (1969), Ruppersberg (1971), Charlson (1969), Kurtz (1972), and Covert *et al.* (1972).

6.6.2. The Ratio between the Extinction Coefficient and the Mass of the Aerosol Particles as a Function of Relative Humidity. Measurements of the extinction or the scattering coefficient as well as visibility observations are widely used as a measure of air pollution by particles, i.e., the mass of atmospheric aerosol particles per unit volume. Experimental and theoretic investigation on this problem are available (e.g., Charlson *et al.*, 1968; Horvath and Charlson, 1969; Horvath, 1969; Charlson, 1969). To clarify some problems related to the indirect determination of air pollution, the humidity dependence of the ratio between the extinction coefficient at the wavelength of light 0.55 μm and the corresponding mass of the aerosol particles per unit volume has been studied by the author (Hänel, 1972b). The most important results of these studies will now be discussed.

Like the total geometric cross section per unit volume [Eq. (6.15)] the total mass M_K of the aerosol particles per unit volume is defined by the sum

$$(6.34) \qquad M_K = \frac{1}{V_K} \sum_{j=1}^{j=N_K} \tfrac{4}{3}\pi r_j^3 \rho_j$$

over N_K particles within the volume V_K of the air. ρ_j is the "bulk" density of the jth particle and r_j its equivalent radius. Results for M_K will be given in gm cm^{-3}, i.e., in the same units as ρ_j.

Using Eqs. (2.30) and (2.45b), we have for the jth particle

$$r_j^3(f)\rho_j(f) = r_{0j}^3 \frac{r_j^3(f)}{r_{0j}^3} \left\{ \rho_w + (\rho_{0j} - \rho_w)\left(\frac{r_j(f)}{r_{0j}}\right)^{-3} \right\}$$

$$= r_{0j}^3 \rho_{0j} \left\{ 1 + \frac{\rho_w}{\rho_{0j}} \left[\left(\frac{r_j(f)}{r_{0j}}\right)^3 - 1 \right] \right\}$$

At relative humidities smaller than 0.9 to 0.95 the approximation (6.1) for $r_j(f)/r_{0j}$ can be used, yielding

$$r_j^3(f)\rho_j(f) \cong r_{0j}^3 \rho_{0j} \left[1 + \mu_j(f)\frac{f}{1-f} \right]$$

Thus the ratio $M_K(f)/M_{K0}$ of the total masses per unit volume at the relative humidity f and in dry state is, taking into account $\rho_{0j} = \rho_0$ and $\mu_j = \bar{\mu}$, i.e., chemical uniformity in dry state,

(6.35)

$$M_{\mathrm{K}}(f)/M_{\mathrm{K}0} = \left(\sum_{j=1}^{j=N_{\mathrm{K}}} \tfrac{4}{3}\pi r_j^3(f)\rho_j(f) \right) \Big/ \left(\sum_{j=1}^{j=N_{\mathrm{K}}} \tfrac{4}{3}\pi r_{0j}^3 \rho_{0j} \right) \cong 1 + \bar{\mu}(f)\,\frac{f}{1-f}$$

when $f \leqq 0.9$ to 0.95. The values of $M_{\mathrm{K}0}(f)/M_{\mathrm{K}0}$ are compiled in Table XX for all aerosol models.

The ratio between the extinction coefficient σ_{E} and the total mass M_{K} of the aerosol particles is given by

(6.36)
$$\sigma_{\mathrm{E}}/M_{\mathrm{K}} = \sum_{j=1}^{j=N_{\mathrm{K}}} \pi r_j^2 \kappa_{\mathrm{E}j} \Big/ \sum_{j=1}^{j=N_{\mathrm{K}}} \tfrac{4}{3}\pi r_j^3 \rho_j$$

following definitions (6.12) and (6.34) for σ_{E} and M_{K}. Based on this equation and on approximations (6.25) and (6.26a) for $\sigma_{\mathrm{E}}(f)/\sigma_{\mathrm{E}0}$ in the short wavelength region and on approximation (6.35) for $M_{\mathrm{K}}(f)/M_{\mathrm{K}0}$ we obtain the twofold ratio

(6.37)
$$R_{\mathrm{EM}} = \left(\frac{\sigma_{\mathrm{E}}(f)}{M_{\mathrm{K}}(f)} \right) \Big/ \left(\frac{\sigma_{\mathrm{E}0}}{M_{\mathrm{K}0}} \right) \simeq \left(\frac{Q_{\mathrm{K}}(f)}{Q_{\mathrm{K}0}} \right) \Big/ \left(\frac{M_{\mathrm{K}}(f)}{M_{\mathrm{K}0}} \right)$$

$$\cong \left(1 + \frac{\rho_0}{\rho_{\mathrm{w}}} \bar{\mu} \frac{f}{1-f} \right)^{2/3} \Big/ \left(1 + \bar{\mu} \frac{f}{1-f} \right)$$

R_{EM} describes how the ratio between the extinction coefficient and the total mass per unit volume has changed from the dry state to one with the relative humidity f considered. The twofold ratio R_{EM} increases with relative humidity when the total geometric cross section per unit volume of air increases more rapidly than the total mass per unit volume of air does. This is true as long as the water uptake per unit mass of dry substance $\bar{\mu}(f)(f/1 - f)$ does not exceed a value close to that of the ratio ρ_0/ρ_{w} of the mean density of dry substance and the density of water (cf. the ratios $Q_{\mathrm{K}}(f)/Q_{\mathrm{K}0}$ from Table XIII with the ratios $M_{\mathrm{K}}(f)/M_{\mathrm{K}0}$ from Table XX). At relative humidities larger than 0.7 to 0.975 the twofold ratio R_{EM} decreases with relative humidity because then the extinction coefficient increases with r_j^2 and the mass with r_j^3. This complex behavior of the twofold ratio R_{EM} during changes of relative humidity is demonstrated within Table XX, where values of R_{EM} at the wavelength of light 0.55 μm are compiled as a function of relative humidity for all aerosol models 1 to 6. This specific wavelength of light has been chosen because of its importance for visibility considerations. The values show that the difference between increasing and decreasing relative humidity is small in each case. Additionally

TABLE XX. Standardized mass M_{K0}/N_{K1} in dry state and ratio $(\sigma_{E0}/M_{K0})_{\lambda=0.55\,\mu m}$ between the extinction coefficient at the wavelength of light 0.55 μm and the total mass per unit volume of air of dry aerosol particles for the aerosol models 1 to 6. Ratio $M_K(f)/M_{K0}$ between the total masses per unit volume of the aerosol particles at the relative humidity f and in dry state. Values of the twofold ratio $R_{EM} = [\sigma_E(f)/M_K(f)]/[\sigma_{E0}/M_{K0}]$ between the extinction coefficients and the total masses at the relative humidity f and in dry state for the wavelength of light $\lambda = 0.55\,\mu m$

Model	1	2	3	4	5	6
M_{K0}/N_{K1} (gm)	1.55×10^{-14}	1.26×10^{-12}	1.16×10^{-13}	2.12×10^{-12}	1.46×10^{-14}	4.17×10^{-14}
σ_{E0}/M_{K0}(cm^2/gm)	1.03×10^4	8.34×10^3	1.48×10^3	4.58×10^3	1.00×10^4	1.49×10^4 $\quad(\lambda = 0.55\,\mu m)$

Model	1	2	3	4	5	6	1	2	3	4	5	6
			$M_K(f)/M_{K0}$						$(R_{EM})_{\lambda=0.55\,\mu m}$			
inc. f												
0.2	1.02	1.00	1.01	1.01	1.01	1.01	1.02	1.00	1.01	1.00	1.01	1.00
0.4	1.06	1.06	1.02	1.01	1.02	1.02	1.06	1.01	1.01	1.01	1.02	1.00
0.6	1.19	1.20	1.06	1.03	1.10	1.06	1.13	1.05	1.04	1.02	1.08	1.01
0.7	1.37	1.50	1.30	1.21	1.25	1.08	1.20	1.07	1.13	1.12	1.16	1.02
0.75	1.47	3.51	1.94	1.39	1.48	1.26	1.23	0.972	1.19	1.15	1.24	1.07
0.8	1.61	4.08	2.38	1.52	1.72	1.59	1.28	0.920	1.21	1.14	1.32	1.09
0.85	1.85	4.97	2.72	1.68	1.99	1.88	1.32	0.851	1.22	1.15	1.37	1.09
0.9	2.28	6.62	3.49	1.94	2.32	2.07	1.37	0.800	1.19	1.13	1.42	1.10
0.95	3.53	11.2	5.53	2.40	4.03	2.93	1.43	0.631	1.13	1.15	1.47	1.09
0.975	6.27	19.8	9.54	3.74	7.44	4.90	1.42	0.568	0.969	1.11	1.45	1.03
0.99	14.9	44.0	20.3	7.25	18.1	11.7	1.27	0.410	0.702	0.854	1.29	0.883
0.995	29.1	87.2	37.8	13.3	35.3	22.6	1.07	0.329	0.593	0.665	1.09	0.711
dec. f												
0.2	1.02	1.05	1.01	1.01	1.02	1.03	1.02	1.04	1.01	1.01	1.01	1.00
0.4	1.10	1.16	1.04	1.03	1.05	1.04	1.08	1.04	1.03	1.02	1.04	1.01
0.6	1.30	1.71	1.14	1.08	1.17	1.08	1.18	1.10	1.08	1.06	1.13	1.02
0.65	1.34	2.72	1.58	1.24	—	1.10	1.19	1.03	1.16	1.12	—	1.02
0.7	1.41	3.19	1.99	1.41	1.39	1.19	1.21	0.987	1.20	1.15	1.22	1.04
0.75	1.68	—	2.15	1.48	1.67	—	—	—	1.21	1.15	1.30	—

in Table XX there are compiled values for the ratio $(\sigma_{E0}/M_{K0})_{\lambda=0.55\,\mu m}$ of the extinction coefficient and the standardized mass M_{K0}/N_{K1} in dry state for all aerosol models. The arithmetic mean of these values is $1.05 \times 10^4\,\text{cm}^2\,\text{gm}^{-1} \pm 0.59 \times 10^4\,\text{cm}^2\,\text{gm}^{-1}$ and thus close to those found experimentally by the authors mentioned in the beginning of this section.

The R_{EQ} values compiled in Table XX demonstrate that the ratio $\sigma_E(f)/M_K(f)_{\lambda=0.55\,\mu m}$ between the extinction coefficient and the total mass per unit volume of air cannot be regarded as independent of relative humidity, unless an error of $\pm 50\%$ at relative humidities smaller than 0.975 can be accepted. Thus the determination of the total mass per unit volume of air by visibility observation or by measurement of the extinction coefficient can, at the best, yield only the order of magnitude.

6.6.3. The Single Scattering Albedo of Aerosol Particles as a Function of Relative Humidity. The discussion of the general circulation of the atmosphere includes the effect of radiative heating and cooling of the air containing aerosol particles (cf. Chandrasekhar, 1950; Yamamoto and Tanaka, 1972; Eschelbach, 1973; Grassl, 1973a, b, 1974). In this context the single scattering albedo ω of atmospheric aerosol particles is a property of immediate interest, which will be discussed now: We now consider a beam of electromagnetic radiation penetrating a volume of atmospheric air containing aerosol particles. Then ω is defined by

$$(6.38) \qquad \omega = 1 - \sigma_A/\sigma_E$$

for each wavelength of radiation. ω is the fraction of the attenuated electromagnetic radiation of the particles due to scattering. $1 - \omega$ is the remaining fraction transformed into another form of energy or of electromagnetic radiation of another wavelength. Due to this physical meaning of the single scattering albedo of the aerosol particles, its dependence upon relative humidity is of interest.

The single scattering albedo of the dry particles

$$(6.39) \qquad \omega_0 = 1 - \sigma_{A0}/\sigma_{E0}$$

is compiled in Table XXI for aerosol models 3, 5, and 6 and wavelengths of radiation from 0.3 to 12.0 μm. These data show that the ω_0 values at the wavelength 9.25 μm are the lowest for each aerosol type, since at this wavelength the imaginary part of the mean complex refractive index attains the largest value.

In Table XXII the ratio of the single scattering albedos of the particles at the relative humidity f and in dry state

$$(6.40) \qquad \frac{\omega(f)}{\omega_0} = \frac{1}{\omega_0} + \left(1 - \frac{1}{\omega_0}\right)\left(\frac{\sigma_A(f)}{\sigma_{A0}}\right)\left(\frac{\sigma_E(f)}{\sigma_{E0}}\right)^{-1}$$

TABLE XXI. Single scattering albedo ω_0 in dry state versus wavelength λ of radiation for the aerosol models 3, 5, and 6

$\lambda(\mu m)$:	0.3	0.55	1.0	1.3	1.65	2.0	2.5	3.0	9.25	10.0	12.0
Model						ω_0					
3	0.628	0.739	0.673	0.604	0.678	0.467	0.465	0.279	0.224	0.438	0.414
5	0.790	0.723	0.632	0.559	0.482	0.427	0.406	0.224	0.163	0.254	0.347
6	0.890	0.877	0.837	0.799	0.756	0.715	0.708	0.534	0.400	0.666	0.666

are compiled. A few examples of direct plots of the single scattering albedo ω as a function of the relative humidity f are given in Fig. 15. Since in most cases the differences between increasing and decreasing relative humidity are very small, only the values for increasing relative humidity are plotted.

The following features have been derived from the results: (1) At all wavelengths of electromagnetic radiation, except 12.0 μm as well as 3.0 and 10.0 μm for the model 6, there is an increase of the single scattering albedo of the particles with increasing relative humidity.

(2) The imaginary part of the complex refractive index of water is either zero or negligibly small compared with the imaginary part of the mean complex refractive index of the dry substance when the wavelength of electromagnetic radiation lies between 0.3 and 2.5 μm. In this case, insertion of the approximations (6.26a) and (6.30) for $\sigma_E(f)/\sigma_{E0}$ and $\sigma_A(f)/\sigma_{A0}$ into Eq. (6.40) gives

$$(6.41) \quad \frac{\omega(f)}{\omega_0} \cong \frac{1}{\omega_0} + \left(1 - \frac{1}{\omega_0}\right) \frac{n_0^2 + 2}{k_0}$$

$$\times \frac{k_w[1 + (\rho_0/\rho_w)\overline{\mu}(f/1 - f)]^{1/3} + (k_0 - k_w)[1 + (\rho_0/\rho_w)\overline{\mu}(f/1 - f)]^{-2/3}}{\{n_w + (n_0 - n_w)/[1 + (\rho_0/\rho_w)\overline{\mu}(f/1 - f)]\}^2 + 2}$$

This formula is a useful approximation for relative humidities up to 0.995. At all the other wavelengths of radiation, where an increase of the single scattering albedo with increasing relative humidity has been found, there are no good approximations of this generality.

(3) At the wavelength 12.0 μm and for model 6 at the wavelengths 3.0 and 10.0 μm, the single scattering albedo of the aerosol particles first decreases with increasing relative humidity and then at the largest relative humidities increases again. This latter increase comes from the tendency of the single scattering albedo to approach a value close to 0.5 (cf. Deirmendjian, 1969), when pure water absorbs electromagnetic radiation and the generalized size parameters of the most significantly absorbing particles attain values close to or larger than one.

TABLE XXII. Ratio $\omega(f)/\omega_0$ of the single scattering albedos at the relative humidity f and in dry state versus wavelength λ of radiation for the aerosol models 3, 5, and 6[a]

Model 3

$\lambda(\mu m)$:	0.3	0.55	1.0	1.3	1.65	2.0	2.5	3.0	9.25	10.0	12.0
inc. f						$\omega(f)/\omega_0$					
0.2	1.00	1.00	1.00	1.00	1.00	1.01	1.01	1.00	1.00	1.00	0.99
0.4	1.01	1.01	1.00	1.01	1.01	1.01	1.01	1.00	1.00	1.00	0.97
0.5	1.02	1.01	1.01	1.01	1.02	1.02	1.02	1.00	1.00	1.00	0.95
0.6	1.03	1.02	1.03	1.02	1.03	1.05	1.04	1.02	1.01	1.00	0.91
0.65	1.06	1.05	1.06	1.06	1.07	1.10	1.09	1.02	1.03	1.00	0.81
0.7	1.10	1.08	1.11	1.12	1.12	1.20	1.18	1.06	1.09	1.01	0.67
0.75	1.19	1.16	1.23	1.28	1.23	1.46	1.41	1.10	1.31	1.01	0.46
0.8	1.24	1.19	1.27	1.34	1.27	1.56	1.51	1.12	1.44	1.02	0.42
0.85	1.27	1.21	1.30	1.38	1.29	1.61	1.57	1.16	1.53	1.02	0.40
0.9	1.30	1.23	1.33	1.44	1.33	1.70	1.66	1.22	1.71	1.04	0.38
0.95	1.37	1.27	1.39	1.51	1.39	1.83	1.80	1.26	2.03	1.08	0.37
0.975	1.41	1.29	1.43	1.57	1.42	1.94	1.91	1.30	2.31	1.12	0.39
0.99	1.44	1.31	1.45	1.61	1.45	2.02	1.99	1.43	2.56	1.17	0.43
0.995	1.50	1.32	1.46	1.63	1.46	2.05	2.03	1.50	2.69	1.20	0.47
dec. f											
0.2	1.01	1.00	1.01	1.01	1.01	1.01	1.01	1.00	1.00	1.00	0.98
0.4	1.02	1.01	1.02	1.01	1.02	1.02	1.03	1.02	1.01	1.00	0.94
0.5	1.03	1.02	1.03	1.03	1.04	1.04	1.05	1.02	1.01	1.00	0.89
0.6	1.06	1.04	1.06	1.06	1.07	1.10	1.09	1.02	1.03	1.00	0.81
0.65	1.15	1.13	1.18	1.20	1.18	1.32	1.30	1.08	1.19	1.00	0.54
0.7	1.20	1.17	1.23	1.28	1.24	1.45	1.42	1.08	1.32	1.01	0.46
0.75	1.22	1.18	1.25	1.31	1.26	1.49	1.46	1.11	1.37	1.01	0.44

Model 5

$\lambda(\mu m)$:	0.3	0.55	1.0	1.3	1.65	2.0	2.5	3.0	9.25	10.0	12.0
inc. f						$\omega(f)/\omega_0$					
0.2	1.00	1.00	1.01	1.01	1.01	1.01	1.01	1.00	1.00	1.00	0.98
0.4	1.01	1.01	1.01	1.01	1.02	1.02	1.02	1.00	1.00	1.00	0.96
0.5	1.01	1.01	1.02	1.02	1.03	1.03	1.03	1.03	1.01	1.00	0.94
0.6	1.03	1.04	1.06	1.07	1.09	1.08	1.08	1.05	1.03	1.01	0.84
0.65	1.05	1.07	1.09	1.12	1.14	1.14	1.14	1.05	1.07	1.03	0.76
0.7	1.06	1.09	1.13	1.16	1.20	1.20	1.20	1.06	1.11	1.04	0.69
0.75	1.10	1.14	1.20	1.26	1.33	1.35	1.34	1.11	1.24	1.10	0.57
0.8	1.12	1.17	1.25	1.33	1.42	1.47	1.43	1.14	1.37	1.15	0.50
0.85	1.14	1.20	1.30	1.39	1.50	1.56	1.56	1.14	1.49	1.19	0.46
0.9	1.16	1.23	1.34	1.44	1.58	1.65	1.65	1.17	1.63	1.23	0.42

TABLE XXII—*Continued*

Model 5-*Continued*

$\lambda(\mu m)$:	0.3	0.55	1.0	1.3	1.65	2.0	2.5	3.0	9.25	10.0	12.0
inc. f						$\omega(f)/\omega_0$					
0.95	1.19	1.29	1.43	1.58	1.77	1.91	1.94	1.29	2.18	1.42	0.39
0.975	1.22	1.33	1.50	1.67	1.90	2.08	2.13	1.39	2.67	1.57	0.40
0.99	1.24	1.35	1.54	1.73	1.99	2.20	2.27	1.48	3.17	1.73	0.43
0.995	1.25	1.36	1.56	1.76	2.03	2.24	2.32	1.56	3.41	1.84	0.48
dec. f											
0.2	1.01	1.01	1.01	1.01	1.01	1.02	1.01	1.01	1.00	1.00	0.97
0.4	1.02	1.02	1.03	1.04	1.05	1.06	1.05	1.02	1.02	1.01	0.91
0.6	1.05	1.07	1.09	1.12	1.15	1.17	1.14	1.05	1.07	1.03	0.75
0.7	1.09	1.12	1.17	1.22	1.28	1.33	1.29	1.11	1.19	1.08	0.61
0.75	1.12	1.17	1.24	1.32	1.40	1.48	1.43	1.14	1.34	1.14	0.51
0.8	1.13	1.20	1.28	1.37	1.48	1.57	1.52	1.15	1.44	1.17	0.47
0.85	1.15	1.22	1.31	1.41	1.53	1.65	1.60	1.17	1.55	1.21	0.44

Model 6

$\lambda(\mu m)$:	0.3	0.55	1.0	1.3	1.65	2.0	2.5	3.0	9.25	10.0	12.0
inc. f						$\omega(f)/\omega_0$					
0.2	1.00	1.00	1.00	1.00	1.00	1.00	1.00	0.99	1.00	1.00	0.99
0.4	1.00	1.00	1.00	1.00	1.00	1.01	1.00	0.97	1.00	1.00	0.96
0.5	1.00	1.00	1.01	1.01	1.00	1.01	1.01	0.96	1.01	0.99	0.93
0.6	1.01	1.01	1.01	1.01	1.01	1.02	1.02	0.90	1.01	0.99	0.88
0.65	1.01	1.01	1.01	1.02	1.01	1.02	1.02	0.89	1.02	0.98	0.85
0.7	1.01	1.01	1.02	1.02	1.02	1.03	1.02	0.88	1.02	0.98	0.84
0.75	1.02	1.03	1.04	1.05	1.02	1.07	1.06	0.75	1.05	0.94	0.63
0.8	1.04	1.05	1.07	1.09	1.06	1.13	1.11	0.66	1.09	0.89	0.45
0.85	1.05	1.06	1.09	1.11	1.11	1.16	1.15	0.63	1.13	0.85	0.37
0.9	1.05	1.07	1.10	1.13	1.14	1.18	1.16	1.61	1.15	0.84	0.34
0.95	1.07	1.09	1.12	1.16	1.16	1.23	1.22	0.60	1.21	0.79	0.28
0.975	1.08	1.11	1.15	1.19	1.20	1.29	1.27	0.61	1.30	0.76	0.24
0.99	1.09	1.12	1.17	1.22	1.25	1.33	1.33	0.64	1.39	0.74	0.25
0.995	1.10	1.13	1.18	1.24	1.29	1.35	1.34	0.68	1.45	0.75	0.27
dec. f											
0.2	1.00	1.00	1.00	1.01	1.01	1.01	1.01	0.96	1.01	0.99	0.94
0.4	1.00	1.01	1.01	1.01	1.01	1.01	1.01	0.93	1.01	0.99	0.91
0.6	1.01	1.01	1.01	1.02	1.02	1.02	1.02	0.88	1.02	0.98	0.85
0.7	1.02	1.02	1.03	1.04	1.05	1.06	1.05	0.79	1.04	0.96	0.70
0.75	1.04	1.04	1.07	1.08	1.10	1.12	1.10	0.68	1.09	0.90	0.48
0.8	1.05	1.06	1.09	1.11	1.14	1.16	1.15	0.63	1.13	0.86	0.38
0.85	1.05	1.07	1.10	1.12	1.15	1.18	1.16	0.62	1.14	0.84	0.35

[a] inc. = increasing, dec. = decreasing.

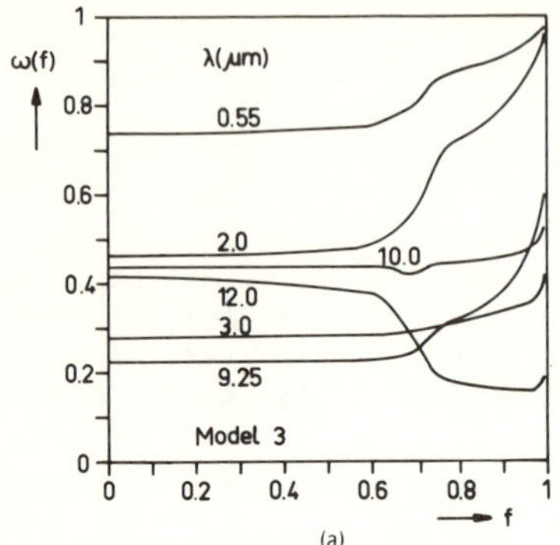

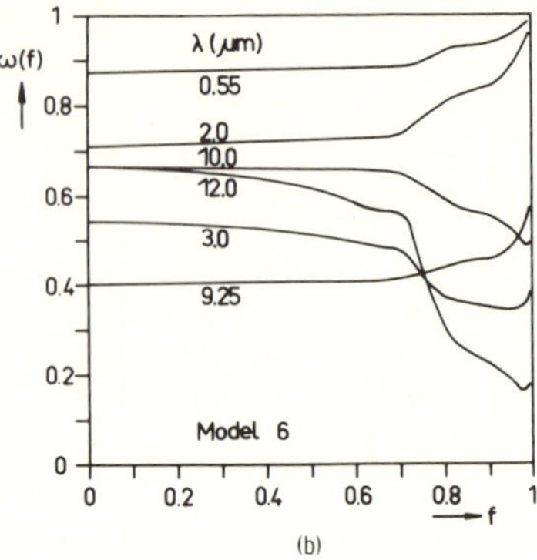

FIG. 15. Single scattering albedo ω as a function of relative humidity f at selected wavelengths λ of light for the aerosol models 3 and 6.

The results indicate that the dependence of the single scattering albedo of atmospheric aerosol particles upon relative humidity cannot be neglected when the relative humidity is larger than about 0.6. Thus a discussion of radiative transfer within the atmosphere must take this effect into account.

7. CONCLUDING REMARKS

In this review equations have been derived and empirical data have been compiled allowing the first realistic model calculations on the humidity dependence of a large variety of properties of atmospheric aerosol particles which are of meteorological interest. These are the size, the mean complex refractive index, and the extinction, the scattering, and the absorption coefficients. All the equations and the numerical results are valid for thermodynamic equilibrium between particles and the surrounding moist air. The formation or the disintegration of a cloud or an unstable fog and the dependent changes of aerosol properties have not been considered. These thermodynamic nonequilibrium cases will be treated in the future.

The application of the measured data is restricted to relative humidities smaller than 0.995, when errors involved are kept small. This is due to the poor knowledge of the surface tension and the structure of the particle's liquid-air interface. Studies are necessary on these subjects because of their immediate impact on cloud and fog formation as well as disintegration. Therefore, more measurements on particles of different aerosol types, including surface tension measurements, are necessary.

Not all the physical properties of atmospheric aerosol particles that are dependent upon relative humidity have been mentioned. Nevertheless, the results show that relative humidity has a large influence on aerosol properties and thus must be taken into account.

Acknowledgments

The author thanks all his colleagues, especially Professor Kurt Bullrich for discussions and the Deutsche Forschungsgemeinschaft (Sonderforschungsbereich "Atmosphärische Spurenstoffe"), Bad Godesberg, Germany, as well as the Air Force Cambridge Research Laboratories, Bedford, Massachusetts, for financial support.

List of Symbols

a Coefficient describing the temperature dependence of the surface tension of pure water

a_w Water activity of a particle's liquid cover with a hypothetical plane surface in absence of dry air

b Coefficient describing the concentration dependence of the surface tension of a solution

f Relative humidity

f_c Critical relative humidity

i Van't Hoff factor of a pure salt

$\bar{i}$ Mean van't Hoff factor of a mixed solute

k Imaginary part of the mean complex refractive index

k_0 Imaginary part of the mean complex refractive index in dry state

k_w Imaginary part of the complex refractive index of pure liquid water

m_0 Mass of the dry substance

m_s Mass of dry salt or dry mixed solute

m_w Mass of water

M_K Total mass of airborne atmospheric aerosol particles per unit volume of air

M_{K0} Total mass of dry airborne atmospheric aerosol particles per unit volume of air

n Real part of the mean complex refractive index

n_0 Real part of the mean complex refractive index in dry state

n_w Real part of the complex refractive index of pure liquid water

N_K Total number of airborne particles within the volume V_K of air

N_{K1} Total number of airborne particles per unit volume

p Total pressure of air

p_a Partial water vapor pressure within the air

p_c Critical partial water vapor pressure over a particle's curved surface

p_r Equilibrium partial water vapor pressure over a particle's curved surface

p_w Equilibrium partial water vapor pressure over a plane surface of water saturated with dry air at the pressure p and the absolute temperature T

Q_K Total geometric cross section of the equivalent spheres of airborne atmospheric aerosol particles per unit volume of air

Q_{K0} Total geometric cross section of the equivalent spheres of dry airborne atmospheric aerosol particles per unit volume of air

r Equivalent radius of a particle, i.e., the radius of a sphere with the same volume

r_c Critical equivalent radius of a particle

r_0 Equivalent radius of a dry particle

R_{AQ} Twofold ratio between the absorption coefficients and the total geometric cross sections of the equivalent spheres of aerosol particles at the relative humidity f and in dry state

R_{EQ} Twofold ratio between the extinction coefficients and the total geometric cross sections of the equivalent spheres of aerosol particles at the relative humidity f and in dry state

R_{EM} Twofold ratio between the extinction coefficients and the total masses of aerosol particles at the relative humidity f and in dry state

R_W Specific gas constant of pure water

T Absolute temperature at the particle's surface

V Volume

V_0 Volume in dry state

V_W Volume of water

α Size parameter

α_A Generalized size parameter of absorption

α_{A0} Generalized size parameter of absorption in dry state

α_E Generalized size parameter of extinction

α_{E0} Generalized size parameter of extinction in dry state

δ Curvature correction term

η Exponential mass increase coefficient of an aerosol particle

η^0 Exponential mass increase coefficient of an aerosol particle at infinite dilution

$\bar{\eta}$ Exponential mass increase coefficient of an aerosol sample

$\bar{\eta}^0$ Exponential mass increase coefficient of an aerosol sample at infinite dilution

η_s Exponential mass increase coefficient of one pure solute

η_s^0 Exponential mass increase coefficient of one pure solute at infinite dilution

$\bar{\eta}_s$ Exponential mass increase coefficient of a mixture of solutes

$\bar{\eta}_s^0$ Exponential mass increase coefficient of a mixture of solutes at infinite dilution

$\bar{\eta}_s^{id}$ Idealized exponential mass increase coefficient of a mixture of solutes

η^{id} Idealized exponential mass increase coefficient of a particle

κ_A Efficiency factor of absorption of a particle

κ_{A0} Efficiency factor of absorption in dry state

κ_E Efficiency factor of extinction of a particle

κ_{E0} Efficiency factor of extinction in dry state

κ_S Efficiency factor of scattering of a particle

κ_{S0} Efficiency factor of scattering in dry state

λ Wavelength of light

μ Linear mass increase coefficient of a particle

$\bar{\mu}$ Linear mass increase coefficient of an aerosol sample

μ_s Linear mass increase coefficient of one pure solute

μ_s^0 Linear mass increase coefficient of one pure solute at infinite dilution

$\bar{\mu}_s$ Linear mass increase coefficient of a mixture of solutes

$\bar{\mu}_s^0$ Linear mass increase coefficient of a mixture of solutes at infinite dilution

$\bar{\mu}_s^{id}$ Idealized linear mass increase coefficient of a mixture of solutes

μ^{id} Idealized linear mass increase coefficient of a particle

ν Number of moles of ions formed from one mole of solute

ρ Mean density

ρ_0 Mean density in dry state

ρ_w Density of pure liquid water

σ Surface tension of the liquid cover of a particle

σ_A Absorption coefficient of airborne aerosol particles

σ_{A0} Absorption coefficient of airborne aerosol particles in dry state

σ_E Extinction coefficient of airborne aerosol particles

σ_{E0} Extinction coefficient of airborne aerosol particles in dry state

σ_{EM} Extinction coefficient of air molecules

σ_S Scattering coefficient of airborne aerosol particles

σ_{S0} Scattering coefficient of airborne aerosol particles in dry state

ϕ Practical osmotic coefficient of one pure solute
$\bar{\phi}$ Practical osmotic coefficient of a mixture of solutes
ω Single scattering albedo of airborne particles
ω_0 Single scattering albedo of dry airborne particles

REFERENCES

Arago, D. F. J., and Biot, I. B. (1806). E.g. *Mém. Acad. Fr.* **7**.

Arons, A. B., and Keith, C. H. (1954). The growth of sea-salt particles by condensation of atmospheric water vapour. *J. Meteorol.* **11**, 173–184.

Barger, W. R., and Garrett, W. D. (1970). Surface active organic material in the marine atmosphere. *J. Geophys. Res.* **75**, 4561–4566.

Bergstrom, R. W. (1973). Extinction and absorption coefficients of the atmospheric aerosol as a function of particle size. *Contrib. Atmos. Phys.* **46**, 223–234.

Blanchard, D. C. (1968). Surface active organic material on airborne salt particles. *Proc. Int. Conf. Cloud Physics*, Toronto, Canada.

Bodmann, O. (1969a). Partielle spezifische Refraktionen von Polymethylmethacrylat und Polystyrol. I. Einfluss verschiedener Lösungsmittel. *Makromol. Chem.* **122**, 196–209.

Bodmann, O. (1969b). Partielle spezifische Refraktionen von Polymethylmethacrylat und Polystyrol. II. Eine Methode zur Endgruppenanalyse. *Makromol. Chem.* **122**, 210–222.

Brunauer, S., Emmett, P. H., and Teller, E. (1938). Adsorption of gases in multimolecular layers. *J. Am. Chem. Soc.* **60**, 309.

Bullrich, K. (1964). Scattered radiation in the atmosphere and the natural aerosol. *Adv. Geophys.* **10**, 99–200.

Bullrich, K., Eschelbach, G., Fischer, K., and Hänel, G. (1972). New aspects of scattering and absorbing properties of atmospheric aerosol particles. *J. Colloid Interface Sci.* **39**, 546–550.

Buma, T. J. (1960). A statistical study between visibility and relative humidity at Leeuwarden. *Bull. Am. Meteorol. Soc.* **41**, 357–360.

Butler, R. D. (1933). Immersion liquids of intermediate refraction. *Am. Mineral.* **18**, 386–401.

Chandrasekhar, S. (1950). "Radiative Transfer." Oxford Univ. Press (Clarendon), London and New York.

Charlson, R. J. (1969). Atmospheric visibility related to aerosol mass concentration. *Environ. Sci. Technol.* **3**, 913–918.

Charlson, R. J., Ahlquist, N. C., and Horvath, H. (1968). On the generality of correlation of atmospheric aerosol mass concentration and light scatter. *Atmos. Envir.* **2**, 455–464.

Chen, C.-S. (1971). Relationship between dynamical and microphysical processes in clouds. Ph.D. Dissertation, University of California, Los Angeles.

Covert, D. S., Charlson, R. J. and Ahlquist, N. C. (1972). A study of the relationship of chemical composition and humidity to light scattering by aerosols. *J. Appl. Meteorol.* **11**, 968–976.

Dalal, N. (1947). Über das Feuchtewachstum von Schwebeteilchen verschiedener Hygroskopizität. Ph.D. Dissertation, University of Heidelberg.

Dale, D., and Gladstone, F. (1858). *Philos. Trans. Roy. Soc. London* **148**, 887.

Dale, D., and Gladstone, F. (1864). *Philos. Trans. Roy. Soc. London* **153**, 317.

Defay, R., Prigogine, I., Bellemans, A., and Everett, D. H. (1966). "Surface Tension and Adsorption." Longmans, Green, New York.

Deirmendjian, D. (1969). "Electromagnetic Scattering on Spherical Polydispersions." Amer. Elsevier, New York.

Dessens, H. (1947). Les noyaux de condensation de l'atmosphère. *Meteorologie* pp. 321–327.

Dieterici, C. (1922). Das Refraktionsvermögen von Flüssigkeiten. *Ann. Phys.* **67**, 337–351.

Dietrici, C. (1923). Über das Refraktionsvolumen sehr verdünnter Lösungen. *Ann. Phys.* **70**, 558–560.

Dietrich, G. (1954). "Ozeanographie." Georg Westermann Verlag, Braunschweig.

Dufour, L., and Defay, R. (1963). "Thermodynamics of clouds." (Int. Geophys. Ser., vol. 6) Academic Press, New York.

Eggleton, A. E. J. (1969). The chemical composition of atmospheric aerosols on Tees-side and its relation to visibility. *Atmos. Environ.* **3**, 355–372.

Eiden, R. (1966). The elliptical polarization of light scattered by a volume of atmospheric air. *Appl. Opt.* **5**, 569–575.

Eiden, R. (1971). Determination of the complex index of refraction of spherical aerosol particles. *Appl. Opt.* **10**, 749–754.

Eiden, R., and Eschelbach, G. (1973). Das atmosphärische Aerosol und seine Bedeutung für den Energiehaushalt der Atmosphäre. *Z. Geophys.* **39**, 189–228.

Eschelbach, G. (1973). Calculs numériques concernant la luminance, le taux de polarisation et les divergences du flux énergétique de la région visible du spectre solaire dans l'atmosphère troublé tenant compte de la diffusion multiple. *Ann. Geophys.* **29**, 329–339.

Fett, W. (1967). Sichtweite und atmosphärisches Aerosol in Berlin-Dahlem. *Contrib. Atmos. Phys.* **40**, 262–278.

Fischer, K. (1971). Bestimmung der Absorption von sichtbarer Strahlung durch Aerosolpartikeln. *Contrib. Atmos. Phys.* **43**, 244–254.

Fischer, K. (1973). Mass absorption coefficient of natural aerosol particles in the 0.4–2.4 μm wavelength interval. *Contrib. Atmos. Phys.* **46**, 89–100.

Fischer, K. (1975). Strahlungsabsorption durch Aerosol in den Spektralbereichen atmosphärischer Transparenz. Ph.D. Dissertation. Nat. Fachbereiche, Joh. Gutenberg-Universität, Mainz.

Fischer, K., and Hänel, G. (1972). Bestimmung physikalischer Eigenschaften atmosphärischer Aerosolteilchen über dem Atlantik. "*Meteor*" *Forschungsergeb, Reihe B*, **8**, 59–62.

Fletcher, N. H. (1962). "The Physics of Rainclouds." Cambridge Univ. Press, London and New York.

Frankenberger, E. (1964). Beziehungen zwischen der Normsichtweite und der relativen Feuchte nach Messungen in Quickborn. *Contrib. Atmos. Phys.* **37**, 183–196.

Frankenberger, E. (1967). Über die Normsicht bei Quickborn/Hol. und den Austausch von trubendem Aerosol. *Ber. Dtsch. Wetterdienstes.* **106** (Bd. 15).

Gaines, G. L. (1966). "Insoluble Monolayers at Liquid-Gas Interfaces." Wiley (Interscience), New York.

Garland, J. A. (1969). Condensation on ammonium sulphate particles and its effect on visibility. *Atmos. Environ.* **3**, 347–354.

Garrett, W. D. (1967a). Stabilization of air bubbles at the air-sea interface by surface active material. *Deep-Sea Res.* **14**, 661–672.

Garrett, W. D. (1967b). The organic chemical composition of the ocean surface. *Deep-Sea Res.* **14**, 221–227.

Gibson, R. E. (1934). The influence of concentration on the compressions of aqueous solutions of certain sulfates and a note on the representation of the compressions of aqueous solutions as a function of pressure. *J. Am. Chem. Soc.* **56**, 4–14.

Gibson, R. E. (1935). The influence of the concentration and nature of the solute on the compressions of certain aqueous solutions. *J. Am. Chem. Soc.* **57**, 284.

Glasstone, S. (1964). "Textbook of Physical Chemistry." Macmillan, London.

Goetz, A., and Pueschel, R. F. (1965). The effect of nucleating particulates on photochemical aerosol formation. *APCA J.* **15**, 90–95.

Goff, J. A., and Gratch, S. (1945). Thermodynamic properties of moist air. *ASHRAE Trans.* **51**, 125.

Grassl, H. (1971). Bestimmung der Grössenverteilung von Wolkenelementen aus spektralen Transmissionsmessungen. *Contrib. Atmos. Phys.* **43**, 255–284.

Grassl, H. (1973a). Separation of atmospheric absorbers in the 8–13 micrometer region. *Contrib. Atmos. Phys.* **46**, 75–88.

Grassl, H. (1973b). Aerosol influence on radiative cooling. *Tellus* **25**, 386–395.

Grassl, H. (1974). Einfluss verschiedener Absorber des Fensterbereiches auf Abkühlungsraten und auf die Bestimmung der Oberflächentemperatur. *Contrib. Atmos. Phys.* **47**, 1–13.

Hänel, G. (1966). Vorbereitende Studien zur Messung des Brechungsindex atmosphärischer Aerosolteilchen mit vorläufigen Messergebnissen des Realteils bei verschiedenen relativen Luftfeuchtigkeiten. Diplomarbeit, Institut für Meteorologie, Johannes Gutenberg-Universität, Mainz.

Hänel, G. (1968). The real part of the mean complex refractive index and the mean density of samples of atmospheric aerosol particles. *Tellus* **20**, 371–379.

Hänel, G. (1969). Messung des Realteils des mittleren Brechungsindex und der mittleren Dichte von Proben atmosphärischer Aerosolteilchen. *Ann. Meteorol.* [NF] **4**, 138–139.

Hänel, G. (1970a). Die Grösse atmosphärischer Aerosolteilchen als Funktion der relativen Luftfeuchtigkeit. *Contrib. Atmos. Phys.* **43**, 119–133.

Hänel, G. (1970b). On the employment of jet impactors for collecting particles of unknown properties. *Atmos. Environ.* **4**, 289–300.

Hänel, G. (1971). New results concerning the dependence of visibility on relative humidity and their significance in a model for visibility forecast. *Contrib. Atmos. Phys.* **44**, 137–167.

Hänel, G. (1972a). Computation of the extinction of visible radiation by atmospheric aerosol particles as a function of the relative humidity, based upon measured properties. *Aerosol Sci.* **3**, 377–386.

Hänel, G. (1972b). The ratio of the extinction coefficient of the mass of atmospheric aerosol particles as a function of the relative humidity. *Aerosol Sci.* **3**, 455–460.

Hänel, G., and Bullrich, K. (1970). Berechnungen der spektralen Strahlungsextinktion an atmosphärischen Aerosolpartikeln mit verschiedenen komplexen Brechungsindices. *Contrib. Atmos. Phys.* **43**, 202–207.

Hänel, G., and Gravenhorst, G. (1974). Jet impactor characteristics versus relative humidity. *Aerosol Sci.* **4**, 47–54.

Harned, H. S., and Owen, B. B. (1958). "The Physical Chemistry of Electrolytic Solutions." Van Nostrand-Reinhold, Princeton, New Jersey.

Harrison, L. P. (1965). Fundamental concepts and definitions relating to humidity. *In* "Humidity and Moisture." (A. Wexler, Ed.). Vol. 3, pp. 3–69. Van Nostrand-Reinhold, Princeton, New Jersey, Chapman & Hall, London.

Heller, W. (1945). The determination of refractive index of colloidal particles by means of a new mixture rule or from measurements of light scattering. *Phys. Rev.* **68**, 5–10.

Heller, W. (1965). Remarks on refractive index mixture rules. *J. Phys. Chem.* **69**, 1123–1129.

Heller, W., and Pugh, T. L. (1957). Experimental investigations on the effect of light scattering upon the refractive index of colloidal particles. *J. Colloid Sci.* **12**, 294–307.

Herbert, F. (1975). A reexamination of the equilibrium conditions in the theory of water drop nucleation. *Tellus* **27**, 406–413.

Hochrainer, D. (1971). A new centrifuge to measure the aerodynamic diameter of aerosol particles in the submicron range. *J. Colloid Sci.* **36**, 191–194.

Horvath, H. (1969). Sichtweite und Streulicht als Mass für die Luftverunreinigung. *Staub-Reinhalt. Luft* **29**, 503–508.

Horvath, H., and Charlson, R. J. (1969). The direct optical measurement of atmospheric air pollution. *J. Am. Ind. Hyg. Assoc.* **30**, 500–509.

Howell, W. E. (1949). The growth of cloud drops in uniformly cooled air. *J. Meteorol.* **6**, 134–149.

Irvine, W. M., and Pollack, J. B. (1968). Infrared optical properties of water and ice spheres. *Icarus* **8**, 324–360.

Jaenicke, R., Junge, C., and Kanter, H. J. (1971). Messungen der Aerosolgrössenverteilung über dem Atlantik. *"Meteor" Forschungsergeb. Reihe B*, **7**, 1–54.

Jarvis, N. L., Garrett, W. D., Scheiman, M. A., and Timmons, C. O. (1967). Surface chemical characterization of surface active material in sea water. *Limnol. Oceanogr.* **12**, 88–96.

Jones, G. and Ray, W. A. (1941). The surface tension of solutions of electrolytes as a function of the concentration. *J. Am. Chem. Soc.* **63**, 288–294.

Junge, C. (1952a). Die Konstitution des atmosphärischen Aerosols. *Ann. Meteorol.* **5**, Beiheft.

Junge, C. (1952b). Das Grössenwachstum der Aitkenkerne. *Ber. Dtsch. Wetterdienstes* **38**, 264–267.

Junge, C. E. (1963). "Air Chemistry and Radioactivity." (Int. Geophys. Ser., Vol. 4) Academic Press, New York.

Junge, C., and McLaren, E. (1970). Relationship of cloud nuclei spectra to aerosol size distribution and composition. *J. Atmos. Sci.* **28**, 382–390.

Kasten, F. (1968). Der Einfluss der Aerosol-Grössenverteilung und ihrer Änderung mit der relativen Feuchte auf die Sichtweite. *Contrib. Atmos. Phys.* **41**, 33–51.

Kasten, F. (1969). Visibility forecast in the phase of pre-condensation. *Tellus* **21**, 631–635.

Keith, C. H., and Arons, A. B. (1954). The growth of sea-salt particles by condensation of atmospheric water vapour. *J. Meteorol.* **11**, 173–184.

Ketseridis, G. (1972). Entwicklung einer Analysenmethode zur Isolierung und Identifizierung des organischen Anteils von Reinluftaerosolteilchen. Ph.D. Dissertation., Joh. Gutenberg-Universität, Mainz.

Kleinschmidt, E. (1935). "Handbuch der meteorologischen Instrumente und ihrer Auswertung." Springer Verlag, Berlin.

Koehler, H. (1936). The nucleus in and the growth of hygroscopic droplets. *Trans. Faraday Soc.* **32**, 1152.

Kornfeld, P. (1970). Numerical solution for condensation of atmospheric vapour on soluble and insoluble nuclei. *J. Atmos. Sci.* **27**, 256–264.

Kortüm, G. (1963). "Einführung in die chemische Thermodynamik." Vandenhoek und Ruprecht, Göttingen.

Kortüm, G. (1966). "Lehrbuch der Elektrochemie." Verlag Chemie, Weinheim.

Koschmieder, H. (1926). Theorie der horizontalen Sichtweite. *Contrib. Atmos. Phys.* **12**, 171.

Kurtz, U. (1972). Zum spektralen Verlauf der Strahlungsextinktion und ihrer Beeinflussung durch die relative Luftfeuchtigkeit. *Meteorol. Rundsch.* **25**, 134–140.

Lorentz, H. A. (1880). *Wied. Ann.* **9**, 641.

Lorenz, L. (1880). *Wied. Ann.* **11**, 70.

Low, R. D. H. (1969). A theoretical study of nineteen condensation nuclei. *J. Rech. Atmos.* **4**, 65–78.

McDonald, J. E. (1953). Erroneous cloud physics application of Raoult's law. *J. Meteorol.* **10**, 68–70.

Mason, B. J. (1963). "The Physics of Clouds." Oxford Univ. Press (Clarendon), London and New York.

Meszaros, A. (1971). On the variation of the size distribution of large and giant atmospheric particles as a function of the relative humidity. *Tellus* **23**, 436–440.

Middleton, W. E. K. (1952). "Vision through the Atmosphere." Univ. of Toronto Press, Toronto.

Mie, G. (1908). Beitrag zur Optik trüber Medien. *Ann. Phys.* **25**, 377–445.

Mordy, W. (1959). Computations of the growth by condensation of a population of cloud droplets. *Tellus* **11**, 16–44.

Müller, G. (1964). "Methoden der Sedimentuntersuchung." E. Schweitzerbart'sche Verlags-Buchhandlung, Stuttgart.

Nakagaki, M., and Heller, W. (1956). Effect of light scattering upon the refractive index of dispersed colloidal spheres. *J. Appl. Phys.* **27**, 975–979.

Neiburger, M., and Chien, C. W. (1960). Computations of the growth of cloud drops by condensation using an electronic digital computer. *In* "Physics of precipitation," pp. 191–210. Monograph No. 5, Amer. Geophys. Union, Washington, D.C.

Neiburger, M., and Wurtele, M. G. (1949). On the nature and size of particles in haze, fog, and stratus of the Los Angeles region. *Chem. Rev.* **44**, 321–335.

Orr, C., Hurd, F. K., and Corbett, W. J. (1958). Aerosol size and relative humidity. *J. Colloid Sci.* **13**, 472–482.

Osipov, L. I. (1962). "Surface Chemistry," Van Nostrand-Reinhold, Princeton, New Jersey, Chapman & Hall, London.

Owens, J. C. (1967). Optical refractive index of air: Dependence on pressure, temperature and composition. *Appl. Opt.* **6**, 51–59.

Paluch, I. R. (1971). A model for cloud droplet growth by condensation in an inhomogeneous medium. *J. Atmos. Sci.* **28**, 629–639.

Penndorf, R. B. (1958). An approximation method to the Mie theory for colloidal spheres. *J. Phys. Chem.* **62**, 1537–1542.

Penndorf, R. B. (1962a). Scattering and extinction coefficients for small absorbing and non-absorbing aerosols. *J. Opt. Soc. Am.* **52**, 896–904.

Penndorf, R. B. (1962b). Scattering and extinction coefficients for small spherical aerosols. *J. Atmos. Sci.* **19**, 193.

Pueschel, R. F., Charlson, R. J., and Ahlquist, N. C. (1969). On the anomalous deliquescence of sea-spray aerosols. *J. Appl. Meteorol.* **8**, 995–998.

Quiney, R. G., and Carswell, A. I. (1972). Laboratory measurements of light scattering by simulated atmospheric aerosols. *Appl. Opt.* **11**, 1611–1618.

Robinson, R. A. (1945). The water activities of lithium chloride solutions up to high concentrations at 25°C. *Trans. Faraday Soc.* **41**, 756–758.

Robinson, R. A. (1952). The osmotic properties of aqueous sodium chloride cesicum chloride mixtures at 25°. *J. Am. Chem. Soc.* **74**, 6035–6036.

Robinson, R. A., and Stokes, R. H. (1945). A thermodynamic study of bivalent metal halides in aqueous solution. Part XV: Double chlorides of uni- and bivalent metals. *Trans. Faraday Soc.* **41**, 752–758.

Robinson, R. A., and Stokes, R. H. (1959). "Electrolyte Solutions." Butterworth, London.

Ruppersberg, G. H. (1971). Die Änderung des maritimen Dunst-Streukoeffizienten mit der relativen Feuchte. *"Meteor" Forschungergeb. Reihe B*, **6**, 37–60.

Stöber, W., Flaschsbart, H., and Hochrainer, D. (1970). Der aerodynamische Durchmesser von Latexaggregaten und Asbestfasern. *Staub*, **30**, 277–285.

Tait, P. G. (1898). "Scientific Papers," Vol. 2. Cambridge University Press, London and New York.

Thudium, J. (1976). A gas pycnometer (microliter) for determining the mean density of atmospheric aerosol particles. *Aerosol Sci.* **7**, to be published.

van de Hulst, R. C. (1957). "Light Scattering by Small Particles." Wiley, New York.

Volz, F. (1954). Die Optik und Meteorologie der atmosphärischen Trübung. *Ber. Dtsch. Wetterdienstes.* **13** [2].

Volz, F. E. (1972). Infrared refractive index of atmospheric aerosol substances. *Appl. Opt.* **11**, 755–759.

Volz, F. E. (1973). Infrared optical constants of ammonium sulfate, Sahara dust, volcanic pumice, and flyash. *Appl. Opt.* **12**, 564–568.

Wall, E. (1942). Zur Physik der Wasserdampfkondensation an Kernen. *Z. Angew. Meteorol.* **59**, 106–125.

Walter, H. (1973). Coagulation and size distribution of condensation aerosols. *Aerosol Sci.* **4**, 1–15.

Whitby, K. T. (1974). Modeling of multimodal aerosol distributions. *In* "Proceedings of the Jahreskongress 1974 der Gesellschaft für Aerosolforschung e.V." Gesellschaft für Aerosolforschung e.V., Bad Soden, Germany.

Winkler, P. (1969). Untersuchungen über das Grössenwachstum natürlicher Aerosolteilchen mit der relativen Feuchte nach einer Wägemethode. *Ann. Meteorol.* [NF] **4**, 134–137.

Winkler, P. (1970). Zusammensetzung und Feuchtewachstum von atmosphärischen Aerosolteilchen. Ph.D. Dissertation., Joh. Gutenberg-Universität, Mainz.

Winkler, P. (1973). The growth of atmospheric aerosol particles as a function of the relative humidity. Part II: An improved concept of mixed nuclei. *Aerosol Sci.* **4**, 373–387.

Winkler, P., and Junge, C. E. (1971). Comments on "Anomalous deliquescence of sea-spray aerosols." *J. Appl. Meteorol.* **10**, 159–163.

Winkler, P., and Junge, C. (1972). The growth of atmospheric aerosol particles as a function of relative humidity. Part I: Method and measurements at different locations. *J. Rech. Atmos.* (Mémorial Henri Dessens), pp. 617–638.

Woodcock, A. H., and Gifford, M. M. (1948). Sampling atmospheric sea-salt nuclei over the ocean. Woods Hole Ocean. Inst. Contrib. No. 458. pp. 177–197.

Yamamoto, G., and Tanaka, M. (1972). Increase of global albedo due to air pollution. *J. Atmos. Sci.* **29**, 1405–1412.

Zebel, G. (1956). Über Wachstum und Wachstumsgeschwindigkeit von Aerosolen aus wasserlöslichen Substanzen in Abhängigkeit von der relativen Luftfeuchtigkeit. *Z. Aerosol-Forsch. Ther.* **5**, 264–288.

Zerull, R. (1973). Mikrowellenanalogieexperimente zur Lichtstreuung an Staubpartikeln. *Bundesminist. Bild. Wiss. Forschungsber. Kernforsch.* **W 73–18**.

Zolotarev, V. M., Mikhailov, B. A., Alperovich, L. I., and Popov, S. I. (1970). Dispersion and absorption of liquid water in the infrared and radio regions of the spectrum. *Opt. Spectrosc.* **31**, 430–432.

A THREE-DIMENSIONAL MODEL FOR THE NUMERICAL SIMULATION OF ESTUARIES

Enrique A. Caponi*

*Institute for Fluid Dynamics and Applied Mathematics
University of Maryland, College Park, Maryland*

1. Introduction

The transition region between the freshwater portion of a river and the open ocean is a biologically rich environment. The effects on its ecologic equilibrium of both human activity and natural phenomena are not fully understood, and new means for the investigation of those effects are a constant need.

Most of the literature on estuarine modeling takes the form of reports to various governmental agencies, and it is difficult to estimate the present state of the art in this subject. However, there have been efforts in the direction of assessing that state at different times. A comprehensive report on numerical and physical approaches to estuarine modeling has been published by Ward and Espey (1971). Orlob (1972) presented a representative survey of the one- and two-dimensional numerical models developed in the United States which had been applied to practical problems at the

---

Present address: Laboratorio de Hidráulica Aplicada/INCYTH, Casilla de Correo 21, 1802 Aeropuerto Ezeiza, Prov. de Buenos Aires, Argentina.

operational level. Trends in the modeling of estuarine and related systems (up to 1973) have been recently summarized (Nihoul, 1975).

The referenced literature shows convincing evidence that tidal elevations and velocities, as well as the distribution of water quality parameters can be adequately predicted in well-mixed systems by one- and two-dimensional models. However, most estuarine flows are fully three-dimensional, either due to the physical processes involved (e.g., the typical salt-driven circulation of coastal plain estuaries) or to the complicated geometry of the basin.

In recent years, several investigators have presented numerical approaches to the three-dimensional description of flows in enclosed and semienclosed natural basins under the action of specific stimuli. Wind-driven three-dimensional circulations have been simulated for shallow, homogeneous lakes (Liggett, 1969) with a layered formulation where convective and horizontal diffusion terms were neglected. Tidal currents in regions with several open boundaries have been computed solving the linearized, primitive equations for a homogeneous fluid with a multilayer model which required the external specification of thermohaline currents (Hamilton *et al.*, 1973). Mixed numerical and analytical models to simulate three-dimensional currents without resorting to a three-dimensional grid system have been developed to obtain the response of the Irish Sea (Heaps, 1973) and of coastal waters (Forristall, 1974) to applied wind stress fields, assuming a homogeneous fluid. Results of the application on Lake Ontario of three-dimensional hydrodynamic models for the prediction of water levels, currents, and temperatures on the basis of prescribed atmospheric conditions have recently been published (Simons, 1974).

A layered model specifically designed to handle as many of the variables of importance for estuarine dynamics as possible has been presented by Leendertse *et al.* (1973). In that report, the response of closed basins (lakes) under the influence of a constant wind stress was studied. Results of the application of that model to several realistic geometries with open boundaries have recently been reported (Leendertse and Liu, 1975).

In this article, a different tidal, fully three-dimensional model for the numerical simulation of estuarine circulations (Caponi and Faller, 1973), which does not invoke the hydrostatic approximation, is developed. The model can describe the effects of tides, river discharges, and wind and atmospheric pressure distributions on the circulation of a variable density fluid contained in irregularly shaped natural basins. In the process of experimenting with this model, Lagrangian markers were introduced. These remain in the present code and are shown to be useful for several applications.

The problem of modeling a continuous system in order to obtain a numerical procedure for prediction is not trivial because of the limitation to keep only a finite quantity of numbers for the description. These values are identified with physical observables and are related among them by so-called finite difference equations. Such identification in the case of geophysical flows can, at most, be loosely stated, and the finite difference equations become essentially empirical relations with adjustable coefficients.

A discussion of some of these problems forms the contents of Section 2. The equations of conservation of mass, momentum, and energy in the continuum description of a fluid are simplified to yield differential equations for sufficiently smooth flows in geophysical scales. The question of the physical meaning of the variables solved for in a numerical simulation is examined as well as the problem of the representation of the influence of the subgrid scales upon the resolved scales. A particular form of spatial averaging is used to obtain a set of three finite difference equations relating the fluxes through faces of computational cells, as well as a companion equation for salt contents. The choice of dependent variables allows the absolute satisfaction of the finite difference analog to $\nabla \cdot \mathbf{v} = 0$. The way in which the subsequent finite differencing of the temporal axis affects the structure of the terms representing subgrid scales is briefly entertained.

In Section 3, a simple model is adopted for the simulation of the subgrid scales, and an algorithm for the independent temporal integration of the three flux equations [based on the Marker-and-Cell method (Harlow and Welch, 1965; Welch *et al.*, 1965)] is discussed. Appropriate boundary conditions for the modeling of estuarine boundaries are obtained from physical considerations, and the stability properties of the method are examined.

The second part of Section 3 briefly describes the organization of the computer program that codifies the model just developed.

Section 4 shows the results obtained for several test cases: tidal, river, wind, and salt-driven circulations in a rotating prismatic basin open to the sea. The capabilities of a built-in feature of the code, that of identifying and following surface tracers, are illustrated in connection with the analysis of tidally averaged circulations.

Section 5 shows results in a complex geometry resembling the Chesapeake Bay for a period of a few simulated days. Changes in the salinity distribution are discussed, and flow patterns at various depths in different stages of the tidal cycle are shown.

It is found that the fully three-dimensional simulation of the hydro-dynamics of an estuary by the method proposed here is feasible. The question of the final usefulness of results generated in this form can only be answered by further experimentation and is dependent, as in all flow simulations, upon the correct parameterization of the subgrid scales.

2. Governing Equations

In this section, a way of obtaining equations for the large-scale description of the hydrodynamic behavior of a mass of fluid of geophysical dimensions is presented.

The first subsection deals with the equations for a sufficiently smooth flow, derived from conservation laws. Simplifying assumptions that are appropriate for smooth flows of geophysical scales are adopted, and the boundary conditions for the resultant equations are stated.

In the second subsection, two approaches to the treatment of flows involving more scales of motion than those resolved are examined. A third approach is adopted that consists of direct integration of the equations stated in the previous subsection over the elementary computational (observational) cells and results in the direct formulation of spatial finite difference equations for suitably defined quantities. The difference between that approach and the usual integral approach is that the resultant continuity equation is free of terms involving the subgrid scales. The questions of temporal finite differencing and of the meaning of the modeling of physical flows by numerical methods is briefly entertained.

2.1. The Continuum Description of a Fluid

2.1.1. Conservation Laws. The equations governing the behavior of a fluid of density ρ follow from the conservation laws for mass, momentum, and energy in a fixed control volume τ_0, bounded by a surface Σ with external normal $\mathbf{d\Sigma}$:

$$(2.1) \qquad \frac{\partial}{\partial t} \int_{\tau_0} \rho \, d^3x = -\oint_\Sigma \rho \mathbf{v} \cdot \mathbf{d\Sigma}$$

$$(2.2) \qquad \frac{\partial}{\partial t} \int_{\tau_0} \rho v_i \, d^3x = -\oint_\Sigma (\rho v_i)\mathbf{v} \cdot \mathbf{d\Sigma} + \int_{\tau_0} F_i \, d^3x$$

$$(2.3) \qquad \frac{\partial}{\partial t} \int_{\tau_0} \frac{1}{2} \rho v^2 \, d^3x = -\oint_\Sigma \left(\frac{1}{2}\rho v^2\right)\mathbf{v} \cdot \mathbf{d\Sigma} + \int_{\tau_0} \mathbf{v} \cdot \mathbf{F} \, d^3x$$

where $\mathbf{v}$ is the momentum per unit mass, and $\mathbf{F}$ is the total force per unit volume.[1] These equations state that the rate of change of the quantity $Q \equiv \int_{\tau_0} q \, d\tau$ is given by the flux of q through the surface plus a volume source (sink) term for q acting within τ_0.

---

[1] Note that Eq. (2.3) is just the first moment of Eq. (2.2) and therefore, it is not an independent equation. It gives rise to an independent equation for a new variable—the internal energy—below, when combined with the first law of thermodynamics.

By allowing $\tau_0 \to 0$ (in the "physical" sense), the equivalent "differential" equations for the continuum description of the fluid are obtained:

$$(2.4) \qquad \frac{\partial \rho}{\partial t} = -\partial_j(\rho v_j)$$

$$(2.5) \qquad \frac{\partial \rho v_i}{\partial t} = -\partial_j(\rho v_i v_j) + F_i$$

$$(2.6) \qquad \frac{\partial}{\partial t}\left(\frac{1}{2}\rho v_l v_l\right) = -\partial_j\left(\frac{1}{2}\rho v_m v_m v_j\right) + v_i F_i$$

The meaning and use of these differential equations is clear for laminar flows, i.e., flows which admit continuous description and measurement, in both space and time. In this case, the volume force acting as a source term in the momentum equation can be adequately approximated by

$$(2.7) \qquad F_i = -\partial_i(p + \rho\psi) + \partial_j \sigma_{ij}$$

where p is the pressure, ψ the geopotential, and

$$(2.8) \qquad \sigma_{ij} = \eta(\partial_j v_i + \partial_i v_j) + \left(\zeta - \frac{2}{3}\eta\right)\partial_m v_m$$

is the viscous stress tensor representing (molecular) dissipative effects. The symbols η and ζ stand for the first and second coefficients of dynamic viscosity.

With the explicit form for $\mathbf{F}$ given by Eq. (2.7), the momentum equation can be rewritten as

$$(2.9) \qquad \frac{\partial \rho v_i}{\partial t} = -\partial_i \Pi_{ij} - \partial_i(\rho\psi)$$

where $\Pi_{ij} = p\,\delta_{ij} + \rho v_i v_j - \sigma_{ij}$ is the total momentum flux density tensor.

The source term for the kinetic energy equation is the work performed within τ_0 by all the present forces (pressure, viscous, and those derived from ψ), and with the help of Eq. (2.7) can be expressed as,

$$(2.10) \qquad \frac{\partial}{\partial t}\int_{\tau_0} \frac{1}{2}\rho v^2\,d^3x = -\oint_\Sigma \left[\left(\frac{1}{2}\rho v^2 + p\right)\delta_{ij} + \sigma_{ij}\right]v_i\,d\Sigma_i$$

$$- \int_{\tau_0} v_i\,\partial_i(\rho\psi)\,d^3x + I$$

where

$$(2.11) \qquad I \equiv \int_{\tau_0} (p\,\partial_i v_i - \sigma_{ij}\,\partial_j v_j)\,d^3x$$

stands for the work of the pressure forces that result in compression of the fluid, and the volume generation of heat per unit time by the viscous forces.

The work performed in a time interval δt by the applied forces on the system that results in compression or expansion, is related to the first term in I:

$$(2.12) \qquad \delta L = \delta t \int_{\tau_0} p\, \partial_i v_i \, d^3x$$

while the change of the total heat contents δQ inside τ_0 during δt is given by

$$(2.13) \qquad \delta Q = \delta t \int_{\tau_0} \sigma_{ij}\, \partial_j v_i \, d^3x + \delta t \oint_\Sigma \kappa(\partial_j T)\, d\Sigma_j$$

where $\sigma_{ij}\, \partial_j v_i$ is the heat generated by friction per unit volume, and the heat flux through the surface is taken as proportional to the temperature gradient. The first law of thermodynamics as applied to the fixed volume τ_0 states that

$$(2.14) \qquad \frac{dU}{dt} = \frac{\delta Q}{\delta t} - \frac{\delta L}{\delta t}$$

where

$$(2.15) \qquad \frac{dU}{dt} = \frac{d}{dt} \int_{\tau_0} \rho e \, d^3x = \int_{\tau_0} \frac{\partial(\rho e)}{\partial t}\, d^3x + \oint_\Sigma \rho e \mathbf{v} \cdot \mathbf{d\Sigma}$$

is the temporal variation of the contents of internal energy $U = \int_{\tau_0} \rho e \, d^3x$ in the fixed volume τ_0.

Combination of Eqs. (2.12) to (2.15), results in an equation for the rate of change of internal energy e in the control volume, as follows

$$(2.16) \qquad \int_{\tau_0} \frac{\partial(\rho e)}{\partial t}\, d^3x + \oint_\Sigma \rho e \mathbf{v} \cdot \mathbf{d\Sigma} = \oint_\Sigma \kappa(\nabla T) \cdot \mathbf{d\Sigma}$$

$$+ \int_{\tau_0} [\sigma_{ij}\, \partial_j v_i - p(\nabla \cdot \mathbf{v})]\, d^3x$$

Finally, combination of Eqs. (2.10), (2.11), and (2.16) results in an equation for the total energy density $\rho(e + 1/2v^2)$:

$$(2.17) \quad \int_{\tau_0} \frac{\partial}{\partial t}\left(\frac{1}{2}\rho v^2 + \rho e\right) d^3x = -\oint_\Sigma \left[\left(\frac{1}{2}\rho v^2 + p + \rho e\right)\mathbf{v} - \kappa\nabla T\right] \cdot \mathbf{d\Sigma}$$

$$+ \oint_\Sigma v_i \sigma_{ij}\, d\Sigma_j - \int_{\tau_0} v_i\, \partial_i(\rho\psi)\, d^3x$$

where $(\frac{1}{2}\rho v^2 + \rho e + p)\mathbf{v} - \mathbf{v} \cdot \boldsymbol{\sigma} - \kappa \nabla T$ is the total energy flux vector. In its differential form, the energy equation is

(2.18)

$$\frac{\partial}{\partial t}\left(\frac{1}{2}\rho v^2 + \rho e\right) = -\nabla\left[\left(\frac{1}{2}\rho v^2 + p + \rho e\right)\mathbf{v} - \kappa\nabla T - \mathbf{v}\cdot\boldsymbol{\sigma}\right] - \mathbf{v}\cdot\nabla(\rho\psi)$$

At this stage, five equations (one each for mass and energy, and three for the momentum), with nine unknowns $[\rho, \mathbf{v}, p, T, e, \eta, \zeta$, all of which are functions of $(\mathbf{x}, t)]$, have been found. Therefore, four additional relations are required. Two of these are provided by empirical relations linking η and ζ to the state variables. The other two are a thermodynamic relation for the internal energy e and an experimentally determined relation for the density ρ in terms of the same state variables.

For a fluid consisting of a single chemical component, such state variables are pressure and temperature. Otherwise, the fluid ought to be thought of as divided up into N sufficiently small volumes, in each of which relations of the type $\rho_1 = \rho_1(p, T)$, $\rho_2 = \rho_2(p, T)$, ..., $\rho_N = \rho_N(p, T)$ hold. These will be parametrically dependent upon the relative concentrations of the several chemical components. For nonreacting substances, it will be possible to relate these relations to those corresponding to the single components that form part of the mixture. Thus, for nonreactive solutes, the equation of state will adopt the form

(2.19) $$\rho = \rho(p, T, \mathbf{C})$$

where $\mathbf{C}$ is an m-dimensional concentration vector, and m is the number of dissolved species.

The addition of m solutes requires the introduction of m equations for the C_i's. These are obtained by stating equations for the conservation of each species in the form

(2.20) $$\int_{\tau_0} \frac{\partial \rho C_i}{\partial t} d^3x = -\oint_\Sigma \rho(C_i\mathbf{v} - D^i\nabla C_i)\cdot \mathbf{d\Sigma} + \int_{\tau_0} \rho r_i\, d^3x$$

or

(2.21) $$\frac{\partial \rho C_i}{\partial t} = -\nabla(\rho C_i\mathbf{v}) + \nabla\cdot(\rho D^i\nabla C_i) + \rho r_i, \qquad i = 1, m$$

where D^i is the diffusion or mass transfer coefficient[2] for the ith species,

---

[2] The general expression for the diffusive flux in Eq. (2.20) is very complicated (cf., for example, Landau and Lifschitz, 1959, p. 224), but in the case of relatively small pressure gradients and low concentrations it can be taken as directly proportional to the concentration gradient. In this same approximation, the extra contribution to the internal energy [Eq. (2.15)] that arises for heterogeneous fluids is neglected.

describing the transfer of solute from one place to another by molecular processes, and r_i is the volume rate of creation or destruction of C_i ($r = 0$ for a conservative substance).

Since Eq. (2.4) for the conservation of total mass remains valid, it can be combined with Eq. (2.21) to give,

$$(2.22) \qquad \frac{\partial C_i}{\partial t} = -\mathbf{v} \cdot \nabla C_i + \frac{1}{\rho} \nabla(\rho D^i \nabla C_i) + r_i, \qquad i = 1, m$$

When C represents concentrations of biological species,[3] for example, terms of the Lotka-Volterra type will have to be added to r to take into account linear and nonlinear effects of the type "fish A eats fish B," etc.:

$$(2.23) \qquad \frac{\partial C_i}{\partial t} = -\mathbf{v} \cdot \nabla C_i + \frac{1}{\rho} \nabla(\rho D^i \nabla C_i) + r_i - \alpha_i C_i$$

$$+ \sum_m \beta_{im} C_m + \sum_m \sum_n \gamma_{imn} C_m C_n + \cdots$$

In this equation, repeated indices do not imply summation.

2.1.2. Simplifying Assumptions. The set of equations presented so far is able to describe more types of phenomena than are relevant to the dynamics of estuaries. Therefore, we restrict our attention to the equations that result from the following approximations:

(a) The influence of all C_i's (except one) on the fluid density is neglected. If S, the salinity, is proportional to the concentration of this solute, then the equation of state reduces to the form $\rho = \rho(p, T, S)$. This approximation uncouples the dynamic equations from those for the other species, which therefore become "passive" pollutants.

(b) For a fluid in its liquid phase, variations of density with pressure are small. This dependence can be neglected when both sound waves are not important to the problem and the mass of fluid under consideration is shallow enough. Thus,

$$(2.24) \qquad \rho = \rho(T, S)$$

In particular, we will only keep for the rest of this discussion the first linear terms in the Taylor expansion of Eq. (2.24):

$$(2.25) \qquad \rho = \rho_0[1 + \alpha(S - S_0) - \beta(T - T_0) + \cdots]$$

---

[3] Since we are still considering laminar flows, Eq. (2.22) can only be applied at this stage to organisms whose independent motions have a negligible influence on the dynamics of the fluid, if Eqs. (2.9) and (2.17) are to remain unchanged.

where

$$\rho_0 = \rho(S_0, T_0) \qquad \text{and} \qquad \alpha \equiv \frac{1}{\rho_0}\frac{\partial \rho}{\partial S}\bigg)_{S=S_0}, \qquad \beta \equiv -\frac{1}{\rho_0}\frac{\partial \rho}{\partial T}\bigg)_{T=T_0}$$

(c) Heat production by viscous dissipation is neglected in the energy equation and the internal energy of the solution is taken as directly proportional to the temperature. Internal energy changes due to compression and expansion are therefore neglected, consistently with assumptions (b) and (d) below.

(d) For a homogeneous fluid at constant temperature, assumption (b) results in the "incompressibility condition"

$$(2.26) \qquad \nabla \cdot \mathbf{v} = 0$$

when applied to the equation for conservation of mass. Extra terms are originated because of temperature and salinity variations. For low concentrations and small temperature changes, it is usual practice to identify the incompressibility equation with the equation for mass conservation. This forms part of the Boussinesq approximation, which we adopt. Consistently, variations in density are neglected in the momentum equations, except when they are multiplied by g.

(e) For fluids on a geophysical scale, the external forces are those derived from the geopotential. This we approximate by gz, where z is the distance from the datum surface. For the description of this rotating system in terms of a coordinate frame rotating with it, a noninertial "force" is required. Its components are given by $-2\rho\varepsilon_{ijk}\Omega_j v_k$, where Ω is the angular velocity and ε_{ijk} is the completely antisymmetric Levy-Civita tensor. Order of magnitude arguments show that for all practical purposes the approximation

$$(2.27) \qquad -2\varepsilon_{ijk}\Omega_j v_k \simeq 2\Omega \sin \lambda \cdot \begin{pmatrix} v \\ -u \\ 0 \end{pmatrix} \equiv \begin{pmatrix} fv \\ -fu \\ 0 \end{pmatrix}$$

is appropriate, where λ is the geographical latitude. Furthermore, in this work we set $f = $ constant, a constant latitude approximation, adequate for the description of regions extending over only a few degrees of latitude.

With these assumptions, the momentum equations simplify:

$$(2.28) \qquad \frac{\partial v_i}{\partial t} = -\partial_j(v_i v_j) - \partial_i \varphi + f \begin{pmatrix} v \\ -u \\ 0 \end{pmatrix}$$

$$-g[1 + \alpha(S - S_0) - \beta(T - T_0)]\,\delta_{i3} + \nu\,\partial_m\,\partial_m v_i$$

where φ is the ratio p/ρ_0. Likewise, the energy equation reduces to

$$(2.29) \qquad \frac{\partial T}{\partial t} = -\partial_i(u_i T) + \chi \partial_m \partial_m T$$

where $\chi = \kappa/\rho C_p$ is the coefficient of thermal diffusivity (or thermometric conductivity). In the same manner,

$$(2.30) \qquad \frac{\partial S}{\partial t} = -\partial_i(v_i S) + D_S \nabla^2 S$$

and the concentration C_j of the jth passive pollutant or species transforms into

(2.31)

$$\frac{\partial C_j}{\partial t} = -\partial_i(v_i C_j) + D^j \nabla^2 C_j + r_j - \alpha_j C_j + \sum_m \beta_{jm} C_m + \sum_m \sum_n \gamma_{jmn} C_m C_n + \cdots$$

The convective terms in Eqs. (2.28) through (2.31) have been written in conservative form with the help of Eq. (2.26) for reasons that will become clear in Section 2.2.3.

The assumption of incompressibility implies that the pressure field is at all times to be consistent with the velocity distribution through the momentum equations. Thus, the field φ is the solution to the Poisson equation obtained by taking the divergence of Eq. (2.28) and applying the appropriate boundary conditions.

For the consideration of problems in geophysical scales, it is convenient to split the pressure field into a hydrostatic and a hydrodynamic part:

$$(2.32) \qquad \varphi = \varphi^h + \varphi^d$$

where the first is chosen to exactly balance the gravitational contribution to the z component of Eq. (2.28)

$$(2.33) \qquad \partial_z \varphi^h \equiv -g[1 + \alpha(S - S_0) - \beta(T - T_0)]$$

Upon integration,

$$(2.34) \qquad \varphi^h(z) = \varphi^h(z = H) + g \int_z^H [1 + \alpha(S - S_0) - \beta(T - T_0)]\, dz$$

$$= \varphi^h(z = H) + g(H - z) + g \cdot I(S, T)$$

where H is the z value of the free surface and

$$(2.35) \qquad I(S, T) \equiv \int_z^H [\alpha(S - S_0) - \beta(T - T_0)]\, dz$$

is the contribution to the hydrostatic pressure due to vertical changes in fluid density.

In that manner, the momentum equations take the form

$$(2.36) \qquad \frac{\partial v_i}{\partial t} = -\partial_j(v_i v_j) - \partial_i \varphi^d + v\,\partial_m\,\partial_m v_i + A_i$$

where

$$A \equiv \begin{pmatrix} fv - g\,\partial_x(H+I) \\ -fu - g\,\partial_y(H+I) \\ 0 \end{pmatrix}$$

Eqs. (2.26), (2.29), (2.30), and (2.36) form a system of six equations for the seven unknowns v, T, S, φ^d, H. The new variable H appeared in Eq. (2.34) as a consequence of incorporating a boundary condition into one of the variables. To solve for H, an integral equation is needed. Alternatively, if the velocity field at the surface is known, the relation

$$(2.37) \qquad w_{surf} = \frac{dH}{dt} = \frac{\partial H}{\partial t} + u_{surf}\frac{\partial H}{\partial x} + v_{surf}\frac{\partial H}{\partial y}$$

provides an equation for $\partial H/\partial t$.

An equation for the pressure deviation φ^d is derived taking the divergence of the momentum equation and explicitly using the incompressibility condition:

$$(2.38) \qquad \nabla^2\varphi^d = -\partial_i\,\partial_j(v_i v_j) + \partial_i A_i$$

The description of n concentrations will require the incorporation of n extra equations of the form of Eq. (2.31). Since these are uncoupled from the hydrodynamic equations, their solution need not be simultaneous to the solution of the latter.

2.1.3. Boundary Conditions. The solution to the dynamic equations for any particular problem requires the specification of an appropriate set of initial and boundary conditions. Boundary conditions for the case of laminar flows are presented below.

For a three-dimensional mass of fluid such as an estuary, certain boundaries may be regarded as fixed in space, at least if their changes occur in characteristic times much longer than those associated with the phenomena under investigation. In this category fall both the rigid bottom and lateral boundaries, and the boundaries of mass, momentum, and energy exchange with other masses of fluid such as the ocean and inflowing rivers. The boundary of interaction with the atmosphere is different in this sense, since its position varies in time in a manner not known *a priori*.

2.1.3.1. Impenetrable boundaries. By definition, no mass flux is allowed through this type of boundary. If $\hat{n}$ denotes the normal to a surface Σ, then

$$(2.39) \qquad v_\perp = \mathbf{v} \cdot \hat{n} \Big|_\Sigma = 0$$

at all times. Thus, $\partial v_\perp / \partial t = 0$, and from the momentum equation (2.36), the corresponding consistent boundary condition for φ^{d} is obtained. That is,

$$(2.40) \qquad \frac{\partial \varphi^{\mathrm{d}}}{\partial n}\Big|_\Sigma = A_\perp + v(\partial_m \, \partial_m v_\perp)\Big|_\Sigma$$

The solution of the momentum equations also requires a condition on the component of $\mathbf{v}$ parallel to the surface Σ. This is provided by the no-slip condition

$$(2.41) \qquad v_\parallel \Big|_\Sigma = 0$$

Conditions (2.40) and (2.41) guarantee that the velocity field vanishes at the fixed wall. The condition of no flux through this kind of boundaries implies

$$(2.42) \qquad \frac{\partial c}{\partial n}\Big|_\Sigma = 0$$

where c stands for salinity, temperature, etc.

2.1.3.2. Lateral boundaries of interaction with other masses of fluid. The specification of $\mathbf{v} = \mathbf{v}(\mathbf{x}, t)$, $S = S(\mathbf{x}, t)$, $T = T(\mathbf{x}, t)$, and $C_m = C_m(\mathbf{x}, t)$ at these boundaries is required for a solution to be possible. If less information is available, the complete information should be built based upon a particular model for such a boundary. In the simulation of systems of geophysical scales, these are doubtlessly the most difficult type of boundaries because conditions specified here are liable to introduce artificial perturbations in the interior. The approach followed in this work is given in Section 3.

2.1.3.3. Surface. This is the boundary through which the fluid interacts with the medium above it. In our case, we are interested in describing a "one-way" interaction mechanism, i.e., the influence of the dynamic atmosphere over the mass of water under consideration. That requires the specification of the temporal-spatial dependence of the applied atmospheric pressure and shear stress on the surface.

In the approximation in which both fluids can be taken as immiscible, the requirement that the interface moves as a single surface is that

$$(2.43) \qquad \mathbf{v}^{(1)}_{\mathrm{surf}} = \mathbf{v}^{(2)}_{\mathrm{surf}}$$

where the superscripts refer to the two fluids. Furthermore, since this is an interaction surface, the forces exerted by these fluids on each other must be equal and opposite:

$$(2.44) \qquad \partial_j(p\,\delta_{ij} - \sigma_{ij})^{(1)} = -\partial_j(p\,\delta_{ij} - \sigma_{ij})^{(2)}$$

Thus, if $\hat{n}(= \hat{n}^{(1)} = -\hat{n}^{(2)})$ is the external normal to the interface, the condition above is equivalent to

$$(2.45) \qquad n_j(p\,\delta_{ij} - \sigma_{ij})^{(1)} - n_j(p\,\delta_{ij} - \sigma_{ij})^{(2)} = 0$$

Hence, in terms of the dynamic pressure φ^d, surface boundary conditions are generated by

$$(2.46) \qquad n_j[\varphi^d\,\delta_{ij} - v(\partial_i v_j + \partial_j v_i)]_{\text{water}} = n_j[\varphi^d_a\,\delta_{ij} - \tau_{ij}]_{\text{air}}$$

Scalar conditions are obtained from the vector Eq. (2.46) by successively multiplying it by the normal to the surface $\hat{n}$, and the independent tangent versors $\hat{m}^{(k)} \equiv (m_1, m_2, m_3)$, $k = 1, 2$:

$$(2.47) \quad n_j[\varphi^d\,\delta_{ij} - v(\partial_i v_j + \partial_j v_i)]_w\, n_i = n_j[\varphi^d_a\,\delta_{ij} - \tau_{ij}]_a\, n_i$$

$$n_j[\varphi^d\,\delta_{ij} - v(\partial_i v_j + \partial_i v_i)]_w\, m_i^{(k)} = n_j[\varphi^d_a\,\delta_{ij} - \tau_{ij}]_a\, m_i^{(k)}, \qquad k = 1, 2$$

Since $n_i\,m_j^{(k)}\,\delta_{ij} = 0$, $n_i\,n_j\,\delta_{ij} = 1$, Eqs. (2.47) yield

$$(2.48) \qquad \{\varphi^d - 2v[n_1^2\,\partial_x u + n_2^2\,\partial_y v + n_3^2\,\partial_z w + n_1 n_2(\partial_x v + \partial_y u)$$

$$+ n_2 n_3(\partial_y w + \partial_z v) + n_1 n_3(\partial_x w + \partial_z u)]\}_{\text{water}}$$

$$= \{\varphi^d_a - \tau_{ij}\,n_i\,n_j\}_{\text{air}}$$

$$(2.49) \qquad [v(\partial_i v_j + \partial_j v_i)]_{\text{water}}\, n_j\, m_i^{(k)} = [\tau_{ij}]_{\text{air}}\, n_j\, m_i^{(k)}, \qquad k = 1, 2$$

In particular, for a flat horizontal interface, $\hat{n} = (0, 0, 1)$, $\hat{m}^{(1)} = (1, 0, 0)$, $m^{(2)} = (0, 1, 0)$, and Eqs. (2.48) and (2.49) reduce to

$$(2.50) \qquad \varphi^d = \{\varphi^d_a + \tau_{33}\}_{\text{air}} + 2v\,\frac{\partial w}{\partial z}$$

$$(2.51) \qquad v\left(\frac{\partial u}{\partial z} + \frac{\partial w}{\partial x}\right) = \tau_{xz} + \tau_{zx}$$

$$(2.52) \qquad v\left(\frac{\partial v}{\partial z} + \frac{\partial w}{\partial x}\right) = \tau_{yz} + \tau_{zy}$$

2.2. Real Flows: Turbulence

The equations stated so far are able to describe the behavior of an incompressible fluid with salt contents, provided the initial and boundary conditions are known. Due to the (classic) intrinsic error generation in any measurement process, the deterministic character of the continuous fluid equations is apparent only in certain kinds of flows: the so-called "laminar flows."

In many laboratory situations and in nature, flows are usually found to present "turbulent" characteristics: the details of a particular, apparently disorganized flow depend very strongly upon variations in the initial and boundary conditions below the experimentally detectable threshold. In such cases, the exact solution to the set of dynamic equations, even if available, is of little practical use. Nevertheless, certain mean quantities related to the instantaneous continuous variables, are expected to evolve independently of the details of the initial conditions and boundary perturbations, in certain cases.

In order to be able to make meaningful predictions, the above mentioned equations must be acted upon by an averaging operator satisfying the conditions: (a) that the resulting equations be sufficiently simple for their solution or conceptual understanding, and (b) that the "mean" quantities related by these resulting equations be experimentally identifiable and measurable.

In order to satisfy the first of these conditions, it is usual practice to define space (and/or time) averages[4] so that for any variable q:

$$(2.53) \qquad q = \bar{q} + q'$$

Subsequent manipulation of the momentum, temperature, and salinity transport equations and incorporation of certain assumptions (see below) yield equations for the mean values of velocity, temperature, and salinity, similar to the equations for the unaveraged quantities, except for an anomalous contribution due to the correlations between the perturbations of the velocity field and the perturbations of the field under consideration (Reynolds' stresses). The problem of accounting for these terms is still unsolved.

---

[4] The simplest kind of averaging, time averaging at a given spatial position (Reynolds), is a particular case of the general form (Monin and Yaglom, 1971):

$$(2.54) \quad \overline{q(x, y, z, t)} = \int\!\!\!\int\!\!\!\int\!\!\!\int_{-\infty}^{\infty} q(x + \xi, y + \eta, z + \zeta, t + \tau)\omega(\xi, \eta, \zeta, \tau)\, d\xi\, d\eta\, d\zeta\, d\tau$$

with the normalization condition for the weighting function ω:

$$(2.55) \qquad \int\!\!\!\int\!\!\!\int\!\!\!\int_{-\infty}^{\infty} \omega(\xi, \eta, \zeta, \tau)\, d\xi\, d\eta\, d\zeta\, d\tau = 1$$

2.2.1. Governing Equations (à la Reynolds). In order to obtain sufficiently simple equations for the mean values of the fluid dynamic variables, the averaging operator must satisfy (Monin and Yaglom, 1971; van Mieghem, 1973) the following relationships:

$$(2.56) \qquad \overline{q_1 + q_2} = \bar{q}_1 + \bar{q}_2$$

$$(2.57) \qquad \overline{aq} = a\bar{q}, \qquad \text{if} \quad a = \text{const.}$$

$$(2.58) \qquad \bar{a} = a, \qquad \text{if} \quad a = \text{const.}$$

$$(2.59) \qquad \overline{\frac{\partial q}{\partial s}} = \frac{\partial \bar{q}}{\partial s}, \qquad \text{where} \quad s \text{ is } x, y, z, \text{ or } t$$

$$(2.60) \qquad \overline{\bar{q}_1 q_2} = \bar{q}_1 \bar{q}_2$$

As a consequence of (2.56) and (2.60), the following properties hold,

$$(2.61a) \qquad \bar{\bar{q}} = \bar{q}$$

$$(2.61b) \qquad \overline{q'} = \overline{q - \bar{q}} = 0$$

$$(2.61c) \qquad \overline{\bar{q}_1 \bar{q}_2} = \bar{q}_1 \bar{q}_2$$

$$(2.61d) \qquad \overline{q_1 q_2} = \bar{q}_1 \bar{q}_2 + \overline{q'_1 q'_2}$$

Properties (2.56) through (2.59) are satisfied by any (running) averaging procedure defined by Eqs. (2.54) and (2.55).

In particular, for the Reynolds' averaging (i.e., $s = t$, $\omega(\xi, \tau) = \omega(\tau)\,\delta(\xi)$ and $\omega(\tau) = 1$ for $|\tau| \le \delta t/2$, $\omega(\tau) = 0$ otherwise), condition (2.59) holds

$$(2.62) \qquad \frac{\partial \bar{q}}{\partial t} = \frac{1}{\Delta t} \frac{\partial}{\partial t} \int_{-\Delta t/2}^{\Delta t/2} q(t + \tau)\, d\tau = \lim_{\varepsilon \to 0} \frac{\bar{q}(t + \varepsilon) - \bar{q}(t)}{\varepsilon}$$

$$= \frac{1}{\Delta t} \lim_{\varepsilon \to 0} \frac{1}{\varepsilon} \left[\int_{\varepsilon - \Delta t/2}^{\varepsilon + \Delta t/2} q(t + \tau)\, d\tau - \int_{-\Delta t/2}^{\Delta t/2} q(t + \tau)\, d\tau \right]$$

$$= \frac{1}{\Delta t} \lim_{\varepsilon \to 0} \int_{-\Delta t/2}^{\Delta t/2} \frac{q(t + \varepsilon + \tau) - q(t + \tau)}{\varepsilon}\, d\tau \equiv \overline{\frac{\partial q}{\partial t}}$$

On the other hand, condition (2.60) is not exactly satisfied:

$$\overline{\bar{q}_1 q_2} = \frac{1}{\Delta t} \int_{-\Delta t/2}^{\Delta t/2} \bar{q}_1(t + \tau) q_2(t + \tau)\, d\tau = \frac{1}{\Delta t^2} \int_{-\Delta t/2}^{\Delta t/2} d\tau$$

$$\times \left[q_2(t + \tau) \int_{-\Delta t/2}^{\Delta t/2} d\tau' q_1(t + \tau + \tau') \right]$$

204 ENRIQUE A. CAPONI

If q_1 is expanded in Taylor's series around $t + \tau'$:

$$q_1(t + \tau + \tau') = q_1(t + \tau') + \tau \left.\frac{\partial q_1}{\partial \tau}\right|_{t+\tau'} + \frac{1}{2} \tau^2 \left.\frac{\partial^2 q_1}{\partial \tau^2}\right|_{t+\tau'} + \cdots$$

and this expansion is substituted into the expression for $\overline{\overline{q_1} q_2}$, we obtain

$$(2.63) \quad \overline{\overline{q_1} q_2} = \frac{1}{\Delta t} \int_{-\Delta t/2}^{\Delta t/2} d\tau \left\{ q_2(t + \tau) \left[\bar{q}_1(t) + \tau \frac{\overline{\partial q_1(t)}}{\partial t} + \frac{1}{2} \tau^2 \frac{\overline{\partial^2 q_1(t)}}{\partial t^2} + \cdots \right] \right\}$$

$$= \bar{q}_1 \bar{q}_2 + \frac{\partial \bar{q}_1}{\partial t} \frac{1}{\Delta t} \int_{-\Delta t/2}^{\Delta t/2} \tau q_2(t + \tau)\, d\tau$$

$$+ \frac{1}{2} \frac{\partial^2 \bar{q}_1}{\partial t} \frac{1}{\Delta t} \int_{-\Delta t/2}^{\Delta t/2} \tau^2 q_2(t + \tau)\, d\tau + \cdots$$

Thus, postulate (2.60) holds only if $\bar{q}_1$ is a constant. Similar considerations apply to all of Eqs. (2.61).

Relation (2.63) also shows that when there is a gap between the characteristic interval associated with the "fluctuations" and that associated with the "mean," an averaging interval that is both much larger than the former and much smaller than the latter can be found, for which the assumption is satisfied to a good approximation. In all other cases, extra terms (besides $\bar{q}_1 \bar{q}_2$) should have to be considered.

When a running average satisfying the Reynolds' postulate is applied to the equations for a homogeneous, incompressible fluid, simple relations ("governing equations") are obtained among the mean quantities:

$$(2.64) \qquad\qquad\qquad \partial_i \bar{v}_i = 0$$

$$(2.65) \qquad\qquad \frac{\partial \bar{v}_i}{\partial t} = -\partial_j(\bar{v}_i \bar{v}_j + \bar{p}\, \delta_{ij} - \bar{\sigma}_{ij}) - \partial_j \overline{v_i' v_j'}$$

For a turbulent flow (or for a large enough averaging interval), the molecular contribution $\bar{\sigma}_{ij}$ is recognized negligible with respect to the correlation terms $\overline{v_i' v_j'}$, and a way of evaluating these must be found if the averaged equations are to be of any use.

By taking moments of the momentum equations and applying the averaging procedure, equations for the time evolution of the correlation terms can be obtained, only as functions of higher order correlations. In this way, a hierarchy of equations is generated, and a solution is possible only if it is closed at some stage. The literature generated by this possibility is very extensive, and its survey is outside the scope of this article. The simplest approach of closing the system at the first stage (Boussinesq's idea of linking the turbulent stresses to the mean values themselves through an "eddy viscosity coefficient"), is the one generally used in the simulation of flows in geophysical scales.

2.2.2. Application to Large-Scale Flows: Continuous Running Averages vs Computational Observational Box Averages and Integro-Interpolation Methods. The meaning of the solutions to the system of equations (2.64) and (2.65) is not precisely defined in the surveyed literature. Part of the problem resides in a confusion between properties of running averages and of averages taken over fixed regions in the space generated by the independent variables. The first approach leads to differential equations and thus requires the adoption of a finite differencing procedure before attempting a numerical solution. The second directly provides spatial difference equations on mean quantities, but as it is shown below, it should be carefully prescribed if the "means" are required to be divergence-free.

The validity of the form used for the last term in the right-hand side of Eq. (2.65) is restricted to flows where property (2.60) is satisfied. Only in these cases, the parameterization of the Reynolds' stresses $\overline{v_i' v_j'}$ in terms of the "mean" quantities remains a valid question. Nevertheless, this has always been the predominant approach for the description of geophysical flows, regardless of the existence of a gap between the characteristic scales and periods associated to the "mean" motions and those associated with the "turbulent" motions. Differential equations for the mean quantities are obtained upon the adoption of a relation

$$(2.66) \qquad \rho \overline{v_i' v_j'} = F_{ij}(\overline{\mathbf{v}}, \mathbf{x}, t, \Delta)$$

between the Reynolds' stress terms, the mean quantities, the position x, and the average interval Δ.

In principle, such procedure would be objectionable only on a conceptual basis. If solutions to the resulting equations are useful for prediction, the conclusion should be that the parameterization used [Eq. (2.66)] is appropriate to the given problem. But it should be noted that in general F_{ij} stands for

$$(2.67) \qquad F_{ij}(\overline{\mathbf{v}}, \mathbf{x}, t, \Delta) \equiv \rho(\overline{v_i v_j} - \overline{v}_i \overline{v}_j)$$

The physical meaning of F_{ij}, related to the rate of mean energy dissipation required for equations (2.64) and (2.65) to describe correctly the relation among mean variables is unchanged.

In practice, however, the solution to Eqs. (2.64) and (2.65) for any moderately complicated configuration requires a numerical approach. This in turn requires discretization of either, or both, the spatial and temporal coordinates to obtain difference equations on grid points separated by characteristic intervals δ. Regardless of the discretization procedure used, if $\delta \ll \Delta$ (i.e., any feature in the mean quantities resulting from a Reynolds' averaging with averaging interval Δ extends over many grid points), Eq. (2.67) can be used for the solution. On the other hand, if $\delta \gtrsim \Delta$, features of the "mean" quantities will not be resolved by the grid; the "mean" quantities are "turbulent" as far as the grid is concerned, and Eq. (2.67), based on the averaging interval Δ, lacks any relevance to this solution.[5] If its use were intended, the energy cascade from large to small scales given by the nonlinear terms would be abruptly interrupted at the smallest resolvable scale δ. The form given by Eq. (2.67) would not account for the total energy dissipation required, and the net result would be a continuous pumping of energy into the largest resolvable wave number ($\kappa = \pi/\delta$). Although this effect results from clear physical reasons, the phenomenon is sometimes called "nonlinear instability."

The stable behavior of the solution can be recuperated by increasing the rate of energy dissipation above the level given by Eq. (2.67). That is accomplished by the use of dissipative schemes, artificial viscosities, etc., which form part of the problem of modeling subgrid motions.

In a paper on this subject, Lilly (1966) explicitly stated the form of the terms that require modeling. To arrive at them, a Reynolds' averaging procedure was used with Δ(spatial) $= \delta$(spatial): the so-called spatial mesh cube average operator:

$$(2.68) \qquad \bar{q}(x, y, z, t) \equiv \frac{1}{\delta^3} \int_{-\delta/2}^{\delta/2} d\xi \int_{-\delta/2}^{\delta/2} d\eta \int_{-\delta/2}^{\delta/2} d\zeta \, q(x + \xi, y + \eta, z + \zeta, t)$$

where δ is the mesh separation distance. Although such choice suggests the direct integration of the differential equations, a Reynolds' running average was employed (Deardorff, 1969). Hence, $\bar{q}$ is a filtered variable, continuous in space and time, and further discretization of the resulting differential equations is required.

Faller (1971) proposed a different procedure whereby averaging operators were defined on each of the spatial temporal axes, resulting in mean values with respect to one or several of the independent variables. These means

---

[5] Similar observations apply for a discrete observational grid. "Turbulence," in its essence, is in the eye of the observer.

were defined only at prespecified points of the grid superposed on the continuous space-time volume. Thus

$$(2.69) \quad q_{1000}(x, y, z, t) \equiv \frac{1}{\delta x} \int_{-\delta x/2}^{\delta x/2} d\xi \, q_{0000}(x + \xi, y, z, t)$$

$$= q_{0000}(x, y, z, t) - q_{2000}(x, y, z, t)$$

$$(2.70) \quad q_{1111}(x, y, z, t) \equiv \frac{1}{\delta x \delta y \delta z \delta t} \int_{-\delta t/2}^{\delta t/2} d\tau \int_{-\delta x/2}^{\delta x/2} d\xi \int_{-\delta y/2}^{\delta y/2} d\eta \int_{-\delta z/2}^{\delta z/2} d\zeta \, q_{0000}$$

$$\times (x + \xi, y + \eta, z + \zeta, t + \tau)$$

etc., where the 0 subscript refers to the unmodified variable, 1 refers to an average along the corresponding axis, and 2 refers to a departure from that average. An average like q_{1000} is defined at all (x, y, z, t) such that $x = idx$ (integer i), while q_{1111} is *only* defined at $x = i\delta x$, $y = j\delta y$, $z = k\delta z$ $t = m\delta t$, with i, j, k, m integers. On the other hand q_{0000}, the unperturbed quantity, is defined at all (x, y, z, t), as is q_{2000}, the deviation of q_{0000} from q_{1000}. (Mixed deviations, like q_{1201} are restricted to $x = i\delta x$, $t = m\delta t$ only, etc.) The observables were regarded to be the space-time averages defined at the centers of the computational (observational) grid (i.e., q_{1111}). Equations for their time evolution were obtained for a system of equations containing the momentum equations in the Boussinesq approximation, the continuity equation, the heat energy equation, and an equation of state. The contributions from the subgrid motions were explicitly written in terms of partial averages and deviations from these. The beauty of the method resides not in the complicated expressions for the terms that require parameterization, but in the fact that, for the momentum equation, a set of finite difference equations in the averages result automatically, without resorting to discretization techniques. When the averages are defined in this way, property (2.60) holds identically, but now the averaging and differentiation operators do not commute unless they refer to different axes; this averaging procedure does not comply with property (2.59). Thus, for the one-dimensional case,

$$(2.71) \quad \left(\frac{\partial q}{\partial t}\right)_{01} \equiv \frac{1}{\delta t} \int_{-\delta t/2}^{\delta t/2} \frac{\partial q}{\partial t} \, dt = \frac{q_{00}(t + \delta t/2) - q_{00}(t - \delta t/2)}{\delta t}$$

and the right-hand side can be readily expressed in terms of the q_{01} values at $t \pm \delta t/2$, and the deviations q_{02} from those. For example (cf. Fig. 1),

$$(2.72) \qquad q_{00}\left(t + \frac{\delta t}{2}\right) = q_{01}\left(t + \frac{\delta t}{2}\right) + q_{02}\left(t + \frac{\delta t}{2}\right)$$

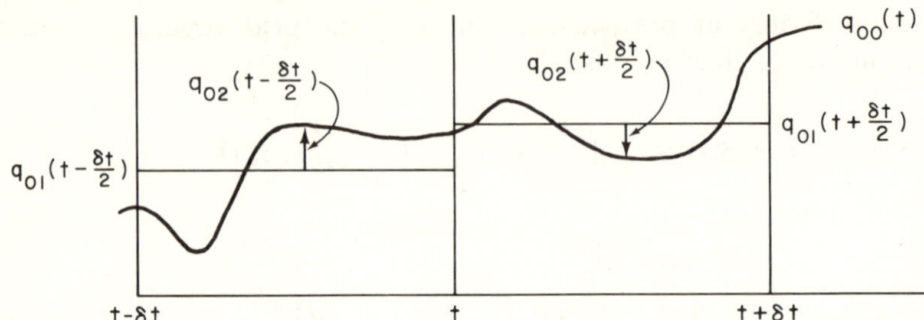

FIG. 1. Means and deviations in Faller's averaging.

The inconvenience with this approach is that a "continuity" equation of the type of Eq. (2.64) is lost for the averaged quantities. In effect,

$$(2.73) \quad \left(\frac{\partial u}{\partial x} + \frac{\partial v}{\partial y} + \frac{\partial w}{\partial z}\right)_{1111}$$

$$= \frac{1}{\delta x}\left[u_{1111}\left(x + \frac{\delta x}{2}, y, z, t\right) - u_{1111}\left(x - \frac{\delta x}{2}, y, z, t\right)\right]$$

$$+ \frac{1}{\delta y}\left[v_{1111}\left(x, y + \frac{\delta y}{2}, z, t\right) - v_{1111}\left(x, y - \frac{\delta y}{2}, z, t\right)\right]$$

$$+ \frac{1}{\delta z}\left[w_{1111}\left(x, y, z + \frac{\delta z}{2}, t\right) - w_{1111}\left(x, y, z - \frac{\delta z}{2}, t\right)\right] + R$$

$$= 0$$

where

$$(2.74) \qquad R \equiv \delta_x(u_{0000} - u_{1112} - u_{1121} - \cdots - u_{2222})$$

$$+ \delta_y(v_{0000} - v_{1112} - v_{1121} - \cdots - v_{2222})$$

$$+ \delta_z(w_{0000} - w_{1112} - w_{1121} - \cdots - w_{2222})$$

$$\neq 0$$

and

$$(2.75) \quad \delta_x q(x, y, z, t) \equiv \frac{1}{\delta x}\left[q\left(x + \frac{\delta x}{2}, y, z, t\right) - q\left(x - \frac{\delta x}{2}, y, z, t\right)\right]$$

A similar approach is followed in so-called integro-interpolation methods, a version of which is given by Romov (1973) for the system of equations for an unsaturated atmosphere. Although the vertical and temporal axes were treated by usual discretization techniques, this followed the direct integration of the equations in the horizontal plane. For a typical computational horizontal cell of side s, whose lower left corner is located at $x_i = is$, $y_j = js$, the following averages are defined (cf. Fig. 2)

$$(2.76) \qquad q_{\overline{i,j}} \equiv \frac{1}{s^2} \int_{y_j}^{y_{j+1}} d\eta \int_{x_i}^{x_{i+1}} d\xi \, q(\xi, \eta)$$

$$(2.77) \qquad q_{\overline{i},j} \equiv \frac{1}{s} \int_{x_i}^{x_{i+1}} q(\xi, y_j) \, d\xi; \qquad q_{i,\overline{j}} \equiv \frac{1}{s} \int_{y_j}^{y_{j+1}} q(x_i, \eta) \, d\eta$$

$$(2.78) \qquad q_{\overline{i}, j+1} \equiv \frac{1}{s} \int_{x_i}^{x_{i+1}} q(\xi, y_{j+1}) \, d\xi; \qquad q_{i+1,\overline{j}} \equiv \frac{1}{s} \int_{y_j}^{y_{j+1}} q(x_{i+1}, \eta) \, d\eta$$

As a consequence, the properties

$$(2.79) \qquad \left(\frac{\partial q}{\partial x}\right)_{\overline{i,j}} = \frac{1}{s} [q_{i+1,\overline{j}} - q_{i,\overline{j}}]$$

$$(2.80) \qquad \left(\frac{\partial q}{\partial y}\right)_{\overline{i,j}} = \frac{1}{s} [q_{\overline{i}, j+1} - q_{\overline{i}, j}]$$

$$(2.81) \qquad \overline{qf} = \bar{q}\bar{f} + \overline{q'f'}$$

(where $q' = q - \bar{q}$, and the overbar on q denotes any of the averages), are satisfied exactly.

Direct integration on a unit horizontal cell provides less equations than the number of variables. For example, application of the procedure to

$$(2.82) \qquad \frac{\partial u}{\partial t} + \frac{\partial u^2}{\partial x} + \frac{\partial uv}{\partial y} = 0$$

results in

$$(2.83) \qquad \frac{\partial u_{\overline{ij}}}{\partial t} + \frac{1}{s} (u^2)_{i+1,\overline{j}} - \frac{1}{s} (u^2)_{i,\overline{j}} + \frac{1}{s} (uv)_{\overline{i}, j+1} - \frac{1}{s} (uv)_{\overline{i}, j} = 0$$

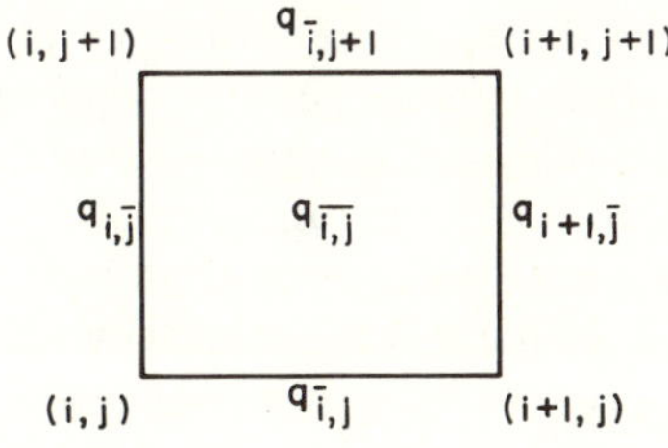

FIG. 2. Location of averaged quantities in Romov's averaging.

The nonlinear contributions $(u^2)_{i+1,\bar{j}}$, etc., are expressed in terms of partial averages (2.77), (2.78) by use of Eq. (2.81). Then the centered and partial averages are related by expanding the unknown functions in powers of x and y, with retention of second order terms.

Since the final equations are prognostic equations for the $q_{\overline{i,j}}$ at every z level, the continuity equation transforms, as in Faller's analysis, into an expression involving unresolved scales. In the next section an alternative choice of dependent variables is proposed that guarantees a simple expression for the incompressibility condition.

2.2.3. The Flux Approach to Turbulent Description. It is now widely recognized (Noh, 1964; Arakawa, 1966; Kurihara and Holloway, 1967; Potter, 1973) that a necessary condition for avoiding certain kinds of instabilities in the numerical solution of the nonlinear flow equations is that the finite difference equations used as discretized versions of the differential equations be written in a "conservative form." In order that the passage from the latter to the former be more natural, the modern literature prefers to refer to the nonlinear terms for incompressible flows as $\partial_x(u^2) + \partial_y(uv) + \partial_z(uw)$, instead of the differentially equivalent $u\,\partial_x u + v\,\partial_y u + w\,\partial_z u$ used traditionally (Lamb, 1932; Phillips, 1956).

In the "running average" approach described in the first part of Section 2.2.2, there is no other reason to prefer one form over the other. For integro-interpolation methods and computational cell averages like those presented in the second part of Section 2.2.2, the nonlinear terms automatically turn out to adopt a conservative form. However, the incompressibility equation for the means becomes dependent on the subgrid scales.

We propose here an interpretation slightly different than that of the second approach, but able to exactly satisfy Eq. (2.64) for properly defined means. The price to pay is a different interpretation for the terms representing the subgrid scales (SGS), but this does not seem to be objectionable.

We start with the integral equations for the incompressibility, momentum, and salinity equations, and integrate them over a unit computational cell centered at $x_i = i\delta x$, $y_j = j\delta y$, $z_k = k\delta z$. Although in what follows the same stages of the previous two approaches are repeated, a new notation is defined to include all the relevant information for this discussion.

We define an operator $\{\ \}_{i,j,k}^{lmn}$ as an average of the quantity between braces over a volume $l\delta x$, $m\delta y$, $n\delta z$ centered at $(x_i = i\delta x, \ y_j = j\delta y, \ z_k = k\delta z)$, where i, j, k, l, m, n are not necessarily all integers. Thus

$$(2.84) \quad \{q\}_{i,j,k}^{111} \equiv \frac{1}{\delta x\,\delta y\,\delta z} \int_{-\delta x/2}^{\delta x/2} d\xi \int_{-\delta y/2}^{\delta y/2} d\eta \int_{-\delta z/2}^{\delta z/2} d\zeta\; q(x_i + \xi,\, y_j + \eta,\, z_k + \zeta,\, t)$$

is the volume average of q over an elementary cell of sides δx, δy, δz, centered at $i\delta x$, $j\delta y$, $k\delta z$. On the other hand,

$$(2.85) \quad \{q\}_{i,\,j,\,k+1/2}^{110} \equiv \frac{1}{\delta x \delta y} \int_{-\delta x/2}^{\delta x/2} d\xi \int_{-\delta y/2}^{\delta y/2} d\eta \; q\left(x_i + \xi, \, y_j + \eta, \, \left(k + \frac{1}{2}\right)\delta z\right)$$

is the surface average of q over the upper face of such unit cell.

The value of q at a general point x, y, z can be expressed in a variety of ways:

$$(2.86) \qquad q(x, y, z, t) = \{q\}_{i,\,j,\,k}^{111} + \overset{111|i,\,j,\,k}{\widehat{q}} \, (x - x_i, \, y - y_j, \, z - z_k)$$

$$= \{q\}_{i,\,j,\,k}^{011} + \overset{011|i,\,j,\,k}{\widehat{q}} \, (x - x_i, \, y - y_j, \, z - z_k)$$

$$= \cdots$$

where the "crowned quantities" are deviations from an average of the type described by the left superscript, centered at x_i, y_j, z_k, and are defined at all points in space. Thus

$$(2.87) \qquad q(x, y, z, t) = \{q\}_{i,\,j,\,k}^{011} + \overset{011|i,\,j,\,k}{\widehat{q}} \, (x - x_i, \, y - y_j, \, z - z_k)$$

$$= \{q\}_{i-1,\,j,\,k}^{011} + \overset{011|i-1,\,j,\,k}{\widehat{q}} \, (x - x_{i-1}, \, y - y_j, \, z - z_k)$$

$$= \cdots$$

Note also that

$$(2.88) \quad \left\{ \overset{r_1'r_2'r_3'}{\left| \frac{\overset{r_1r_2r_3|i,\,j,\,k}{\widehat{q}}}{q} \right.} \right\}_{i',\,j',\,k'} = 0, \qquad \text{iif} \quad \begin{pmatrix} r_1 \\ r_2 \\ r_3 \end{pmatrix} = \begin{pmatrix} r_1' \\ r_2' \\ r_3' \end{pmatrix} \quad \text{and} \quad \begin{pmatrix} i \\ j \\ k \end{pmatrix} = \begin{pmatrix} i' \\ j' \\ k' \end{pmatrix}$$

From these definitions, the incompressibility condition, integrated over the elementary cell

$$(2.89) \qquad 0 = \int_{-\delta x/2}^{\delta x/2} d\xi \int_{-\delta y/2}^{\delta y/2} d\eta \int_{-\delta z/2}^{\delta z/2} d\zeta \, [\partial_l v_l(x_i + \xi, \, y_j + \eta, \, z_k + \zeta)]$$

$$\equiv \delta x \cdot \delta y \cdot \delta z \cdot \{\nabla \cdot \mathbf{v}\}_{i,\,j,\,k}^{111}$$

results in

$$\delta y \cdot \delta z \cdot [\{u\}_{i+1/2,\,j,\,k}^{011} - \{u\}_{i-1/2,\,j,\,k}^{011}] + \delta x \cdot \delta z \cdot [\{v\}_{i,\,j+1/2,\,k}^{101} - \{v\}_{i,\,j+1/2,\,k}^{101}]$$

$$+ \delta y \cdot \delta z \cdot [\{w\}_{i,\,j,\,k+1/2}^{110} - \{w\}_{i,\,j,\,k-1/2}^{110}] = 0$$

or, dividing by the cell's volume,

$$(2.90) \qquad \delta_x\{u\}^{011}_{i,j,k} + \delta_y\{v\}^{101}_{i,j,k} + \delta_z\{w\}^{110}_{i,j,k} = 0$$

where the δ_i operators have been defined by Eq. (2.75).

The same procedure applied on the first component of the momentum equation obtained in Section 2.1.2 results in

$$(2.91) \qquad \frac{\partial}{\partial t}\{u\}^{111}_{i,j,k} = -\delta_x[\{u^2\}^{011}_{i,j,k}] - \delta_y[(uv)^{101}_{i,j,k}] - \delta_z[(uw)^{110}_{i,j,k}]$$

$$- g\delta_x[\{H\}^{01}_{i,j}] - g\delta_x[\{I\}^{011}_{i,j,k}] - \delta_x[\{\varphi\}^{011}_{i,j,k}]$$

$$+ f\{v\}^{111}_{i,j,k} + \nu\{\nabla^2 u\}^{111}_{i,j,k}$$

(where φ now stands for the ratio of dynamic pressure to density) with similar expressions for the other components. (Note that only two sub- and superscripts are given for H, since it is independent of z, thus $\{H\}^{111} = \{H\}^{110} = \{H\}^{11}$ for a cell within the fluid.) The salinity equation is transformed into

$$(2.92) \qquad \frac{\partial}{\partial t}\{S\}^{111}_{i,j,k} = -\delta_x[\{uS\}^{011}_{i,j,k}] - \delta_y[\{vS\}^{101}_{i,j,k}] - \delta_z[\{wS\}^{110}_{i,j,k}]$$

$$+ D^S\{\nabla^2 S\}^{111}_{i,j,k}$$

At this stage we propose to depart from previous analyses and to consider the average fluxes $\{u\}^{011}_{i+1/2,j,k}$, $\{v\}^{101}_{j,j+1/2,k}$, $\{w\}^{110}_{i,j,k+1/2}$ (for all i, j, k in the net), rather than the volume averages $\{q\}^{111}_{i,j,k}$, as the dependent flow variables. The other dependent variables are the volume averages $\{\varphi\}^{111}_{i,j,k}$ and $\{S\}^{111}_{i,j,k}$. This choice of dependent variables lets Eq. (2.90) stand without further modification. It also simplifies the application of boundary conditions, as will be shown in the next chapter.

In order to obtain the same number of equations as of variables, Eqs. (2.91) and (2.92) must be modified on several accounts. In that direction, Eq. (2.91) is rewritten for an elementary computational volume centered at the center of the cell's face with normal $(1, 0, 0)$ (cf. Fig. 3)

$$(2.93)$$

$$\frac{\partial}{\partial t}\{u\}^{111}_{i+1/2,j,k} = \frac{\{u^2\}^{011}_{i,j,k} - \{u^2\}^{011}_{i+1,j,k}}{\delta x} + \frac{\{uv\}^{101}_{i+1/2,j-1/2,k} - \{uv\}^{101}_{i+1/2,j+1/2,k}}{\delta y}$$

$$+ \frac{\{uw\}^{110}_{i+1/2,j,k-1/2} - \{uw\}^{110}_{i+1/2,j,k+1/2}}{\delta z} + f\{v\}^{111}_{i+1/2,j,k}$$

$$+ (g/\delta x)[\{H\}^{01}_{i,j} + \{I\}^{011}_{i,j,k} - \{H\}^{01}_{i+1,j} - \{I\}^{011}_{i+1,j,k}]$$

$$+ (g/\delta x)[\{\varphi\}^{011}_{i,j,k} - \{\varphi\}^{011}_{i+1,j,k}]$$

$$+ \nu\{\nabla^2 u\}^{111}_{i+1/2,j,k}$$

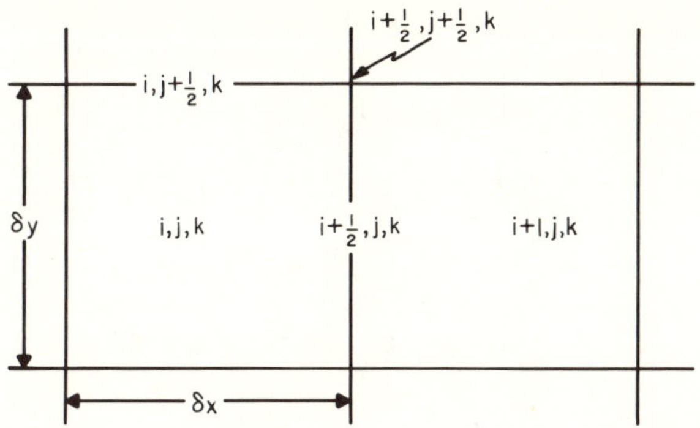

FIG. 3. Cross section of the computational cell by a plane normal to the z axis.

Since

$$u(x_{i+1/2} + \xi, y_j + \eta, z_k + \zeta) \equiv \{u\}^{111}_{i+1/2,\,j,\,k} + \overline{u}^{111|i+\frac{1}{2},\,j,\,k}(\xi, \eta, \zeta)$$

$$\equiv \{u\}^{011}_{i+1/2,\,j,\,k} + \overline{u}^{011|i+\frac{1}{2},\,j,\,k}(\xi, \eta, \zeta)$$

the volume and face averages are related by

$$(2.94) \qquad \{u\}^{111}_{i+1/2,\,j,\,k} = \{u\}^{011}_{i+1/2,\,j,\,k} + \left\{ \overline{u}^{011|i+\frac{1}{2},\,j,\,k} \right\}^{111}_{i+1/2,\,j,\,k}$$

Similar expressions hold for φ, H, and I.

The nonlinear fluxes in Eq. (2.93) are treated as follows:

$$(2.95) \qquad \{u^2\}^{011}_{i,\,j,\,k} = \left\{ \left(\{u\}^{011}_{i-1/2,\,j,\,k} + \overline{u}^{011|i-\frac{1}{2},\,j,\,k} \right) \right.$$

$$\left. \times \left(\{u\}^{011}_{i+1/2,\,j,\,k} + \overline{u}^{011|i+\frac{1}{2},\,j,\,k} \right) \right\}^{011}_{i,\,j,\,k}$$

$$= \{u\}^{011}_{i-1/2,\,j,\,k} \cdot \{u\}^{011}_{i+1/2,\,j,\,k}$$

$$+ \{u\}^{011}_{i-1/2,\,j,\,k} \cdot \left\{ \overline{u}^{011|i+\frac{1}{2},\,j,\,k} \right\}^{011}_{i,\,j,\,k}$$

$$+ \{u\}^{011}_{i+1/2,\,j,\,k} \cdot \left\{ \overline{u}^{011|i-\frac{1}{2},\,j,\,k} \right\}^{011}_{i,\,j,\,k}$$

$$+ \left\{ \overline{u}^{011|i-\frac{1}{2},\,j,\,k} \cdot \overline{u}^{011|i+\frac{1}{2},\,j,\,k} \right\}^{011}_{i,\,j,\,k}$$

(2.96)

$$
\begin{aligned}
\{uv\}^{101}_{i+1/2,\,j+1/2,\,k} = \ &\frac{1}{4}\left(\{u\}^{011}_{i+1/2,\,j,\,k} + \{u\}^{011}_{i+1/2,\,j+1,\,k}\right) \\[4pt]
&\times \left(\{v\}^{101}_{i,\,j+1/2,\,k} + \{v\}^{101}_{i+1,\,j+1/2,\,k}\right) \\[4pt]
&+ \frac{1}{4}\left(\{u\}^{011}_{i+1/2,\,j,\,k} + \{u\}^{011}_{i+1/2,\,j+1,\,k}\right) \\[4pt]
&\times \left\{\frac{101|i,\,j+\frac{1}{2},\,k}{v} + \frac{101|i+1,\,j+\frac{1}{2},\,k}{v}\right\}^{101}_{i+1/2,\,j+1/2,\,k} \\[4pt]
&+ \frac{1}{4}\left(\{v\}^{101}_{i,\,j+1/2,\,k} + \{v\}^{101}_{i+1,\,j+1/2,\,k}\right) \\[4pt]
&\times \left\{\frac{011|i+\frac{1}{2},\,j,\,k}{u} + \frac{011|i+\frac{1}{2},\,j+1,\,k}{u}\right\}^{101}_{i+1/2,\,j+1/2,\,k} \\[4pt]
&+ \frac{1}{4}\left\{\left(\frac{011|i+\frac{1}{2},\,j,\,k}{u} + \frac{011|i+\frac{1}{2},\,j+1,\,k}{u}\right)\right. \\[4pt]
&\times \left.\left(\frac{101|i,\,j+\frac{1}{2},\,k}{v} + \frac{101|i+1,\,j+\frac{1}{2},\,k}{v}\right)\right\}^{101}_{i+1,\,j+1/2,\,k}
\end{aligned}
$$

with similar expressions for the $\{uw\}^{110}$ terms

Equation (2.93) is then transformed into:

$$
\begin{aligned}
(2.97)\qquad \frac{\partial}{\partial t}\{u\}^{011}_{i+1/2,\,j,\,k} = \ &\{u\}^{011}_{i+1/2,\,j,\,k}\cdot\frac{\{u\}^{011}_{i-1/2,\,j,\,k} - \{u\}^{011}_{i+3/2,\,j,\,k}}{\delta x} \\[6pt]
&- \frac{1}{4\delta y}\left[\left(\{u\}^{011}_{i+1/2,\,j,\,k} + \{u\}^{011}_{i+1/2,\,j+1,\,k}\right)\right. \\[4pt]
&\times \left(\{v\}^{101}_{i,\,j+1/2,\,k} + \{v\}^{101}_{i+1,\,j+1/2,\,k}\right) \\[4pt]
&- \left(\{u\}^{011}_{i+1/2,\,j,\,k} + \{u\}^{011}_{i+1/2,\,j-1,\,k}\right) \\[4pt]
&\left.\{v\}^{101}_{i,\,j-1/2,\,k} + \{v\}^{101}_{i+1,\,j-1/2,\,k}\right] \\[6pt]
&- \frac{1}{4\delta z}\left[\left(\{u\}^{011}_{i+1/2,\,j,\,k} + \{u\}^{011}_{i+1/2,\,j,\,k+1}\right)\right. \\[4pt]
&\times \left(\{w\}^{110}_{i,\,j,\,k+1/2} + \{w\}^{110}_{i+1,\,j,\,k+1/2}\right) \\[4pt]
&- \left(\{u\}^{011}_{i+1/2,\,j,\,k} + \{u\}^{011}_{i+1/2,\,j,\,k-1}\right) \\[4pt]
&\left.\times \left(\{w\}^{110}_{i,\,j,\,k-1/2} + \{w\}^{110}_{i+1,\,j,\,k+1/2}\right)\right]
\end{aligned}
$$

$$+ \frac{f}{4}\left[\{v\}^{101}_{i,\,j+1/2,\,k} + \{v\}^{101}_{i+1,\,j+1/2,\,k}\right.$$

$$+ \{v\}^{101}_{i,\,j-1/2,\,k} + \{v\}^{101}_{i+1,\,j-1/2,\,k}]$$

$$+ \frac{1}{\delta x}\left[g\{H\}^{11}_{i,\,j} + \{gI + \varphi\}^{111}_{i,\,j,\,k} - g\{H\}^{11}_{i+1,\,j}\right.$$

$$- \{gI + \varphi\}^{111}_{i+1,\,j,\,k}]$$

$$+ R^{u}_{i+1/2,\,j,\,k}$$

where

(2.98)

$$R^{u}_{i+1/2,\,j,\,k} = -\frac{\partial}{\partial t}\left\{\frac{011|i+\frac12,\,j,\,k}{u}\right\}^{111}_{1+1/2,\,j,\,k}$$

$$+ \frac{1}{\delta x}\left[\{u\}^{011}_{i-1/2,\,j,\,k}\cdot\left\{\frac{011|i+\frac12,\,j,\,k}{u}\right\}^{011}_{i,\,j,\,k}\right.$$

$$- \{u\}^{011}_{i+3/2,\,j,\,k}\cdot\left\{\frac{011|i+\frac12,\,j,\,k}{u}\right\}^{011}_{i+1,\,j,\,k}$$

$$+ \{u\}^{011}_{i+1/2,\,j,\,k}\cdot\left[\left\{\frac{011|i-\frac12,\,j,\,k}{u}\right\}^{011}_{i,\,j,\,k}\right.$$

$$- \left\{\frac{011|i+\frac12,\,j,\,k}{u}\right\}^{011}_{i+1,\,j,\,k}\right]$$

$$+ \left\{\frac{011|i+\frac12,\,j,\,k}{u}\cdot\frac{011|i-\frac12,\,j,\,k}{u}\right\}^{011}_{i,\,j,\,k}$$

$$- \left\{\frac{011|i+\frac12,\,j,\,k}{u}\cdot\frac{011|i+\frac12,\,j,\,k}{u}\right\}^{011}_{i+1,\,j,\,k}\right]$$

$$+ \frac{1}{4\delta y}\left[(\{u\}^{011}_{i+1/2,\,j-1,\,k} + \{u\}^{011}_{i+1/2,\,j,\,k})\right.$$

$$\times\left\{\frac{101|i,\,j-\frac12,\,k}{v} + \frac{101|i+1,\,j-\frac12,\,k}{v}\right\}^{101}_{i-1/2,\,j-1/2,\,k}$$

$$- (\{u\}^{011}_{i+1/2,\,j,\,k} + \{u\}^{011}_{i+1/2,\,j+1,\,k})$$

$$\times\left\{\frac{101|i,\,j+\frac12,\,k}{v} + \frac{101|i+1,\,j+\frac12,\,k}{v}\right\}^{101}_{i+1/2,\,j+1/2,\,k}$$

$$+ (\{v\}^{101}_{i,\,j-1/2,\,k} + \{v\}^{101}_{i+1,\,j-1/2,\,k})$$

$$\times\left\{\frac{011|i+\frac12,\,j-1,\,k}{u} + \frac{011|i+\frac12,\,j,\,k}{u}\right\}^{101}_{i+1/2,\,j-1/2,\,k}$$

$$-\left(\{v\}^{101}_{i,j+1/2,k} + \{v\}^{101}_{i+1,j+1/2,k}\right)$$

$$\times\left\{\overline{u}^{\,011|i+1/2,j,k} + \overline{u}^{\,011|i+1/2,j+1,k}\right\}^{101}_{i+1/2,j+1/2,k}$$

$$+\left\{\left(\overline{u}^{\,011|i+1/2,j-1,k} + \overline{u}^{\,011|i+1/2,j,k}\right)\right.$$

$$\times\left.\left(\overline{v}^{\,101|i,j-1/2,k} + \overline{v}^{\,101|i+1,j-1/2,k}\right)\right\}^{101}_{i+1/2,j-1/2,k}$$

$$-\left\{\left(\overline{u}^{\,011|i+1/2,j,k} + \overline{u}^{\,011|i+1/2,j+1,k}\right)\right.$$

$$\times\left.\left(\overline{v}^{\,101|i,j+1/2,k} + \overline{v}^{\,101|i+1,j+1/2,k}\right)\right\}^{101}_{i+1/2,j+1/2,k}$$

$$+\frac{1}{4\delta z}\left\{\sim \text{terms involving similar quantities in } u \text{ and } w\right\}$$

$$+\frac{f}{4}\left\{\overline{v}^{\,101|i,j+1/2,k} + \overline{v}^{\,101|i+1,j+1/2,k} + \overline{v}^{\,101|i,1,j-1/2,k}\right.$$

$$+\left.\overline{v}^{\,101|i+1,j-1/2,k}\right\}^{101}_{i+1/2,j,k}$$

$$-\frac{g}{\delta x}\left[\left\{\overline{H}^{\,11|i,j}\right\}^{01}_{i,j,k} + \left\{\overline{I}^{\,111|i,j,k}\right\}^{011}_{i,j,k} - \left\{\overline{H}^{\,11|i+1,j}\right\}^{01}_{i+1,j}\right.$$

$$-\left.\left\{\overline{I}^{\,111|i+1,j,k}\right\}^{011}_{i+1,j,k}\right]$$

$$-\frac{1}{\delta x}\left\{\left\{\overline{\varphi}^{\,111|i,j,k}\right\}^{011}_{i,j,k} - \left\{\overline{\varphi}^{\,111|i+1,j,k}\right\}^{011}_{i+1,j,k}\right\}$$

$$+\nu\{\nabla^2 u\}^{111}_{i+1/2,j,k}$$

In Eqs. (2.97) and (2.98) the only face averages for the velocity components that are included are of the form $\{u\}^{011}_{i+\alpha/2,j,k}$, $\{v\}^{101}_{i,j+\beta/2,k}$, $\{w\}^{110}_{i,j,k+\gamma/2}$, with α, β, γ odd integers (cf. Fig. 3), which represent unit fluxes through cell's faces perpendicular to the x, y, and z axes, respectively. Hence, curly brackets and superscripts can be dropped for these means (and kept for all others) in order to simplify the notation:

$$(2.99)\qquad \{u\}^{011}_{i+\alpha/2,j+\beta,k+\gamma} \rightarrow u_{i+\alpha/2,j+\beta,k+\gamma}$$

$$\{v\}^{101}_{i+\alpha,j+\beta/2,k+\gamma} \rightarrow v_{i+\alpha,j+\beta/2,k+\gamma}$$

$$\{w\}^{110}_{i+\alpha,j+\beta,k+\gamma/2} \rightarrow w_{i+\alpha,j+\beta,k+\gamma/2}$$

Likewise, since volume averages have been taken as the primordial ones for φ and S, we denote

(2.100)
$$\{\psi\}^{1\,1\,1}_{i+\alpha,\,j+\beta,\,k+\gamma} \to \psi_{i+\alpha,\,j+\beta,\,k+\gamma}$$

where ψ stands for φ, S, or I. Also

(2.101)
$$\{H\}^{1\,1}_{i+\alpha,\,j+\beta} \to H_{i+\alpha,\,j+\beta}$$

The continuity and momentum equations, compactly written in terms of the face averages, are

(2.102)
$$\delta_x u_{i,\,j,\,k} + \delta_y v_{i,\,j,\,k} + \delta_z w_{i,\,j,\,k} \equiv 0$$

(2.103)
$$\frac{\partial u_{i+1/2,\,j,\,k}}{\partial t} = -[\delta_x(u^2) + \delta_y(uv) + \delta_z(uw)]_{i+1/2,\,j,\,k}$$
$$- \{\delta_x[\varphi + g(H + I)]\}_{i+1/2,\,i,\,k}$$
$$+ \frac{f}{4}\left(v_{i,\,j+1/2,\,k} + v_{i+1,\,j+1/2,\,k} + v_{i,\,j-1/2,\,k} + v_{i+1,\,j-1/2,\,k}\right)$$
$$+ R^u_{i+1/2,\,j,\,k}$$

(2.104)
$$\frac{\partial v_{i,\,j+1/2,\,k}}{\partial t} = -[\delta_x(uv) + \delta_y(v^2) + \delta_z(vw)]_{i,\,j+1/2,\,k}$$
$$- \{\delta_y[\varphi + g(H + I)]\}_{i,\,j+1/2,\,k}$$
$$- \frac{f}{4}\left(u_{i+1/2,\,j,\,k} + u_{i-1/2,\,j,\,k} + u_{i-1/2,\,j+1,\,k} + u_{i+1/2,\,j+1,\,k}\right)$$
$$+ R^v_{i,\,j+1/2,\,k}$$

(2.105)
$$\frac{\partial w_{i,\,j,\,k+1/2}}{\partial t} = -[\delta_x(uw) + \delta_y(vw) + \delta_z(w^2)]_{i,\,j,\,k+1/2}$$
$$- [\delta_z\,\varphi]_{i,\,j,\,k+1/2} + R^w_{i,\,j,\,k+1/2}$$

where

(2.106)
$$(u^2)_{i,\,j,\,k} \equiv u_{i-1/2,\,j,\,k} \cdot u_{i+1/2,\,j\,k}$$

(2.107)
$$(uv)_{i+1/2,\,j+1/2,\,k} \equiv \tfrac{1}{4}(u_{i+1/2,\,j,\,k} + u_{i+1/2,\,j+1,\,k}) \cdot (v_{i,\,j+1/2,\,k} + v_{i+1,\,j+1/2,\,k})$$

The salinity equation takes the form

$$(2.108) \qquad \frac{\partial S_{i,j,k}}{\partial t} = -[\delta_x(uS) + \delta_y(vS) + \delta_z(wS)]_{i,j,k} + R^S_{i,j,j}$$

where, if the complete expression for $R^S_{i,j,k}$ could be used,[6]

$$(2.109) \qquad (uS)_{i+1/2,j,k} = \tfrac{1}{2}u_{i+1/2,j,k} \cdot (S_{i,j,k} + S_{i+1,j,k})$$

Equations (2.102) to (2.108) are partially ready for their numerical solution. The procedure employed automatically provided a set of finite difference equations (in space), and now a form must be found for their temporal discretization.

Consistency would require a similar treatment of the temporal axis, as was done in the previously described approach by Faller (1971). However, for simplicity we will argue differently. Equations (2.103), (2.104), (2.105), and (2.108) can be compactly written at grid point i as

$$(2.110) \qquad \frac{\partial q^l(\mathbf{x}_i, t)}{\partial t} = F^l(\mathbf{q}, \mathbf{x}_i, t) + R^l(\mathbf{q}, \mathbf{x}_i, t), \qquad l = 1, 2, 3$$

where $\mathbf{x}_i$ stands for $[(i + \tfrac{1}{2})\delta x, \, j\delta y, \, k\delta z]$, $[i\delta x, \, (j + \tfrac{1}{2})\delta y, \, k\delta z]$, and $[i\delta x, \, j\delta y, \, (k + \tfrac{1}{2})\delta z]$ if q^l stands for u, v, w when $l = 1, 2, 3$, respectively, and $x_i = [i\delta x, \, j\delta y, \, k\delta z]$ if $q^l = S$.

Direct temporal integration from $t = n\delta t$ to $(n + 1)\delta t$ gives

$$(2.111) \qquad \frac{q^l(\mathbf{x}_i, t + \delta t) - q^l(\mathbf{x}_i, t)}{\delta t} = F^l(\mathbf{q}, \mathbf{x}_i, t + \tau) + R^l(\mathbf{q}, \mathbf{x}_i, t + \tau)$$

where $0 \leq \tau \leq \delta t$. The evaluation of F^l and R^l at $t + \tau$ requires the adoption of a prescription for their computation at times t not on the set of discrete temporal axis points. Such specification could only be given if F^l and R^l were already known as a function of time.

Let us simply consider the case where a running temporal average operator is applied to Eq. (2.110). If the symbol $\langle \, \rangle$ denotes temporally averaged quantities, Eq. (2.110) transforms into

$$(2.112) \qquad \frac{\partial \langle q^l \rangle}{\partial t} = \langle F^l \rangle + \langle R^l \rangle$$

If $\langle q^l \rangle$, $\langle F^l \rangle$ and $\langle R^l \rangle$ are available only at specified grid points of the temporal axis (but defined for all t), then their values at any time can be written in terms of the available values at t_n and a deviation from them

$$(2.113) \qquad \langle P^l(t_n + \tau) \rangle = \langle P^l(t_n) \rangle + \overline{P^l(\tau)}^{\,t|n}$$

---

[6] Note that $R^S_{i,j,k}$ does not include the temporal derivative of the volume average of a perturbation quantity, as do the R terms for u, v, and w [cf. Eq. (2.98)].

where P stands for q, F, or R.[7] Thus, direct integration of Eq. (2.112) yields

$$(2.114) \qquad \langle q^l(\mathbf{x}_i, t_{n+1}) \rangle = \langle q^l(\mathbf{x}_i, t_n) \rangle + \delta t \cdot \langle F^l(\mathbf{q}, \mathbf{x}_i, t_n) \rangle + \mathscr{R}^l$$

where

$$(2.115) \quad \mathscr{R}^l \equiv \delta t \cdot \langle R^l(\mathbf{q}, \mathbf{x}_i, t_n) \rangle + \int_0^{\delta t} d\tau \left[\overline{F^l(\mathbf{q}, \mathbf{x}_i, \tau)}^{\,t|n} + \overline{R^l(\mathbf{q}, \mathbf{x}_i, \tau)}^{\,t|n} \right]$$

Therefore, the adoption of the spatial averaging mentioned above and the simple temporal discretization (2.113) on the running temporal averages, result in a system of finite difference equations of the form (2.114) and (2.115).

Different ways have been proposed for the temporal discretization, leading to different finite difference equations and consequently different expressions for the effect $\mathscr{R}^l$ on the unresolved scales. It is important to stress that the correctness of the predictions obtained with any such scheme is, in the last analysis, dependent upon the correct modeling of the influence of the subgrid motions.

The modeling of $\mathscr{R}^l$ from first principles is, in general, hopeless. The solution envisioned at the present time relies in a systematic program of trial and error experimentation.

Concluding Comment. One of the reasons for this discussion was the clarification of the meaning of the required boundary conditions for the finite difference equations chosen and of the meaning of the variables for which such system solves. Another motivation was the confusion generated in this author's mind in his early encounters with numerical simulations by the use of highly sophisticated numerical schemes used for the solution of flow problems in geophysical scales (e.g., Marchuk *et al.*, 1973). Such confusion disappeared after the realization that there are *no* differential equations applicable to the "mean" variables, unless very crude simplifying assumptions are made for the representation of the subgrid scales ("turbulence"). Then, only puzzlement remained as to the relevance of the use of elegant schemes for the numerical integration of finite difference equations consistent with and convergent to differential equations part of whose spatial terms are fictional.

The crude integration method used here has been chosen because of its simplicity and because of the belief that successful predictions will not come as a result of improved "finite differencing," but as a result of an improved parameterization of the unresolved scales.

---

[7] As defined, F and R are continuous functions of time. Therefore, the mean value theorem of integral calculus is applicable.

     ENRIQUE A. CAPONI

3. THE NUMERICAL MODEL

In the first subsection, a very simplified model for the representation of the subgrid scales is adopted, and the technique employed for the solution of the set of finite difference equations, as well as the boundary conditions used, are explained.

The second subsection briefly describes the computational organization of the computer program that codifies the model.

[It is hereby warned that the word "velocity" (in the x_i direction) is used many times in this chapter as a shorthand for the longer expression "average flux through (the corresponding) computational cell face."]

3.1. The Solution Technique

As a first approximation to the fully three-dimensional simulation of estuarine flows, we have adopted for this work one of the simplest forms for the modeling of the $\mathscr{R}^l$ terms in the dynamic equations (2.115):

$$(3.1) \qquad \mathscr{R}^l(x, y, z, t) = v_1^l[\langle q^l(x + \delta x, y, z, t)\rangle - 2\langle q^l(x, y, z, t)\rangle$$
$$+ \langle q^l(x - \delta x, y, z, t)\rangle]$$
$$+ v_2^l[\langle q^l(x, y + \delta y, z, t)\rangle - 2\langle q^l(x, y, z, t)\rangle$$
$$+ \langle q^l(x, y - \delta y, z, t)\rangle]$$
$$+ v_3^l[\langle q^l(x, y, z + \delta z, t)\rangle - 2\langle q^l(x, y, z, t)\rangle$$
$$+ \langle q^l(x, y, z - \delta z, t)\rangle]$$

with the further simplifying assumption

$$(3.2) \qquad\qquad v_j^u = v_j^v = v_j^w = v_j^s = v_j$$

for each value of j. These are loosely referred to as "eddy coefficients of viscosity."

Assumptions (3.1) and (3.2) are about the most primitive way of accounting for the dissipation of energy transferred from scales resolvable by the grid to the subgrid scales (SGS), and they are unrealistic to the extent that they transform the dynamic equations into finite difference advection-diffusion equations with the same dissipation coefficient for all scales.

In view of the form of the $\mathscr{R}^l$ terms, slightly more complicated expressions for (3.1) and (3.2) seem hard to justify from first principles, but may prove empirically useful.

In any event, the formulation of the model, as well as the method of solution used for the system of finite difference equations, does not depend on the coefficients v_j being constant. Their expression in terms of mean variables will only require a slight increase in storage and computation time.

The functional dependence (3.1) is explicitly incorporated in the present version of the code, especially in the form in which the surface boundary conditions are imposed. The method of solution, however, is not dependent on it, and the replacement of (3.1) by some other relation should not be a difficult task.

Since the dynamic equations are now entirely written in terms of the $\langle q^l(\mathbf{x}_i, t_n) \rangle$, the notation is simplified by the relabeling:

$$\langle q^l(x_\alpha, y_\beta, z_\gamma, t_n) \rangle \to q^{l^n}_{\alpha, \beta, \gamma}$$

where $x_\alpha = \alpha \delta x$, $y_\beta = \beta \delta y$, $z_\gamma = \gamma \delta z$, $t_n = n \delta t$. The subscripts α, β, γ measure the distance from the origin in units of δx, δy, δz. The temporal notation is given in terms of the integer n, assuming a constant value of δt; this is not a requirement, but an assumption made here for notational simplicity.

3.1.1. Finite Difference Equations: Nondimensional Expression and Numerical Solution. Equations (2.114) and (2.115), with the use of (3.1), are already written in finite difference form. They are nondimensionalized by transforming independent and dependent variables as follows:

$$(3.3) \qquad \begin{aligned} t &= \tau_0 \, t_{\mathrm{nd}} & u &= U_0 \, u_{\mathrm{nd}} \\ x &= X_0 \, x_{\mathrm{nd}} & v &= V_0 \, v_{\mathrm{nd}} \\ y &= Y_0 \, y_{\mathrm{nd}} & w &= W_0 \, w_{\mathrm{nd}} \\ z &= Z_0 \, z_{\mathrm{nd}} & \varphi &= \Phi_0 \, \varphi_{\mathrm{nd}} \\ & & H &= Z_0 \, H_{\mathrm{nd}} \end{aligned}$$

where the subscript nd denotes nondimensional quantities, and the subscript 0 the nondimensionalizing unit. In what follows, the subscript nd is dropped, and all variables are to be interpreted as nondimensional, unless stated otherwise.

In order to simplify implementation on a regular, prismatic grid of elementary cell sides δx, δy, δz, the following values for the normalizing quantities have been chosen:

$$(3.4) \qquad \begin{aligned} X_0 &= \delta x \\ Y_0 &= \delta y \\ Z_0 &= \delta z \\ U_0 &= \delta x / \tau_0 \\ V_0 &= \delta y / \tau_0 \\ W_0 &= \delta z / \tau_0 \end{aligned}$$

Furthermore, we adopt $\tau_0 = 1$ day, and $\Phi_0 = (Z_0/\tau_0)^2$. The set of finite difference equations becomes:

$$(3.5) \qquad D^n_{i,j,k} = u^n_{i+1/2,j,k} - u^n_{i-1/2,j,k} + v^n_{i,j+1/2,k} - v^n_{i,j-1/2,k}$$
$$+ w^n_{i,j,k+1/2} - w^n_{i,j,k-1/2} \equiv 0$$

$$(3.6) \quad u^{n+1}_{i+1/2,j,k} = u^n_{i+1/2,j,k} + \delta t \cdot \left[DU_{i+1/2,j,k} + \left(\frac{Z_0}{X_0}\right)^2 (\varphi_{i,j,k} - \varphi_{i+1,j,k}) \right]$$

$$(3.7) \quad v^{n+1}_{i,j+1/2,k} = v^n_{i,j+1/2,k} + \delta y \cdot \left[DV_{i,j+1/2,k} + \left(\frac{Z_0}{Y_0}\right)^2 (\varphi_{i,j,k} - \varphi_{i,j+1,k}) \right]$$

$$(3.8) \quad w^{n+1}_{i,j,k+1/2} = w^n_{i,j,k+1/2} + \delta t \cdot [DW_{i,j,k+1/2} + \varphi_{i,j,k} - \varphi_{i,j,k+1}]$$

$$(3.9) \qquad S^{n+1}_{i,j,k} = S^n_{i,j,k} + \delta t \cdot DS_{i,j,k}$$

The expressions DU, DV, and DW contain all contributions not included in the dynamic pressure φ:

(3.10)

$$DU_{i+1/2,j,k} = (u^2)_{i,j,k} - (u^2)_{i+1,j,k} + (uv)_{i+1/2,j-1/2,k} - (uv)_{i+1/2,j+1/2,k}$$
$$+ (uw)_{i+1/2,j,k-1/2} - (uw)_{i+1/2,j,k+1/2} + f^u(v)_{i+1/2,j,k}$$
$$+ \frac{gZ_0 \tau_0^2}{X_0^2} [(H + \alpha I_k)_{i,j} - (H + \alpha I_k)_{i+1,j}]$$
$$+ A_1 (u_{i+3/2,j,k} - 2u_{i+1/2,j,k} + u_{i-1/2,j,k})$$
$$+ A_2 (u_{i+1/2,j+1,k} - 2u_{i+1/2,j,k} + u_{i+1/2,j-1,k})$$
$$+ A_3 (u_{i+1/2,j,k+1} - 2u_{i+1/2,j,k} + u_{i+1/2,j,k-1})$$

(3.11)

$$DV_{i,j+1/2,k} = (uv)_{i-1/2,j+1/2,k} - (uv)_{i+1/2,j+1/2,k} + (v^2)_{i,j,k} - (v^2)_{i,j+1,k}$$
$$+ (vw)_{i,j+1/2,k-1/2} - (vw)_{i,j+1/2,k+1/2} - f^v(u)_{i,j+1/2,k}$$
$$+ \frac{gZ_0 \tau_0^2}{Y_0^2} [(H + \alpha I_k)_{i,j} - (H + \alpha I_k)_{i,j+1}]$$
$$+ A_1 (v_{i+1,j+1/2,k} - 2v_{i,j+1/2,k} + v_{i-1,j+1/2,k})$$
$$+ A_2 (v_{i,j+3/2,k} - 2v_{i,j+1/2,k} + v_{i,j-1/2,k})$$
$$+ A_3 (v_{i,j+1/2,k+1} - 2v_{i,j+1/2,k} + v_{i,j+1/2,k-1})$$

$$(3.12) \quad DW_{i,j,k+1/2} = (uw)_{i-1/2,j,k+1/2} - (uw)_{i+1/2,j,k+1/2}$$
$$+ (vw)_{i,j-1/2,k+1/2} - (vw)_{i,j+1/2,k+1/2}$$
$$+ (w^2)_{i,j,k} - (w^2)_{i,j,k+1}$$
$$+ \varphi_{i,j,k} - \varphi_{i,j,k+1}$$
$$+ A_1(w_{i+1,j,k+1/2} - 2w_{i,j,k+1/2} + w_{i-1,j,k+1/2})$$
$$+ A_2(w_{i,j+1,k+1/2} - 2w_{i,j,k+1/2} + w_{i,j-1,k+1/2})$$
$$+ A_3(w_{i,j,k+3/2} - 2w_{i,j,k+1/2} + w_{i,j,k-1/2})$$

$$(3.13) \quad DS_{i,j,k} = (uS)_{i-1/2,j,k} - (uS)_{i+1/2,j,k}$$
$$+ (vS)_{i,j-1/2,k} - (vS)_{i,j+1/2,k}$$
$$+ (wS)_{i,j,k-1/2} - (wS)_{i,j,k+1/2}$$
$$+ A_1(S_{i+1,j,k} - 2S_{i,j,k} + S_{i-1,j,k})$$
$$+ A_2(S_{i,j+1,k} - 2S_{i,j,k} + S_{i,j-1,k})$$
$$+ A_3(S_{i,j,k+1} - 2S_{i,j,k} + S_{i,j,k-1})$$

where

$$(3.14) \qquad A_1 = \frac{v_1 \tau_0}{X_0^2}, \qquad A_2 = \frac{v_2 \tau_0}{Y_0^2}, \qquad A_3 = \frac{v_3 \tau_0}{Z_0^2}$$

and

$$(3.15) \qquad f^u = 4\pi \frac{Y_0}{X_0} \sin \lambda, \qquad f^v = 4\pi \frac{X_0}{Y_0} \sin \lambda$$

with λ standing for the geographical latitude. For the nonlinear and Coriolis terms in the "acceleration terms" [Eqs. (3.10) to (3.12)], the following notation was also used:

$$(3.16)$$

$$(u^2)_{i,j,k} = u_{i-1/2,j,k} \cdot u_{i+1/2,j,k}, \qquad (v^2)_{i,j,k} = v_{i,j-1/2,k} \cdot v_{i,j+1/2,k},$$
$$(w^2)_{i,j,k} = w_{i,j,k-1/2} \cdot w_{i,j,k+1/2}$$
$$(uv)_{i+1/2,j+1/2,k} = \tfrac{1}{4}(u_{i+1/2,j,k} + u_{i+1/2,j+1,k}) \cdot (v_{i,j+1/2,k} + v_{i+1,j+1/2,k})$$
$$(uw)_{i+1/2,j,k+1/2} = \tfrac{1}{4}(u_{i+1/2,j,k} + u_{i+1/2,j,k+1}) \cdot (w_{i,j,k+1/2} + w_{i+1,j,k+1/2})$$
$$(vw)_{i,j+1/2,k+1/2} = \tfrac{1}{4}(v_{i,j+1/2,k} + v_{i,j+1/2,k+1}) \cdot (w_{i,j,k+1/2} + w_{i,j+1,k+1/2})$$
$$(u)_{i,j+1/2,k} = \tfrac{1}{4}(u_{i+1/2,j,k} + u_{i+1/2,j+1,k} + u_{i-1/2,j,k} + u_{i-1/2,j+1,k})$$
$$(v)_{i+1/2,j,k} = \tfrac{1}{4}(v_{i,j+1/2,k} + v_{i+1,j+1/2,k} + v_{i,j-1/2,k} + v_{i+1,j-1/2,k})$$

The nonlinear terms in the salinity equation (3.13) are written in a "donor cell" form, instead of the centered form used for the momentum terms, to insure positiveness of $S_{i,j,k}$:

$$(3.17) \qquad (uS)_{i+1/2,\,j,\,k} = u_{i+1/2,\,j,\,k} \cdot \begin{cases} S_{i,\,j,\,k} & \text{if} \quad u_{i+1/2,\,j,\,k} \geq 0 \\ S_{i+1,\,j,\,k} & \text{if} \quad u_{i+1/2,\,j,\,k} < 0 \end{cases}$$

The variable S stands for the total salt contents in cell i, j, k divided by the cell volume $X_0 Y_0 Z_0$. For a homogeneous fluid of salinity S_0, $S_{i,\,j,\,k} = S_0$ for a computational cell filled with fluid, but $S_{i,\,j,\,k} = f S_0$ (where f is the fractional volume occupied by fluid), for a partially filled cell.

The $(I_k)_{i,\,j}$ terms in the horizontal acceleration terms [Eqs. (3.10) and (3.11)] are proportional to the salt contents in fluid column (i, j) from k to the surface:

$$(3.18) \qquad (I_k)_{i,\,j} = \sum_{k'=k}^{\hat{}} S_{i,\,j,\,k'}$$

The location of the normalized variables with respect to the computational cell are shown in Fig. 4.

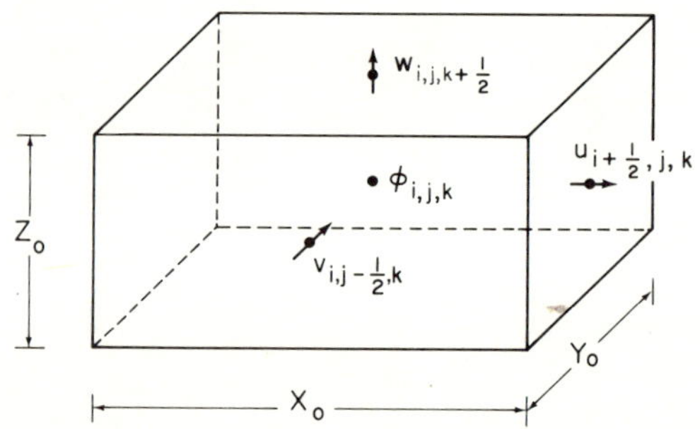

Fig. 4. Arrangement of variables in a computational cell.

(Note: that with the choice of normalizing units given above, the nondimensional computational cell's volume is 1, and the nondimensional areas of each of its faces is also 1. Hence, the nondimensional u, v, w stand both for mean and for total nondimensional mass fluxes between completely filled cells.)

The equations presented so far, in the limit of a two-dimensional, nonrotating, homogeneous fluid in whose description the total pressure φ— rather than φ^H and φ^d—is retained, reduce to a system of finite difference equations that can be successfully solved by the Marker-and-Cell (MAC) Method, due to Harlow and Welch (1965). The method of solution presented here is based on theirs with the necessary changes to include rotation effects and properly attack problems where (Z_0/Y_0), $(Z_0/X_0) \ll 1$. Some of the improvements and variations proposed by several users of this technique have been also incorporated, as will be detailed below in the appropriate sections. [For example, we note here that the form given in Eqs. (3.16) for the (u^2), (v^2), and (w^2) expressions is the ZIP form used by Amsden and Harlow (1970), rather than the more cumbersome relation used in the original MAC formulation.]

In order to update the values of $u_{i+1/2,\,j,\,k}$, $v_{i,\,j+1/2,\,k}$, $w_{i,\,j,\,k+1/2}$, and $S_{i,\,j,\,k}$ from $t = n\delta t$ to $t = (n+1)\delta t$, the values of $\varphi_{i,\,j,\,k}^n$ and $H_{i,\,j}^n$ are required. The obtention of the $H_{i,\,j}^n$ values is regarded here as a boundary problem, and the algorithm is discussed in the section on boundary conditions.

An equation for $\varphi_{i,\,j,\,k}$ at a given time level n is obtained by adding the differences between each of the momentum equations [Eqs. (3.6), (3.7), and (3.8)] evaluated at the opposite faces of the computational cells where they are defined. Incorporating (3.5) to the result and reordering terms, an equation for φ follows:

(3.19)

$$\varphi_{i,\,j,\,k} = \frac{1}{2[1 + (Z_0/Y_0)^2 + (Z_0/X_0)^2]} \left\{ \varphi_{i,\,j,\,k+1} + \varphi_{i,\,j,\,k-1} \right.$$

$$+ \left(\frac{Z_0}{X_0}\right)^2 (\varphi_{i+1,\,j,\,k} + \varphi_{i-1,\,j,\,k})$$

$$+ \left(\frac{Z_0}{Y_0}\right)^2 (\varphi_{i,\,j+1,\,k} + \varphi_{i,\,j-1,\,k})$$

$$\left. + \mathbb{R}_{i,\,j,\,k} \right\}$$

where the source term is

(3.20)
$$\mathbb{R}_{i,\,j,\,k} = \mathbb{R}'_{i,\,j,\,k} + \frac{D_{i,\,j,\,k}^{n+1} - D_{i,\,j,\,k}^n}{\delta t}$$

and

(3.21)
$$\mathbb{R}'_{i,\,j,\,k} = DU_{i-1/2,\,j,\,k} - DU_{i+1/2,\,j,\,k}$$
$$+ DV_{i,\,j-1/2,\,k} - DV_{i,\,j+1/2,\,k}$$
$$+ DW_{i,\,j,\,k-1/2} - DW_{i,\,j,\,k+1/2}$$

The procedure employed is the finite differences equivalent of taking the divergence of the momentum equation (2.36) to obtain the Poisson equation (2.38) for φ^d. Thus, Eq. (3.19) is referred to as a finite differences Poisson equation.

Values of φ on both sides of Eq. (3.19) are to be taken at the same time level n. Direct methods for an efficient solution of elliptic-type equations of the form of Eq. (3.19) have in the last years been developed for the two-dimensional case over a rectangle (Christiansen and Hockney, 1971; Hockney, 1972; Dorr, 1970). Some (Busbee *et al.*, 1970) have been extended to higher dimensions and to L-shaped regions. Roache (1971) has proposed one other algorithm able to tackle higher dimensions and irregular boundaries, but requiring the same class of linear boundary conditions along them. Therefore, none of these methods have been seriously explored here.

Fortunately, the method presently used does not require an accurate solution of Eq. (3.19). In effect, in order to insure mass conservation at successive times, while dynamically updating the three components of the momentum equation, a very fine convergence criterion for the iterative procedure that determines the φ values would be necessary. To avoid this necessity, Welch *et al.* (1965) proposed keeping the D terms in the source term $\mathbb{R}$ [Eq. (3.20)], explicitly computing D at the present time level, and setting $D^{n+1} = 0$. Thus,

$$(3.22) \qquad \mathbb{R}^n_{i,j,k} = \mathbb{R}'^n_{i,j,k} - \frac{D^n_{i,j,k}}{\delta t}$$

This technique, later on extended by Hirt and Harlow (1967) to other problems requiring iterative solution, is equivalent to the introduction of an extra corrective field ψ, whose "finite difference gradient" in the u-v-w equations compensates for errors in the solution of φ—and other round-off errors present in the computation of the right-hand side of the momentum equations—that give rise to a divergent velocity field. The lumping of this extra field with the pressure field φ is computationally sound. Such a technique has been successfully used in other incompressible flow simulations (Deardorff, 1969; Williams, 1969).

Thus, the only criteria influencing the choice of iterative method to be used in the solution of the equation for φ are simplicity of implementation and economy of computation. The first one is fulfilled by the method used here, while the second has not been tested against possible alternatives (Douglas, 1962; Hadjidimos, 1970; Yanenko, 1971).

The actual solution of Eq. (3.19) is performed here using the point successive over-relaxation (SOR) method, whose convergence is faster than that of either the Jacobi or Gauss-Seidel methods (cf., for example, Carnahan *et al.*, 1969) and whose implementation is easier and more economical. Thus, if h denotes the iteration level and the sweep over i, j, k

space is done in the direction of increasing indices, the algorithm prescribes
the computation of a temporary value

$$(3.23) \quad \varphi_{i,j,k}^{(h+1)} = \frac{1}{2[1 + (Z_0/X_0)^2 + (Z_0/Y_0)^2]} \{\varphi_{i,j,k-1}^{h+1} + \varphi_{i,j,k+1}^{h} + (Z_0/X_0)^2$$
$$\times (\varphi_{i-1,j,k}^{h+1} + \varphi_{i+1,j,k}^{h}) + (Z_0/Y_0)^2(\varphi_{i,j-1,k}^{h+1} + \varphi_{i,j+1,k}^{h})$$
$$+ \mathbb{R}_{i,j,k}\}$$

and the evaluation, prior to consideration of the next grid point, of the
final value

$$(3.24) \qquad\qquad \varphi_{i,j,k}^{h+1} = \varphi_{i,j,k}^{h} + \alpha(\varphi_{i,j,k}^{(h+1)} - \varphi_{i,j,k}^{h})$$

The relaxation parameter α determines the rate of convergence, which is
faster for $1 < \alpha < 2$. For simple geometries it is possible to determine
analytically on optimum value for α to minimize the number of iterations
needed to satisfy the convergence criterion. For the geometry of our test
cases, it was experimentally found that $1.6 < \alpha < 1.8$ produced a resultant
field within the specified bounds in the least number of iterations. The
value $\alpha = 1.7$ was adopted throughout.

This procedure converges and is absolutely stable (Carnahan *et al.*, 1969)
for the Neumann, Dirichlet, or mixed problem, for a region entirely bounded
by rigid boundaries. Stability problems may arise as a consequence of the
implementation of some type of boundary conditions at the surface
(Nichols and Hirt, 1971). This problem will be discussed in Section 3.1.5.

3.1.2. Boundary Conditions for the Dynamic Equations. The volume of
fluid is confined to some arbitrarily shaped "container." There are various
kinds of physical boundaries to model: (a) the bed of the river estuary,
i.e., coastline and bottom, (b) the inflowing rivers, (c) the water-air interface,
and (d) the opening(s) to the sea.

All these represent the interaction of the system being simulated with
other systems. The prescription of boundary conditions for the computed
variables at each of these regions means the adoption of a model either
for the external system, or for the interaction, or, most likely, for both.

In the tradition of the MAC method, cells located outside of the fluid
and separated from it by faces representing boundaries of type a, b, or d are
labeled BND, IN, and OUT, respectively. Cells not belonging to any of the
previous categories are labeled EMP or FUL according to their being empty
or containing fluid and not being adjacent to an EMP cell. Cells containing
fluid and adjacent to an EMP cell are labeled SUR. In most of our applications, the set of SUR cells is almost always identical to the set of cells in
which the surface is located.

A simplifying assumption made here is that all the boundaries—except the surface—fall on faces of computational cells. Some procedures have been proposed (Viecelli, 1969; Chan and Street, 1970a) in two-dimensional versions of the MAC method by which boundaries falling at other places could be handled.

Since the u, v, and w values related by Eqs. (2.97) are average fluxes through determined faces of the computational volumes, it would not be difficult to adapt them to piecewise plane boundaries not coincident with the prismatic cell boundaries. This is equivalent to a net of computational cells not all of which are equal. This procedure involves further storage requirements for each of these cells' dimensions, and the appearance of quantities related to these in the nondimensional equations (3.5) to (3.13). The modeling of the boundary conditions for u, v, and w will suffer the same kind of difficulties in their realistic modeling as they do now (cf. below). A difficult part may be the pressure boundary condition on a sloping bottom, but Viecelli's (1969) method may include useful clues for this problem. The solution will be relatively insensitive to lateral pressure boundary conditions in those cases where the normal to the boundary has a zero component in the k direction. (This will be true for most cases of geophysical flows, where $Z_0 \ll X_0, Y_0$.)

Although the possibility of irregularly shaped cells is conceivable, its implementation may prove worthwhile only for cases where the irregular faces have a normal with at least one of its components equal to 0.

It must be noticed that we are aware of difficulties found when using irregular mesh systems (Crowder and Dalton, 1971). However, such problems would not arise if a correct representation of the SGS were available.[8] Furthermore, they do not need to arise in general when solving differential equations numerically. For example, the belief that the use of nonuniform grid intervals invariably causes reflections has been found by Browning *et al.* (1973) to be not necessarily justified for the numerical solution of linear equations. Also, Kálnay (1972) has presented a method for the construction of certain stretched coordinates which she reports to have successfully used in two-dimensional numerical models of the atmosphere of Venus.

3.1.2.1. No-flow boundaries. No transport through this type of boundaries means that the flux through it is zero at any point. In particular, the average normal flux, represented by one of u, v, or w of Eqs. (3.6) to (3.8), vanishes. For our model of regular prismatic cells, this condition is exempli-

---

[8] For a rough representation of SGS, as is the likely case for immediate applications, the irregular grid configuration itself could be modeled by appropriately modifying the SGS representation in regular cells near the boundary.

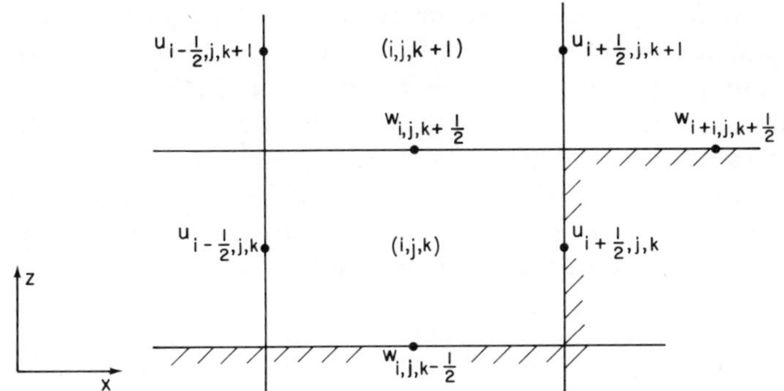

FIG. 5. Cross section of a corner configuration.

fied with reference to a cross section of a typical corner configuration shown in Fig. 5:

$$(3.25) \qquad w^n_{i,\,j,\,k-1/2} = u^n_{i+1/2,\,j,\,k} = w^n_{i+1,\,j,\,k+1/2} = 0$$

for all values of n. The condition is a result of the integration of Eq. (2.39) over a cell's face, and its strict enforcement is required if the whole system is to conserve mass.

For u and v at a lateral boundary to remain zero at successive time steps, the accelerating terms [(3.10), (3.11)] must vanish. This imposes the further condition:[9]

$$(3.26) \qquad (H + \alpha I_k)_\mathrm{B} = (H + \alpha I_k)_\mathrm{F}$$

where the subscript B stands for the (i, j) values of a boundary column, and F for those of the adjacent interior column. Likewise, no-salt flux through the boundary implies

$$(3.27) \qquad S_\mathrm{B} = S_\mathrm{F}$$

The combination of (3.26) and (3.27) results in

$$(3.28) \qquad H_\mathrm{B} = H_\mathrm{F}$$

Note that H and S need not be single valued at boundary cells (corner configurations).

<hr>

[9] Condition (3.26) is actually a simplification of that obtained by setting Eqs. (3.10)–(3.11) equal to zero at lateral boundaries. The full equations also involve the particular type of boundary conditions chosen for flows parallel to the boundary (see below). However, the conditions stated are good in all cases where $gZ_0 \tau_0^2/X_0^2 \gg \mathrm{Max}(f^u, f^v, A_1, A_2, A_3)$. In our examples, the left-hand side of this inequality is $\gtrsim O(10^3)$, while that of the right-hand side is less than 10, and the approximation is good.

There is another condition to be set at this type of boundary whose actual formulation is open to debate, in the same sense and for the same reasons that the representation of SGS everywhere in the volume is open to debate. The problem is that of the correct incorporation of the mean motion energy drainage that the existence of a boundary layer represents. In the case of Eqs. (3.6) to (3.12), the rate of dissipation of mean momentum due to the subgrid scales is everywhere represented by terms proportional to second finite differences. In particular, for the $u_{i-1/2,\,j,\,k}$ value of Fig. 5, the term

$$(3.29)\qquad A_3[(u_{i-1/2,\,j,\,k+1} - u_{i-1/2,\,j,\,k}) + (u_{i-1/2,\,j,\,k-1} - u_{i-1/2,\,j,\,k})]$$

$$= A_3(\tau^u_{k,\,k+1} - \tau^u_{k-1,\,k})$$

$$= \frac{\tau_0^2}{\rho Z_0 X_0}\left(\tau^{u(d)}_{k,\,k+1} - \tau^{u(d)}_{k-1,\,k}\right)$$

is to be evaluated, where X_0, Y_0, Z_0, and A_3 are given by Eqs. (3.4), (III.14). The quantity $u_{i-1/2,\,j,\,k-1}$ is outside of the net, and consequently, its value cannot be provided by the dynamic equation.

It is common practice in two-dimensional models of systems related to ours (Hansen, 1956; Orlob, 1972; Marchuk *et al.*, 1973) to represent the equivalent of $\tau^{u(d)}_{k-1,\,k}$ in Eq. (3.29)—the dimensional "bottom boundary frictional term"—by an expression of the form

$$(3.30)\qquad\qquad\qquad \tau^{u(d)}_{k-1,\,k} = \rho g\,\frac{u(u^2 + v^2)^{1/2}}{C_{\mathrm{h}}^2}$$

where C_{h} is the so-called Chezy coefficient, a dimensional number. Its use in two-dimensional cases, where only equations for the vertically integrated fluxes are available, is convenient, as it conglomerates in a single expression the mean energy drain due to the interaction of unresolved horizontal layers, including the "bottom friction." The value of C_{h} is experimentally determined so that the model is tuned to the system (C_{h} is said to depend, not only on some suitably defined Reynolds' number, but also on something called the "boundary roughness").

The use of Eq. (3.22) has both been proposed (Prichard, 1971) and implemented (Leendertse *et al.*, 1973; Hamilton *et al.*, 1973) for the bottom layer of three-dimensional models. However, we do not feel this to be a convenient practice on two counts. The first is that there is no conceptual difference between the meaning of $\tau^u_{k-1,\,k}$ for the bottom, or for any other cell; in all cases it embodies the frictionlike effects on the resolved quantities by the subgrid scales. The second is that it is as convenient to either let v_3—or the nondimensional A_3 in (3.29)—assume different values near the boundaries, imposing at the same time a no-slip condition (for the mean "velocities") at the boundary, or to let A_3 to be determined by the same relation as it is in the rest of the volume, but specifying an "empirical" law for the needed $u_{i-1/2, j, k-1}$. In terms of this Chezy coefficient, the two options described are (for a flow along x, for simplicity):

$$(3.31) \qquad u_{i-1/2, j, k-1} = -u_{i-1/2, j, k}; \qquad A_3 = \frac{1}{2} \frac{gX_0}{Z_0 C_h^2} |u_{i-1/2, j, k}|$$

or

$$(3.32) \qquad u_{i-1/2, j, k-1} = \left[1 - \frac{gX_0}{A_3 Z_0 C_h^2} |u_{i-1/2, j, k}|\right] u_{i-1/2, j, k}$$

These relations are also valid if A_3 is everywhere defined in terms of the mean variables, instead of being a constant as has been taken in the present tentative version. Notice that when $C_h \to \infty$, Eq. (3.32) gives the "free-slip" condition for mean fluxes:

$$(3.33) \qquad u_{i-1/2, j, k-1} = u_{i-1/2, j, k}$$

while for $C_h^2 = 1/2g\,\mathcal{Re}$, with $\mathcal{Re} = (X_0/A_3 Z_0)|u_{i-1/2, j, k}|$ a conveniently defined computational Reynolds' number, Eq. (3.31) reduces to the "no-slip" condition

$$(3.34) \qquad u_{i-1/2, j, k-1} = -u_{i-1/2, j, k}$$

with A_3 given by the same prescription as in the interior. The same arguments apply to both bottom and lateral impenetrable boundaries.

Consistent with a choice of constant A_1, A_2 used in the examples of Sections 4 and 5, the no-slip condition [Eq. (3.34)] is imposed on the u and v components at the bottom and lateral boundaries. On the other hand, consistency with the boundary condition on H at no flux boundaries, requires the adoption of the free-slip condition [Eq. (3.33)] for the w component at lateral walls.

A short comment on the ZIP form is in order here. Formal evaluation of $DU_{i+1/2, j, k}$ (for the case of Fig. 5), involves the computation of $(u^2)_{i+1, j, k} - (u^2)_{i, j, k}$. By using the ZIP form, that expression is written as $u_{i+1/2, j, k}(u_{i+3/2, j, k} - u_{i-1/2, j, k})$. At an impenetrable boundary, this is identically zero because $u_{i+1/2, j, k} = 0$. The form originally used in MAC consisted in setting $(u^2)_{i, j, k} = 1/4(u_{i-1/2, j, k} + u_{i+1/2, j, k})^2$. Its disadvantage was apparent in early versions of MAC where $DU_{i+1/2, j, k}$ had to be evaluated in order to obtain $R_{i, j, k}$, the source term for the Poisson equation for φ. Consistency required $DU_{i+1/2, j, k}$ to be zero when formally computed, and the use of that form for $(u^2)_{i, j, k}$ imposed the condition that $|u_{i+3/2, j, k}| = |u_{i-1/2, j, k}|$. The ZIP form gives automatically a zero contribution on a wall, and no computational boundary condition is required for the normal flux inside the wall. This argument, though, is not relevant for us because in this version DU is never computed at a boundary wall, due to the adoption of an idea by Easton (cf. Section 3.1.3).

The overall technique seems to handle corners properly and no special treatment is deemed necessary.

3.1.2.2. Inflowing rivers. Cells labeled IN indicate that through the common face between them and internal cells, the river discharge, i.e., $u_{i+1/2, j, k}$ in Fig. 6, is specified as a function of time.

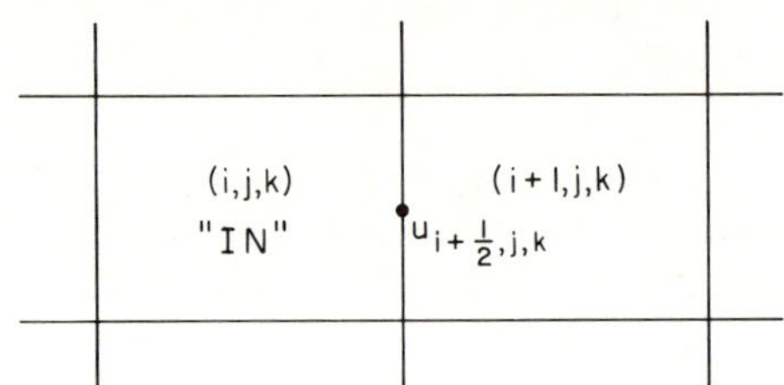

FIG. 6. A typical IN cell at (i, j, k).

In the present implementation, there is a choice of two prescriptions for such an updating:

(a) $u^{n+1}_{i+1/2, j, k} = u^{n}_{i+1/2, j, k}$, regardless of the height distribution in the neighborhood.

(b) Constant flux, with the assumption of surface elevation at the IN column equal to that in the adjacent interior column. (This involves also an assumption of flux values being independent of k.)

Differences between both procedures are going to be apparent (mainly locally) only in cases where the incoming flows are dominant in the circulation compared to tides. In such cases, a more elaborate boundary condition may be justified.

Salinity values can be fixed at these cells or specified by suitable modification of the code. The test cases, which included only one of these IN cells, followed the first approach.

In the application to the Chesapeake Bay though, the same procedure is dangerous because of the inclusion of several rivers. If the salinity values at the different IN cells (as well as at the OUT cells) were specified, these should be self-consistent with the whole representation of the system given by the model if extraneous boundary-related effects propagating to the interior are to be avoided.

One possibility is to fix the value of

$$(3.35) \qquad \Delta S = S_{\text{IN}+1} - S_{\text{IN}} = \text{const.}$$

where the subscript IN designates the IN cell located at (i, j, k) and IN + 1 its neighbor inside the fluid, at either $(i + 1, j, k)$, $(i, j + 1, k)$,

$(i - 1, j, k)$, or $(i, j - 1, k)$. Prescription (3.35) seems inadequate, especially if IN fluxes are to be varied in time, because large influxes may drive S_{IN} to negative values.

The parameterization of S_{IN} should be made in terms of S_{IN+1}, the river discharge, its cross section, and the "mixing coefficients." We are not aware of previous efforts in the direction of the incorporation of a dynamic boundary condition for the determination of S_{IN} in numerical models.

A suitable simplification for this parameterization seems to have to be based upon a relationship of the type

$$(3.36) \qquad S_{IN} = \lambda S_{IN+1}, \qquad 0 \leq \lambda \leq 1$$

where $\lambda = \lambda(u_{IN}, A_1, A_2, A_3, X_0, Y_0, Z_0)$. A general analysis of the consequences of (3.36) for our three-dimensional model is not within our possibilities, but some insight on them can be gained by looking at the dependence of S_{IN} on λ in a steady state, one-dimensional string of cells, as shown in Fig. 7. For simplicity, we assume

$$(3.37) \qquad u_i = u_{IN}, \qquad i = 1, n$$

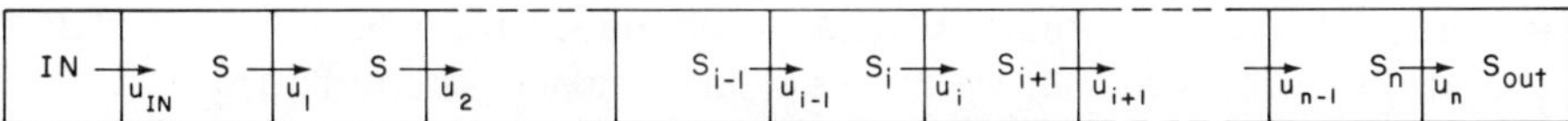

FIG. 7. One-dimensional model for the consideration of some consequences of adopting a dynamic boundary condition [Eq. (3.36)] for the salinity at IN cells.

Thus, the steady-state condition (for the adopted donor cell scheme)

$$(3.38) \qquad DS_i = u_{IN}(S_{i-1} - S_i) + A_1(S_{i+1} - 2S_i + S_{i-1}) = 0$$

implies the recursion relation

$$(3.39) \qquad (R_e + 1)S_{i-1} - (R_e + 2)S_i + S_{i+1} = 0, \qquad 1 < i \leq n$$

where $R_e = u_{IN}/A_1$ is the Reynolds' number of the problem. The value S_{OUT} is obtained by

$$(3.40) \qquad S_{OUT} = \frac{1}{2}(S_{OUT-1} + S_F)$$

where S_F is a specified fixed value appropriate to the "open ocean" (cf. below, in the discussion of OUT boundary conditions).

Equations (3.39) and (3.40) uniquely determine a value for S_{IN} in terms of S_F, λ, and R_e through a relationship whose complexity increases with n. In particular, for $n = 4$

$$(3.41) \quad S_{IN} = \frac{\lambda S_F/2}{(R_e + \tfrac{3}{2})[(R_e + 2)^2 - \lambda(R_e + 1)(R_e + 2) - (R_e + 1)] \atop \qquad\qquad\qquad - (R_e + 1)[R_e + 2 - \lambda(R_e + 1)]}$$

with limits

$$(3.42) \qquad\qquad S_{IN} \xrightarrow[R_e \to 0]{} \frac{\lambda S_F}{5 - 4\lambda}; \qquad 0 \le \lambda < 1$$

and

$$(3.43) \qquad\qquad S_{IN} \xrightarrow[R_e \gg 2]{} \frac{\lambda S_F}{2(1 - \lambda)R_e^3} \xrightarrow[R_e \to \infty]{} 0; \qquad 0 \le \lambda < 1$$

For $\lambda = 1$, Eq. (3.41) reduces to

$$(3.44) \qquad\qquad\qquad S_{IN} = S_F$$

independent of R_e, as expected.

The parameter λ should depend on the cross section and the value of R_e *at* the IN cell, and is to be estimated from experimental steady state values for each IN location in particular. A functional relation between λ and R_e can also be proposed, but we have not investigated its consequences.

3.1.2.3. The "free" surface. In this section we will deal only with the determination of the location of this moving boundary, its temporal variation and the implementation of boundary conditions on the velocity field. Surface conditions on the pressure and salinity fields may introduce stability problems of their own. These are treated in Section 3.1.5.

The precise location of the surface within a computational cell is possible only if it has been defined at $t = 0$, and a mass conservative updating scheme can be found. In the original two-dimensional MAC Method (Harlow and Welch, 1965; Welch *et al.*, 1965), and modifications to it (Amsden and Harlow, 1970; Chan and Street, 1970a), numerical experiments were initiated with a distribution of markers with (x, z) coordinates falling on the prescribed surface for $t = 0$. At every time step, the position of each marker was kinematically updated by a bilinear interpolation[10] from the velocity field in its neighborhood. The updated surface is then defined by joining with straight segments adjacent markers (Fig. 8). Markers were destroyed when moved into BND or OUT cells, and created at a given

---

[10] Chan and Street (1970a, b) proposed a higher order interpolation scheme. However, it is easy to see that only a bilinear form (trilinear for three dimensions) can guarantee conservation of mass.

rate at IN-type cells. However, no mention was made of special provisions to maintain a uniform number of markers along x. These markers have proven to be extremely useful in providing graphical information, as shown by the work of the group at Los Alamos Scientific Laboratory (University of California).

The use of markers for the determination of the position of the surface was intended in this three-dimensional case, but the project was abandoned after many tries when the nature of the difficulties became clear. These difficulties involve both the problem of trajectory errors and that of the impossibility of building an algorithm whose application leads to particle distributions satisfying the most elementary symmetry requirement. The first problem could, in principle, be corrected for (Forester, 1973). On the other hand, the appearance of manifestations of the second difficulty, valid when there is a net flow through the system, can only be delayed to later times by uneconomically increasing the total number of particles allowed in the system.

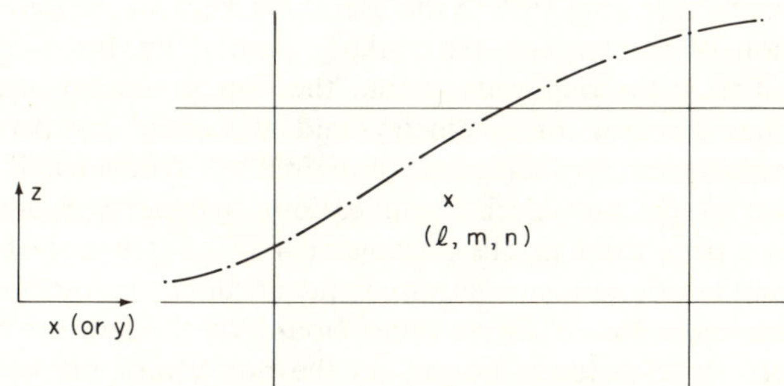

FIG. 8. The position of the surface as determined by the surface markers.

As originally proposed, the markers' positions are updated explicitly, according to

$$(3.45) \qquad \mathbf{x}^{n+1} = \mathbf{x}^n + \Delta t \cdot (\mathbf{u})^n$$

where $\mathbf{x}$ is the position vector of the marker under consideration, and $(\mathbf{u})$ is the linearly interpolated velocity at the marker's position. Forester (1973) showed experimentally that when this technique is used in the simulation of steady (unsteady) flows with curved streamlines (pathlines), there is a strong tendency for the marker particles to drift away from such lines. The trajectory errors can be significantly reduced by substantially reducing the time step. He successfully showed the possibility of reducing such errors with a semiimplicit scheme solved in an iterative form.

The fact that the surface elevation value H enters into our formulation for the dynamic equations, makes explicit the influence of the markers' positions in the dynamics. When values of (Z_0/X_0) are O(1), the splitting of the pressure field into two parts is not necessary, and in such case the markers' positions influence the dynamics indirectly, through the boundary conditions on φ at the surface. When very simple versions for these conditions (see Section 3.1.5) are implemented, the statement (Welch *et al.*, 1965) that the motion of the particles does not affect the dynamics, is approximately correct. It is definitely not correct when more elaborate boundary conditions on φ and the velocity field are imposed at the surface, as those used by Chan and Street (1970a, b) and by Nichols and Hirt (1971).

Thus, for our applications, markers would directly enter into the dynamics, if they were to be used for the determination of the surface elevation values. The procedure becomes dangerous and extremely complicated, involving the continuous creation of new markers in surface areas which became covered by fewer particles than desired and the destruction of markers in overpopulated regions, so as not to exceed allocated storage.

There is one further objection to the use of markers for the determination of the position of the surface, particularly painful for three-dimensional simulations. This is the requirement that the flow generated inside a container that has a plane of symmetry and is excited by perturbations symmetric with respect to such plane, must remain symmetric if the initial conditions are so. (In our specific applications, symmetry of acting forces also requires setting the Coriolis parameter $f = 0$.) It is easy to see that if the following restrictions are imposed: (a) an upper bound on the total number of markers allowed, (b) an upper bound on the number of markers per column (i, j), (c) a lower bound on the number of markers in each of the four quadrants of a column; then symmetry can only be conserved if the particular geometry is built in into the procedures for creation and destruction of markers, a highly undesirable feature. Requirements (a) and (b) are due to the finiteness of available storage, while requirement (c) is imposed in order to obtain an adequate distribution to fit a plane by least squares which would determine a value of H at the center of the column.

The previous reasons, tied to the corresponding complexity of programming and excessively long running times required, make the use of markers for the determination of the surface in the three-dimensional case only marginally suitable for some kind of problems. The only example found in the literature (Nichols and Hirt, 1973) of a three-dimensional flow computation with a free surface using the general principles of the MAC Method updates the surface by the finite differences analog to

$$(3.46) \qquad \frac{\partial H}{\partial t} = w - u \frac{\partial H}{\partial x} - v \frac{\partial H}{\partial y}$$

Markers have been left in this code, and the user may choose to use them or not for tracing purposes.

The method employed here to update the position of the surface is slightly different and guarantees total mass conservation. The value of H assigned to the center of each column is interpreted as the average of the surface elevation in that column, consistently with the discussion of Section 2.2.3. The difference between such values at successive time steps is associated with a change in the volume of fluid in that column by $(H_{i,j}^{n+1} - H_{i,j}^{n}) \cdot \delta x \cdot \delta y$. For mass to be conserved, this volume has to be balanced by the amount of fluid that entered such column during δt, i.e., the sum of fluxes through the (i, j) column's walls.

For the nondimensional quantities, the condition is

$$(3.47) \quad H_{i,j}^{n+1} = H_{i,j}^{n} - \delta t \left[\sum_k \hat{u}_{i+1/2,\,j,\,k}^{n} - \sum_k \hat{u}_{i-1/2,\,j,\,k}^{n} \right.$$

$$\left. + \sum_k \hat{v}_{i,\,j+1/2,\,k}^{n} - \sum_k \hat{v}_{i,\,j-1/2,\,k}^{n} \right]$$

where, referring to Fig. 9,

$$(3.48)$$

$$- \sum_k \hat{u}_{i+1/2,\,j,\,k} + \sum_k \hat{u}_{i-1/2,\,j,\,k} = u_1 + u_2 + \delta_1 u_3 - u_5 - u_6 - u_7 - \delta_2 u_8$$

The procedure has a direct application and requires no interpolation of the velocity field.

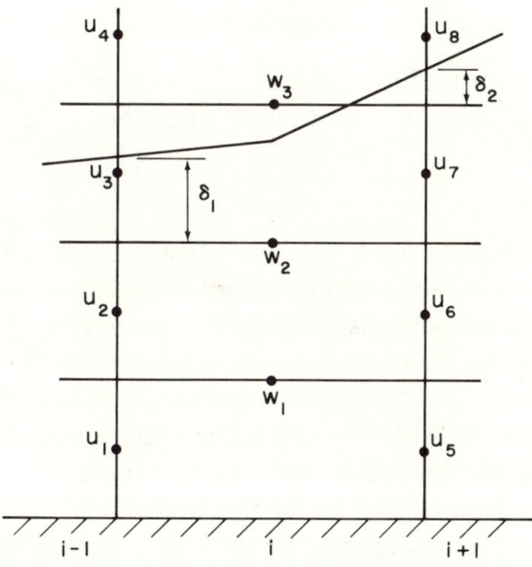

FIG. 9. Two-dimensional cross section to illustrate the algorithm used to update the surface.

[Note that Eqs. (3.47) and (3.48) defined an algorithm that absolutely conserves the total mass in the system only if the δ_l values are computed from the H_{ij}^n values and held fixed until all the $H_{i,j}^{n+1}$ have been determined. The code does this through the use of auxiliary arrays $HI_{i,j} = (H_{i,j} + H_{i+1,j})/2$ and $HJ_{i,j} = (H_{i,j} + H_{i,j+1})/2$].

For the application of Eqs. (3.47)–(3.48), as well as for the updating of flux values at locations near the surface (i.e., u_3 and u_7 in Fig. 9), Eqs. (3.6) to (3.13) require the specification of flux values at positions which are not in the fluid (i.e., u_4, u_8, and w_3 in Fig. 9), for both the computation of the nonlinear terms and the evaluation of the terms standing for the chosen representation of the subgrid scales.

In formal analogy to the derivation leading to Eqs. (2.51) and (2.52) in the horizontal surface approximation, conditions on the jump in u and v across the surface are imposed:

$$(3.49) \qquad u_+ = u_- + \tau_x$$

$$(3.50) \qquad v_+ = v_- + \tau_y$$

where subscripts $+$ and $-$ indicate outside and inside the fluid (i.e., at both sides of the surface), and $\tau_{x(y)}$ are nondimensional quantities that stand for

$$(3.51) \qquad \tau_x = \frac{\tau_0^2}{\rho Z_0 X_0 A_3} \tau_x^d$$

$$(3.52) \qquad \tau_y = \frac{\tau_0^2}{\rho Z_0 Y_0 A_3} \tau_y^d$$

It is customary to express the dimensional surface drags τ_x^d, τ_y^d at an air-water boundary in terms of the mean wind speed u specified at some height z above the surface

$$(3.53) \qquad \tau_i = \rho C(z) u_i^2(z)$$

where ρ is the air density, and C is the shear stress, resistance, drag, or friction coefficient. Its value is parametrically dependent on z, the height at which u is given. Several methods are used to obtain estimates of the friction coefficient (cf., e.g., Chambers *et al.*, 1970, Chapter 2).

This kind of relations have been incorporated into three-dimensional models (Leendertse *et al.*, 1973). It should be understood that the coefficient C must also depend on the size of the computational cell.

The use of Eqs. (3.49)–(3.50) for the extrapolation of tangential fluxes across the surface can be interpreted as the continuation of the fluid in that region subject to the condition that Δu assumes a specified value across the surface. Consistently with this interpretation, vertical fluxes w at places above the surface are determined so that continuity ($D_{i,j,k} = 0$) is exactly valid in these SUR and EMP cells.

3.1.2.4. Openings to the sea. To fully specify in a nonambiguous way conditions at a lateral open boundary it is necessary to specify in it the full flux, pressure, salinity, and height fields as a function of time. For practical reasons, namely, lack of appropriate data, such an approach will be possible to implement only in the far future. (Naturally, the same arguments apply to boundaries representing tributaries.)

In order to generate this boundary data and be able to run representative examples, we have adopted a model for the "sea as seen by the estuary" which has been found to provide reasonable results.

It was found by experimentation that transverse fluxes, leading to mass transfer between adjacent OUT cells, should be allowed to exist in this region. This implies that the surface elevation has to be allowed to vary across the row of consecutive OUT cells representing the opening to the sea, consistently with the transverse flows.

Consider a row of OUT columns along a "y wall," i.e., a wall at constant x. The average height $\bar{H}$ in such a row of columns will change in Δt by the mean of the expression

$$(3.54) \qquad \Delta H^* = \Delta t[u_1 + u_2 + \delta_1 u_3 - u_4 - u_5 - \delta_2 u_6]$$

$$= \Delta t\left[\sum_k \hat{u}_{i-1/2,\,j,\,k} - \sum_k \hat{u}_{i+1/2,\,j,\,k}\right]$$

averaged over the consecutive OUT cells. If $\Delta\bar{H}/\Delta t$ is specified, $\sum_k \hat{u}_{i+1/2,\,j,\,k}$ can be evaluated so as to be consistent with the present values of $u_{i-1/2,\,j,\,k}$ and the prescribed increase in height. Lateral fluxes (the v values) will give extra contributions to each column's H^*, but no net contribution to $\bar{H}$; they are going to be responsible for the slope of the surface at this region.

We make the assumption that

$$(3.55) \qquad u_{i+1/2,\,j,\,k} = u_{i-1/2,\,j,\,k} + \Delta u_{i,\,j}$$

where $\Delta u_{i,\,j}$ is independent of k. If this expression is substituted into Eq. (3.54), for the case represented in Fig. 10, we obtain,

$$(3.56) \qquad \Delta H^* = \Delta t \cdot [\delta_1 u_3 - \delta_2 u_6 - \hat{H}\Delta u_{i,\,j}]$$

where $\hat{H} = 2 + (\delta_1 + \delta_2)/2$, and u_3 is to be given as a function of u_2, as per Eq. (3.49). [Note, that if u_6 were above the surface, its value should be related to that of u_5 by the surface condition (3.49), instead of being related to u_3 through (3.55). The programming respects this and makes sure that the correct factor, as obtained substituting in Eq. (3.56), is computed instead of $\hat{H}$.]

Thus, if $\Delta\bar{H}/\Delta t$ is specified at all times at a row of OUT cells, $\Delta u_{i,j}$ is obtained from (3.56) and the $u_{i+1/2,j,k}$'s from Eq. (3.55). The transverse velocities are dynamically updated using the old values of the velocity field,[11] and the w components are obtained by continuity.

The final value of H at each OUT cell is obtained through the same procedure that updates H everywhere in the interior [cf. Eq. (3.47)], after the velocity field has been extrapolated outside of the surface according to Eqs. (3.49) and (3.50).

The procedure places no arbitrary restriction on the flow through the OUT cells. The total flux through these cells, integrated over an integer number of tidal cycles, equals the total influx through rivers in that time.

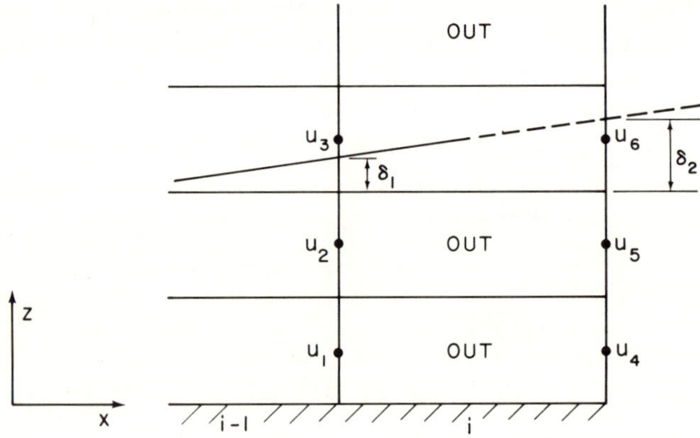

FIG. 10. Longitudinal section of one of the columns in a row of adjacent OUT columns.

Due to the way in which longitudinal velocities at OUT cells are built, it is impossible to use Eq. (3.13) to dynamically update S at OUT cells, specifying fixed values outside, "in the big ocean." The reason is that since $u_{i+1/2,j,k}$ is not dynamically updated, it will not be affected by the saline acceleration proportional to $(I_k)_{i,j} - (I_k)_{i+1,j}$. Thus, the integrated flux $\sum_k u_{i+1/2,j,k}$ will not be able to carry into the system as much salt as required for dynamic equilibrium, and a net leak of salt from the basin results.

We think that it is possible to make $(\Delta u)_{i,j}$ dependent upon k through the existing stratification; if this were done properly, salt variations could

---

[11] The updating prescription at these locations is simplified to avoid making assumptions on v "in the sea"; thus, for example, the terms $\delta_x(uv)$ and $\delta_x^2 v$ are not included. They can be added to specify different boundary conditions.

be computed dynamically at these locations. This possibility has not been investigated as of yet, but a simplified prescription has been adopted:

$$(3.57) \qquad S_{i,j,k} = \frac{1}{2}(S_{i-1,j,k} + S_{i+1,j,k})$$

for (i, j, k) an OUT cell. In the present version of the code, "the ocean" is assumed well mixed, i.e., $S_{i+1,j,k}$ [the S_F of Eq. (3.40)], independent of k.

3.1.3. Boundary Conditions for the Poisson Equation. Boundary values for the evaluation of Eq. (3.23) are required at cells which have a neighboring BND, OUT, IN, or SUR cell. The problem of this last type of cells will be treated together with the salinity surface conditions in Section 3.1.5.

In order to guarantee that φ be consistent with the velocity field and the boundary conditions imposed on it, relations for the φ values at the boundary cell have to be imposed through an examination of Eqs. (3.6) through (3.12). In the original versions of MAC, φ was explicitly evaluated at boundary cells and these values used in the solution of interior points by Eq. (3.23). The procedure is cumbersome and special provisions must be made for corner configurations, where the boundary pressure is not singly valued [e.g., cell $(i + 1, j, k)$ in Fig. 5].

However, the explicit dependence on such boundary values can be eliminated and the procedure simplified by a trick first used by Easton (1972). Consider cell (i, j, k), whose neighbor $(i + 1, j, k)$ is of type BND, OUT, or IN. From Eq. (3.6),

$$(3.58) \qquad \varphi_{i+1,j,k} = \varphi_{i,j,k} + \left(\frac{X_0}{Z_0}\right)^2 \left[DU_{i+1/2,j,k} - \left(\frac{\delta u}{\delta t}\right)_{i+1/2,j,k} \right]$$

where $(\delta u/\delta t)_{i+1/2,j,k}$ stands for $(u^{n+1}_{i+1/2,j,k} - u^n_{i+1/2,j,k})/\delta t$.

By direct substitution of Eq. (3.58) into Eq. (3.23), we obtain,

(3.59)

$$\varphi^{(h+1)}_{i,j,k} = \frac{1}{2[1 + (Z_0/Y_0)^2 + (Z_0/X_0)^2]} \Big\{ \varphi^{h+1}_{i,j,k-1} + \varphi^k_{i,j,k+1} + (Z_0/X_0)^2 $$

$$\times (\varphi^{h+1}_{i-1,j,k} + \varphi^h_{i,j,k})$$

$$+ (Z_0/Y_0)^2(\varphi^{h+1}_{i,j-1,k} + \varphi^h_{i,j+1,k})$$

$$+ \mathbb{R}_{i,j,k} + DU_{i+1/2,j,k}$$

$$- \left(\frac{\delta u}{\delta t}\right)_{i+1/2,j,k} \Big\}$$

Thus, cell (i, j, k) is relaxed in the same way as any cell surrounded by like cells, except that:

(a) the φ value at the neighboring boundary cell is replaced by the value at the cell under consideration;

(b) the total source term is formally equivalent to that given by (3.20), (3.21), except that the externally specified $(\delta u/\delta t)_{i+1/2, j, k}$ is used in place of $DU_{i+1/2, j, k}$. (This is a serious advantage because the computation of $DU_{i+1/2, j, k}$ requires the explicit use of the velocity boundary conditions, a nontrivial exercise in corner configurations.)

For impenetrable walls, $(\delta u/\delta t)_{i+1/2, j, k} = 0$. For IN boundaries, that value is a specified function of time. In this code, OUT cells are treated somewhat differently. We allow the relaxation routine to operate also on OUT cells, and use Eq. (3.58) to compensate for our lack of knowledge at positions outside the grid. The values of $(\delta u/\delta t)_{\text{OUT}+1/2}$ used are actually lagged values, computed in the same section that updates the values of $u_{\text{OUT}+1/2}$ in the manner explained in Sec. 3.1.2.4.

3.1.4. Stability Considerations. Equations (3.6) to (3.8) can be analyzed for stability by the usual harmonic analysis on the linearized equations. As a result of its application to that system of equations, the limitations on the nondimensional δt

$$(3.60) \qquad \delta t < \frac{1}{\text{Max}(|u|, |v|, |w|)}$$

$$(3.61) \qquad \delta t < \frac{1}{2} \frac{1}{A_1 + A_2 + A_3}$$

and

$$(3.62) \qquad A_1 > \frac{1}{2} u^2 \delta t$$

$$A_2 > \frac{1}{2} v^2 \delta t$$

$$A_3 > \frac{1}{2} w^2 \delta t$$

follow. Conditions (3.60) and (3.61) are also conditions for numerical accuracy: the first one states that the fluid must not be allowed to flow across more than one computational cell in one time step, while the second one states that the diffusion of momentum during one time step must also be less than one cell width. Condition (3.62) shows the system to be unstable for zero "eddy viscosity" (i.e., $A_1 = A_2 = A_3 = 0$).

These conditions are a consequence of the application of the von Neuman condition: a linear numerical scheme is stable if the amplification of any Fourier component of the initial data satisfies

$$(3.63) \qquad |\lambda_i|_{\max} \le 1$$

where λ_i are the eigenvalues of the amplification matrix for the numerical scheme.

Such a condition is sufficient (Richtmeyer and Morton, 1957) for stability of *linear* difference schemes with analytical initial data, and it is generally thought that its application to the linearized form of a nonlinear difference scheme also provides sufficient conditions for stability of the nonlinear scheme.

Counter examples to the assumption of validity of this condition for nonlinear equations were presented by Hirt (1968), who found computational instabilities in cases regarded stable by the von Neuman analysis on the linearized equations. He also introduced an alternative way of analyzing finite difference equations, able to predict these instabilities.[12]

Hirt's heuristic analysis consists in only one more step of the analysis method regularly used (Richtmeyer and Morton, 1957; Godunov and Ryaben'ky, 1964) for the determination of consistency conditions for a finite difference scheme. The finite difference equation is first expanded in a Taylor series around a typical grid point. For the simple forward time stepping used, expansion to first order in δt of Eqs. (3.6) to (3.8) (keeping only the nonlinear and dissipative terms), results in

(3.64)

$$\frac{\partial u}{\partial t} + \frac{1}{2}\delta t \frac{\partial^2 u}{\partial t^2} + \cdots = -\frac{\partial u^2}{\partial x} - \frac{\partial uv}{\partial y} - \frac{\partial uw}{\partial z} + \cdots$$

$$+ \left[A_1 + O(\delta x^3)\right]\frac{\partial^2 u}{\partial x^2} + \left(A_2 - \frac{1}{4}\delta y^2 \frac{\partial v}{\partial y} + \cdots\right)\frac{\partial^2 u}{\partial y^2}$$

$$+ \left(A_3 - \frac{1}{4}\delta z^2 \frac{\partial w}{\partial z} + \cdots\right)\frac{\partial^2 u}{\partial z^2} + \cdots$$

and similar equations for the other two components. Note that the coefficient multiplying $\partial^2 u/\partial x^2$ is unaffected by spatial contributions from the nonlinear terms up to second order in the spatial increments, while A_2 and A_3 are. This is a consequence of the use of the ZIP form.

From a region of influence argument applied to the hyperbolic equation (3.64), condition (3.61) follows. More interesting is how the equivalent to

---

[12] A similar approach for stability analysis has been developed by Ferziger (1973), and the connection between the heuristic stability theory and the Fourier method was recently established by Warming and Hyett (1974) for linear finite difference equations.

condition (3.62) is reached. From Eq. (3.64) and its v and w companions, estimates for the second temporal derivatives (to zero order in δt) can be obtained

$$\frac{\partial^2 u}{\partial t^2} \simeq \frac{\partial}{\partial t}\left[-\frac{\partial u^2}{\partial x} - \frac{\partial uv}{\partial y} - \frac{\partial uw}{\partial z} + A_1 \frac{\partial^2 u}{\partial x^2} + A_2 \frac{\partial^2 u}{\partial y^2} + A_3 \frac{\partial^2 u}{\partial z^2} \right]$$

and evaluated using the same equations to zero order in δt and δx, δy, δz. In the assumption of sufficiently smooth u, v, w—where higher order spatial derivatives can be neglected—substitution of the quantities $\partial^2 u/\partial t^2$, $\partial^2 v/\partial t^2$, $\partial^2 w/\partial t^2$ in terms of the spatial derivatives into the u, v, w equations of type (3.64), shows that the original finite difference expressions approximate differential equations of the form

$$(3.65) \quad \frac{\partial u}{\partial t} = -u' \frac{\partial u}{\partial x} - v' \frac{\partial u}{\partial y} - w' \frac{\partial u}{\partial z}$$

$$+ A_1' \frac{\partial^2 u}{\partial x^2} + A_2' \frac{\partial^2 u}{\partial y^2} + A_3' \frac{\partial^2 u}{\partial z^2}$$

$$+ \text{ higher order terms}$$

where

$$(3.66) \quad u' = u\left(1 + \frac{3}{2}\,\delta t\,\frac{\partial u}{\partial x} + \cdots\right)$$

$$(3.67) \quad v' = v\left(1 + \frac{3}{2}\,\delta t\,\frac{\partial u}{\partial x} + 2\delta t\,\frac{\partial v}{\partial y} + \delta t\,\frac{\partial w}{\partial z}\right)$$

$$+ \delta t\left[u\,\frac{\partial v}{\partial x} + w\,\frac{\partial v}{\partial z} - \frac{1}{2}\left(A_1 \frac{\partial^2 v}{\partial x^2} + A_2 \frac{\partial^2 v}{\partial y^2} + A_3 \frac{\partial^2 v}{\partial z^2}\right)\right]$$

$$(3.68) \quad w' = w\left(1 + \frac{3}{2}\,\delta t\,\frac{\partial u}{\partial x} + \delta t\,\frac{\partial v}{\partial y} + 2\delta t\,\frac{\partial w}{\partial z}\right)$$

$$+ \delta t\left[v\,\frac{\partial w}{\partial y} + u\,\frac{\partial w}{\partial x} - \frac{1}{2}\left(A_1 \frac{\partial^2 w}{\partial x^2} + A_2 \frac{\partial^2 w}{\partial y^2} + A_3 \frac{\partial^2 w}{\partial z^2}\right)\right]$$

$$(3.69) \quad A_1' = A_1\left(1 + \frac{1}{2}\,\delta t\,\frac{\partial u}{\partial x}\right) - \frac{1}{2}\,u^2\delta t + \cdots$$

$$(3.70) \quad A_2' = A_2\left(1 + \frac{1}{2}\,\delta t\,\frac{\partial u}{\partial x}\right) - \frac{1}{2}\,v^2\delta t - \frac{1}{4}\,\delta y^2\,\frac{\partial v}{\partial y} + \cdots$$

$$(3.71) \quad A_3' = A_3\left(1 + \frac{1}{2}\,\delta t\,\frac{\partial u}{\partial x}\right) - \frac{1}{2}\,w^2\delta t - \frac{1}{4}\,\delta z^2\,\frac{\partial w}{\partial z} + \cdots$$

The first observation is that a necessary condition for the "effective" u', v', w' to have the same sign as u, v, w, is that

$$(3.72) \qquad \delta t < \frac{1}{2}\,\mathrm{Min}\left(\frac{1}{|\partial u/\partial x|},\ \frac{1}{|\partial v/\partial y|},\ \frac{1}{|\partial w/\partial z|}\right)$$

This condition is not given by the linear analysis.

A condition of the type of Eq. (3.62) is obtained from the examination of Eqs. (3.69)–(3.71) for A'_1, A'_2, A'_3. Hirt's physical reasoning is that these quantities in Eq. (3.65) represent energy dissipation if positive, but will act as energy sources if negative. Thus, in the limit of

$$\delta t \ll \mathrm{Min}\left(\frac{1}{|\partial u/\partial x|},\ \frac{1}{|\partial v/\partial y|},\ \frac{1}{|\partial w/\partial z|}\right)$$

the condition of $A'_i > 0$, $i = 1, 2, 3$, becomes

$$(3.73) \qquad A_1 > \tfrac{1}{2}u^2\delta t + \cdots$$

$$(3.74) \qquad A_2 > \tfrac{1}{2}v^2\delta t + \tfrac{1}{4}\delta y^2\,\frac{\partial v}{\partial y} + \cdots$$

$$(3.75) \qquad A_3 > \tfrac{1}{2}w^2\delta t + \tfrac{1}{4}\delta z^2\,\frac{\partial w}{\partial z} + \cdots$$

where the leading terms are those of Eqs. (3.62).

It should be noted that the analysis has by necessity been very much simplified and there is no guarantee that the fulfillment of conditions (3.60), (3.61), (3.72), and (3.73) to (3.75) is either necessary or sufficient for stability of the numerical solution to the whole set of finite difference equations. This is not a consequence of Hirt's method but of the complexity of the system.

For the geometry used in the test cases (see Section 4), and $f^u = f^v = 0$, we have experimentally verified that the numerical solution to Eqs. (3.6) to (3.8) has an unstable behavior for $A_1 = A_2 = A_3 = 0$. On the other hand, if the leading terms in (3.73)–(3.75) are accounted for by locally modifying the A_i values, a stable numerical solution is obtained. The extra contributions to A'_1, A'_2, A'_3 are grossly accounted for by locally modifying them in the form:

$$(3.76)$$

$$(A_1)_{i+1/2,\,j,\,k} = A_1 + \tfrac{1}{2}\delta t(u_{i+1/2,\,j,\,k})^2$$

$$(3.77)$$

$$(A_2)_{i+1/2,\,j,\,k} = A_2 + \tfrac{1}{32}\delta t$$
$$\times\,(v_{i,\,j+1/2,\,k} + v_{i+1,\,j+1/2,\,k} + v_{i,\,j-1/2,\,k} + v_{i+1,\,j-1/2,\,k})^2$$
$$+\,\tfrac{1}{8}(v_{i,\,j+1/2,\,k} + v_{i+1,\,j+1/2,\,k} - v_{i,\,j-1/2,\,k} - v_{i+1,\,j-1/2,\,k})$$

246

ENRIQUE A. CAPONI

(3.78)

$$(A_3)_{i+1/2,\,j,\,k} = A_3 + \tfrac{1}{32}\delta t$$
$$\times \left(w_{i,\,j,\,k+1/2} + w_{i+1,\,j,\,k+1/2} + w_{i,\,j,\,k-1/2} + w_{i+1,\,j,\,k-1/2}\right)^2$$
$$+ \tfrac{1}{8}\left(w_{i,\,j,\,k+1/2} + w_{i+1,\,j,\,k+1/2} - w_{i,\,j,\,k-1/2} - w_{i+1,\,j,\,k-1/2}\right)$$

for the u equation. Similar expressions are used for the local modification of A_1, A_2, and A_3 in the v and w equations.

The forward time differencing is also responsible for another source of computational growth of mean energy involving the representation of the Coriolis effect. We consider the simplified scheme

(3.79)

$$u_{i+1/2,\,j,\,k}^{n+1} = u_{i+1/2,\,j,\,k}^{n}$$
$$+ \frac{\delta t}{4}\, f^u\left(v_{i,\,j+1/2,\,k}^{n} + v_{i+1,\,j+1/2,\,k}^{n} + v_{i,\,j-1/2,\,k}^{n} + v_{i+1,\,j-1/2,\,k}^{n}\right)$$

(3.80)

$$v_{i,\,j+1/2,\,k}^{n+1} = v_{i,\,j+1/2,\,k}^{n}$$
$$- \frac{\delta t}{4}\, f^v\left(u_{i+1/2,\,j,\,k}^{n} + u_{i+1/2,\,j+1,\,k}^{n} + u_{i-1/2,\,j,\,k}^{n} + u_{i-1/2,\,j+1,\,k}^{n}\right)$$

for the u and v components, and perform a Hirt analysis on the linear combinations of these equations that lead to equations for $u_{i,\,j,\,k} \equiv \tfrac{1}{2}(u_{i+1/2,\,j,\,k} + u_{i-1/2,\,j,\,k})$ and $v_{i,\,j,\,k} \equiv \tfrac{1}{2}(v_{i,\,j+1/2,\,k} + v_{i,\,j-1/2,\,k})$. To first order in δt and second in the spatial coordinates, Eqs. (3.79) and (3.80) are seen to simulate the behavior of the differential equations

(3.81)
$$\frac{\partial u}{\partial t} = f^u v + \frac{1}{2}\,\delta t f^u f^v\left(u + 3\delta x^2\,\frac{\partial^2 u}{\partial x^2} + 3\delta y^2\,\frac{\partial^2 u}{\partial y^2}\right)$$
$$+ f^u\left(2\delta x^2\,\frac{\partial^2 v}{\partial x^2} + \delta y^2\,\frac{\partial^2 v}{\partial y^2}\right)$$

(3.82)
$$\frac{\partial v}{\partial t} = -f^v u + \frac{1}{2}\,\delta t f^u f^v\left(v + 3\delta x^2\,\frac{\partial^2 v}{\partial x^2} + 3\delta y^2\,\frac{\partial^2 v}{\partial y^2}\right)$$
$$+ f^v\left(\delta x^2\,\frac{\partial^2 u}{\partial x^2} + 2\delta y^2\,\frac{\partial^2 u}{\partial y^2}\right)$$

The terms in δx and δy are due to the choice of a staggered grid and are stabilizing. On the other hand, the terms $\tfrac{1}{2}\delta t f^u f^v u$ in Eq. (3.81) and $\tfrac{1}{2}\delta t f^u f^v v$ in Eq. (3.82) give rise to an exponential energy growth with growth rate $\gamma = \tfrac{1}{2}f^u f^v \delta t$. This has been experimentally observed in steady state runs made with $A_1 = A_2 = A_3 = 0$, but incorporating the local modifications given by Eqs. (3.76) to (3.78). A "normal" behavior was recovered by modifying the form of Eqs. (3.79) and (3.80) to:

(3.83)

$$u_{i+1/2, j, k}^{n+1} = \left(1 - \frac{f^u f^v}{2} \delta t\right) u_{i+1/2, j, k}^n$$

$$+ \frac{\delta t}{4} f^u\left(v_{i, j+1/2, k}^n + v_{i+1, j+1/2, k}^n + v_{i, j-1/2, k}^n + v_{i+1, j-1/2, k}^n\right)$$

(3.84)

$$v_{i, j+1/2, k}^{n+1} = \left(1 - \frac{f^u f^v}{2} \delta t\right) v_{i, j+1/2, k}^n$$

$$- \frac{\delta t}{4} f^v\left(u_{i+1/2, j, k}^n + u_{i+1/2, j+1, k}^n + u_{i-1/2, j, k}^n + u_{i-1/2, j+1, k}^n\right)$$

Due to density variations being allowed, for (internal) gravity waves not to grow exponentially, a further condition must be imposed:

(3.85)
$$c < \frac{\Delta}{\delta t}$$

on the dimensional quantities c, Δ, and δt, the wave speed, the grid spacing in the direction of propagation, and the value of the time increment. Physically, condition (3.85) means that information propagating with speed c should travel less than Δ during δt if the grid is to resolve it. The fastest gravity waves are those happening at the surface of discontinuity for the density, i.e., the surface of the fluid. For our shallow water waves, propagating in the xy plane, this condition limits the nondimensional δt to:

(3.86)
$$\delta t < \frac{Y_0}{Z_0} \left[\frac{g\tau_0^2}{Z_0} \cdot \left(1 + \frac{Y_0}{X_0}\right)^2 H_m\right]^{-1/2}$$

where H_m is the maximum depth.

Equations (3.60) and (3.85) are both required for stability when using an explicit integration scheme. They are referred to as Courant or CFL conditions after a similar one derived by Courant, Friedrichs, and Lewy (1928) for the linear wave equation.

A Hirt-type analysis on the chosen donor cell scheme for the salinity equation shows it to be stable if

(3.87)
$$|u| > u^2 \delta t$$

and this is satisfied if the accuracy condition (3.60) is satisfied, as already noticed by Daly and Pracht (1968). It is also possible to see that for FUL cells the donor cell scheme guarantees positive values of S as long as condition (3.87) is satisfied. The alternative form of space centering the convective terms for the salinity equation is not only conditionally stable (in the same sense as the momentum equations are), but cannot guarantee positiveness of S, even though it is able to conserve S. Local corrections suggested by a heuristic analysis are quite complicated to implement given the different spatial arrangement of the variables. Piacsek and Williams (1970) suggested an "absolutely conserving" form for the convective terms. This guaranteed that S^2 be conserved in all the volume, but not S. Furthermore, that formulation may also lead to local negative values of S.

3.1.5. Surface Conditions for the Pressure and Salinity Fields. The free surface capability originates the requirement that δt satisfy condition (3.85) to avoid instabilities. There is another cause for numerical instability associated with the surface, and this is due to the way in which the boundary condition on the dynamic pressure φ is imposed. This is the first problem treated in this section. Then, we examine some problems associated with the updating of S values at cells containing the surface, and the way in which they have been solved. Surface boundary conditions on u, v, and w have been examined previously in Section 3.1.2.3.

3.1.5.1. In formal analogy to the derivation leading to Eq. (2.50) the value of φ at the surface is to satisfy (in the horizontal surface approximation)

$$(3.88) \qquad \varphi_s = \varphi_a + A_3(w_+ - w_-)$$

where $w_{+(-)}$ indicate values of w in the next grid position just above (just below) the surface, and φ_a is an applied dynamic pressure value, chosen to parameterize the SGS.

The original versions of MAC assumed $\varphi_s = \varphi_a$. Problems were soon recognized (Hirt and Shannon, 1968) for cases where A_3 was not negligible in some sense (viscous bore problem), and a finite difference analog to Eq. (2.50) was proposed to determine the value of φ at the center of the cell where the surface was located. The modification led to improved results. The application of $\varphi_s = \varphi_a$ at the center of surface cells was later reported (Chan and Street, 1970a,b) to lead to very irregular surface profiles, particularly when using low viscosities. The application of the condition *at* the location of the surface improved the results.

A method that at the same time used the complete specification for φ_s and assigned this value to the actual position of the surface was then proposed by Nichols and Hirt (1971).

A difficulty associated with the first and third approaches mentioned above is that the orientation of the surface was to be determined in order to apply the normal stress condition. Even though such modifications were thought for and applied to two-dimensional simulations, they seemed to have proved implementable only by approximating the surface orientation as being parallel, perpendicular, or at 45° with the horizontal.

The difficulties are much worse for the three-dimensional case; but for systems where the horizontal grid sizes (X_0, Y_0) are much larger than the vertical grid size Z_0, and surface elevation changes do not take more vertical cells than horizontal cells representing typical wave lengths, the horizontal surface approximation proposed in (3.88) seems appropriate.

Our implementation of Eq. (3.88) follows the one proposed by Nichols and Hirt (1971) and consists in interpolating or extrapolating the value of φ at the center of the cell where the surface lies.

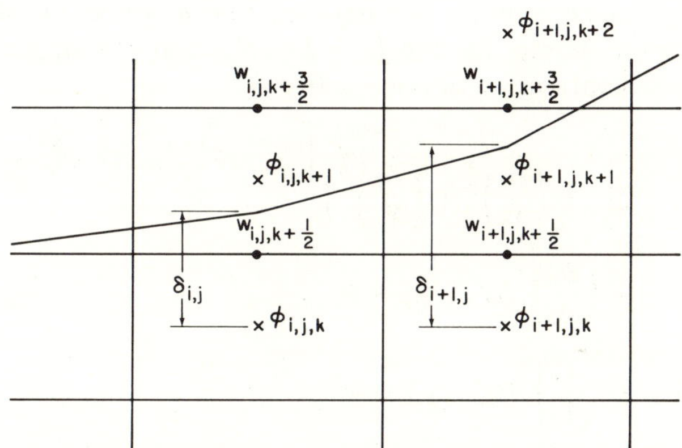

FIG. 11. Application of the surface boundary condition on the dynamic pressure.

With reference to Fig. 11, for the two configurations shown:

$$(3.89) \qquad \varphi_{i,j}^{\text{surf}} = \varphi_{i,j,k} + \delta_{i,j}(\varphi_{i,j,k+1} - \varphi_{i,j,k})$$

$$(3.90) \qquad \varphi_{i+1,j}^{\text{surf}} = \varphi_{i+1,j,k+1} + \delta_{i+1,j}(\varphi_{i+1,j,k+1} - \varphi_{i+1,j,k})$$

where the left-hand side is, according to (3.89):

$$(3.91) \qquad \varphi_{m,j}^{\text{surf}} = \varphi_{a_{m,j}} + A_3(w_{m,j,k+3/2} - w_{m,j,k+1/2}), \qquad m = i, i+1$$

In this example, $w_{m,j,k+1/2}$ is dynamically updated through Eq. (3.8) while $w_{m,j,k+3/2}$ is determined by continuity, after the u and v values have been extrapolated outside the surface where needed.

If φ_F denotes the value of φ at the uppermost FUL cell, φ_S its value *at* the surface given by Eq. (3.88), and φ_E the consistent φ value one grid space above the place where φ_F is defined, the linear interpolation relation among them [Eqs. (3.89) and (3.90)] can be used to obtain

$$(3.92) \qquad \varphi_E = \eta_{i,j}\varphi_S + (1 - \eta_{i,j})\varphi_F$$

where $\eta_{i,j} = \delta_{i,j}^{-1}$.

The procedure is then to iteratively solve φ in the form shown in Section 3.1.3, up to the value φ_F. When solving for φ_F, Eq. (3.92) is to be used to evaluate the value of φ at the neighboring cell above.

With the definitions given above, δ is always larger than one-half the vertical cell's dimension. In Nichols and Hirt (1971), this was not the case, and it was found that the pressure iteration diverged when $\delta < 0.5$. To free the procedure from this instability, Nichols and Hirt (1971) allowed α, the relaxation parameter in Eq. (3.24), to be a function of δ in such a way that φ_F was actually underrelaxed.

In our case, where the orientation of the surface is considered horizontal, it is simple to make implicit the iteration for φ values at the surface. If the subscript F stands for the (i, j, k) values at the uppermost FUL cell under consideration, we obtain φ_F from:

$$(3.93) \quad \varphi_F^{h+1} = \frac{1}{2\left[1 + \left(\dfrac{Z_0}{Y_0}\right)^2 + \left(\dfrac{Z_0}{X_0}\right)^2\right]}\left\{\mathbb{R}_F + \varphi_{i,j,k-1}^{h+1} + \varphi_{i,j,k+1}^{h+1}\right.$$

$$+ \left(\frac{Z_0}{X_0}\right)^2 (\varphi_{i+1,j,k}^{h} + \varphi_{i-1,j,k}^{h+1})$$

$$\left. + \left(\frac{Z_0}{Y_0}\right)^2 (\varphi_{i,j+1,k}^{h} + \varphi_{i,j-1,k}^{h+1})\right\}$$

where φ_F^{h+1} is the final value of φ_F for the $(h+1)$st iteration, and $\varphi_{i,j,k+1}^{h+1}$ is given in terms of φ_F^{h+1} by Eq. (3.88). Thus,

$$(3.94) \qquad \varphi_F^{h+1} = \frac{1}{1 - \dfrac{1 - \eta_{i,j}}{2[1 + (Z_0/Y_0)^2 + (Z_0/X_0)^2]}} \varphi_F^{(h+1)}$$

where

$$(3.95) \qquad \varphi_F^{(h+1)} = \frac{1}{2\left[1 + \left(\dfrac{Z_0}{Y_0}\right)^2 + \left(\dfrac{Z_0}{X_0}\right)^2\right]}\left\{\mathbb{R}_F + \varphi_{i,j,k-1}^{h+1} + \eta_{i,j}\varphi_S\right.$$

$$+ \left(\frac{Z_0}{X_0}\right)^2 (\varphi_{i+1,j,k}^{h} + \varphi_{i-1,j,k}^{h+1})$$

$$\left. + \left(\frac{Z_0}{Y_0}\right)^2 (\varphi_{i,j+1,k}^{h} + \varphi_{i,j-1,k}^{h+1})\right\}$$

The implementation is straightforward, requiring that at the beginning of the procedure all φ_E's are set equal to $(\eta\varphi_S)_{i,\,j}$. In this way, Eq. (3.95) is the same as Eq. (3.23) for regular cells. Thus, for any cell, once the $\varphi^{(h+1)}$ value has been computed, a test is made to decide whether Eq. (3.24) or Eq. (3.94) should be used and the final value φ^{h+1} is obtained.

Once the convergence criterium has been met for the pressure iteration, the final values for the φ_E's are built from their temporary ones and the final φ_F's, according to Eq. (3.92).

3.1.5.2. We have previously mentioned that the variable S stands for mean salinity times the fractional cell's volume occupied by the fluid. The reason is essentially one of consistency with the form in which average fluxes are extrapolated outside of the surface. Thus, the application of these boundary conditions constitute the implicit understanding that the fluid is regarded to continue through the surface, and the specification of a surface tangential stress becomes the requirement that at some level, $\delta u/\delta k$ and $\delta v/\delta k$ assume specified values. The determination of w values by continuity outside of the surface is also consistent with this interpretation, as is the algorithm (columnwise mass conservation) by which surface elevations are updated. In this section, we will use S to denote average salinity, and σ to denote total salt contents in a cell with $X_0 = Y_0 = Z_0 = 1$. Thus, considering for simplicity a two-dimensional case, and making reference to Fig. 12:

$$(3.96) \qquad \sigma_{i,\,k=2} = S_{i,\,k=2}$$

$$\sigma_{i,\,k=3} = S_{i,\,k=3} \cdot \delta_i$$

The nondimensional convective flux between surface cells at i and at $i + 1$ (cf. Fig. 12) is to be interpreted as:

$$(3.97) \qquad F^c_{i+1/2,\,3} = (u\delta)_{i+1/2,\,3}(S)_{i+1/2,\,3}$$

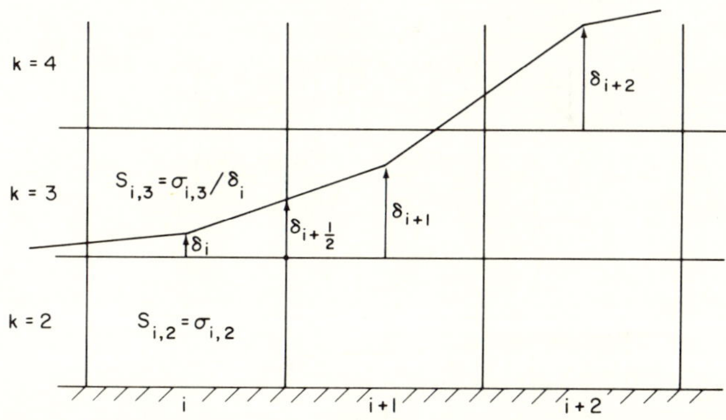

FIG. 12. Relation between salinity S and salt contents σ in partially and in completely filled cells.

Thus, the total flux between such surface cells is given by

$$(3.98) \quad F_{i+1/2,\,3} = \delta_{i+1/2}[u_{i+1/2,\,3}(S)_{i+1/2,\,3} + A_1(S_{i,\,3} - S_{i+1,\,3})]$$

while the total flux between FUL cells is given by

$$(3.99) \quad F_{i+1/2,\,2} = u_{i+1/2,\,2}(S)_{i+1/2,\,2} + A_1(S_{i,\,2} - S_{i+1,\,2})$$

In both, Eqs. (3.98) and (3.99),

$$(3.100) \qquad (S)_{i+1/2,\,k} = \begin{cases} S_{i,\,k} & \text{if} \quad u_{i+1/2,\,k} \geq 0 \\ S_{i+1,\,k} & \text{if} \quad u_{i+1/2,\,k} < 0 \end{cases}$$

The sum of these saline fluxes over the faces of a computational cell where the mass flux is not zero, gives the rate of change of total salt contents σ in that cell.

This algorithm is able to preserve homogeneity. To simplify notation, we now consider the configuration represented in Fig. 13. We can compactly write,

$$(3.101) \qquad F_{i+1/2} = \delta_{i+1/2}[u_{i+1/2}(S)_{i+1/2} - A_1(S_{i+1} - S_i)]$$

where

$$S_i = (\sigma_i/\delta_i), \qquad S_{i+1} = (\sigma_{i+1}/\delta_{i+1})$$

It is easy to show that homogeneity is conserved in the steady state. In effect, for $\delta H/\delta t$ to vanish in all cells, the condition:

$$(3.102) \qquad u_{i-1/2}\,\delta_{i-1/2} = u_{i+1/2}\,\delta_{i+1/2} = u_{i+3/2}\,\delta_{i+3/2}$$

must be satisfied.

The rate of change of salt contents in cell i will be given by

$$\frac{\Delta\sigma_i}{\Delta t} = \frac{\Delta(S_i\,\delta_i)}{\Delta t} = F_{i-1/2} - F_{i+1/2}$$

$$= \delta_{i-1/2}\left[u_{i-1/2}\,\frac{\sigma_{i-1}}{\delta_{i-1}} - A_1\left(\frac{\sigma_i}{\delta_i} - \frac{\sigma_{i-1}}{\delta_{i-1}}\right)\right]$$

$$- \delta_{i+1/2}\left[u_{i+1/2}\,\frac{\sigma_i}{\delta_i} - A_1\left(\frac{\sigma_{i+1}}{\delta_{i+1}} - \frac{\sigma_i}{\delta_i}\right)\right]$$

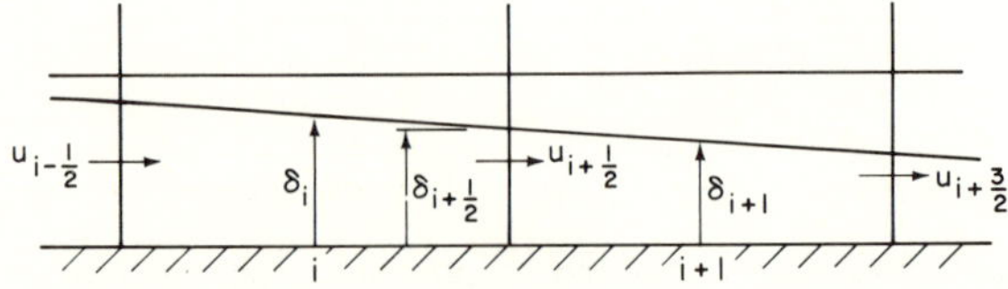

Fig. 13. Simple configuration to illustrate conservation of homogeneity in the steady state.

For a homogeneous fluid,

$$\frac{\sigma_{i-1}}{\delta_{i-1}} = \frac{\sigma_i}{\delta_i} = \frac{\sigma_{i+1}}{\delta_{i+1}} = \cdots$$

and the dissipative fluxes do not contribute to $\Delta\sigma_i/\Delta t$. If furthermore the steady state condition (3.102) is satisfied, then $\Delta\sigma_i/\Delta t = 0$, thus conserving homogeneity.

If the mass fluxes are not equal (unsteady case), then the rate of change of the surface will be given by

$$(3.103) \qquad \frac{\Delta(\delta_i)}{\Delta t} = \frac{\Delta H_i}{\Delta t} = u_{i-1/2}\,\delta_{i-1/2} - u_{i+1/2}\,\delta_{i+1/2}$$

The rate of change for the salt contents is again only convective (homogeneous initial state):

$$(3.104) \qquad \frac{\Delta\sigma_i}{\Delta t} = \frac{\sigma_i}{\delta_i}\left(u_{i-1/2}\,\delta_{i-1/2} - u_{i+1/2}\,\delta_{i+1/2}\right) \equiv S_i\,\frac{\Delta\delta_i}{\Delta t}$$

and consequently, the fluid remains homogeneous. The algorithm also guarantees no saline flux across the surface.

In the present version, interchanges of mass and salt are computed independently, by alternatively switching one of them off when the other is on and vice versa. This procedure is not necessary, and it is only a consequence of having added the salinity algorithm onto the previously developed homogeneous code, desiring to introduce as few changes as possible in the preexisting routines. We will shortly change over to an algorithm that will compute mass and salt fluxes simultaneously, thus avoiding most of the problems for which we had to correct.

Regardless of whether or not the mass and salt fluxes are computed simultaneously, the computation of the latter may require special treatment at surface cells. We refer to Fig. 14 to illustrate this problem. In that

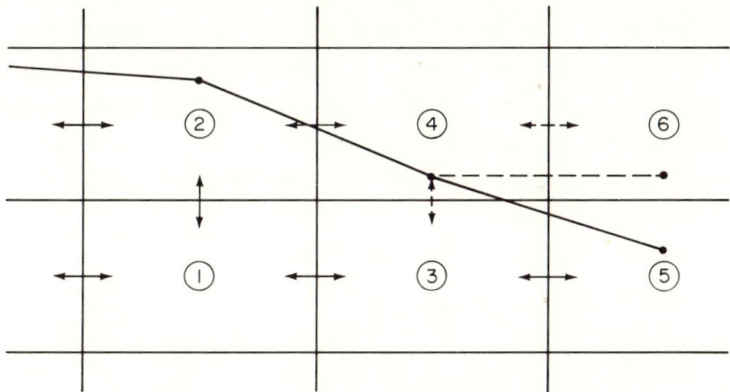

FIG. 14. Computed saline fluxes.

figure, double arrows indicate faces through which salt fluxes can be computed without extra effort. When the surface trace in the cross section shown in that figure follows the broken line, the corresponding saline fluxes between cells 4 and 6 and between 4 and 3 can be computed and they will, together with the exchange between cells 4 and 2, determine $\Delta\sigma_4/\Delta t$.

However, in the case of the solid line surface, there is no exchange between cells 4 and 6, and the flux between cells 3 and 4 depends on the value of the cross section between these last two cells. In order to avoid this further complication, we arbitrarily set to zero the saline fluxes between cells 3 and 4 during the process of salinity updating. In order to avoid further problems when this is the case, cells 3 and 4 are homogenized with respect to salinity after their values are updated. The procedure conserves total salt contents.

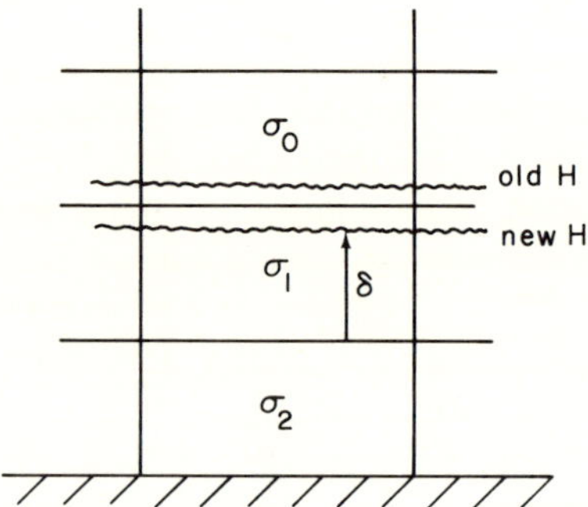

FIG. 15. A possible configuration when H and S are updated independently.

In order to avoid artificial effects at the surface due to the alternative updating of the H and S fields, two kinds of modifications to the old values at the top cells may be required before computing the saline flows:

(a) If the surface has just crossed a cell's top face, moving downward, values are as shown in Fig. 15. In the new configuration, the σ value above the surface (σ_0) should be zero. Setting $\sigma_0 = 0$ would mean a numerical loss of salt. Consequently, we use the procedure

$$\sigma_1 = \sigma_1 + \sigma_0$$

$$\sigma_0 = 0$$

to guarantee overall conservation.

(b) In all cases, when $\delta < \Delta Z/2$, the salinity at the two top cells with fluid is arbitrarily homogenized (cf. Fig. 16):

$$\sigma_S \equiv \sigma_1 + \sigma_0$$

$$\sigma_1 = \sigma_S/(1 + \delta)$$

$$\sigma_0 = \delta \cdot \sigma_S$$

This has to be done in order to avoid artificially large values for $S_0 = \sigma_0/\delta$ in cases where δ becomes very small. Note that both (a) and (b) are able to locally destroy homogeneity. However, total salt conservation, a superseding consideration, is satisfied.

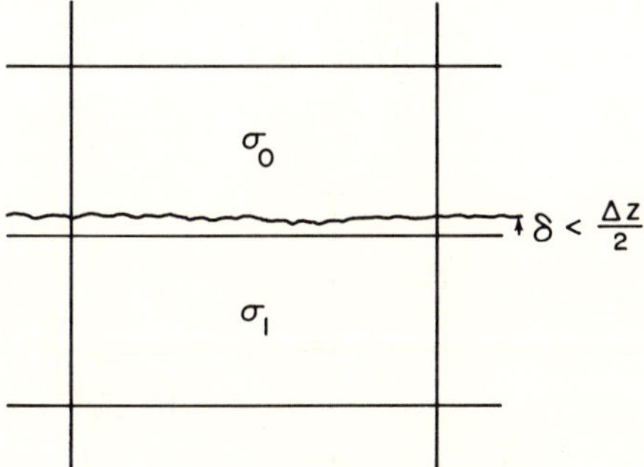

FIG. 16. Configuration requiring homogenization of two top cells with the procedure presently used.

There are cases where algorithm (3.98) can apparently lead to negative surface salinities even when complying with $\Delta t < 1/(\partial u/\partial x)$.

Consider the situation shown in Fig. 17. The change in salt contents $\Delta\sigma_1$ during Δt will be given by

$$\Delta\sigma_1 = -\Delta t \cdot F^S_{3/2} = -\Delta t \cdot \delta_{3/2}\left[u\frac{\sigma_1}{\delta_1} - A_1\left(\frac{\sigma_2}{\delta_2} - \frac{\sigma_1}{\delta_1}\right)\right]$$

The convective flux alone will account for a change:

$$-\Delta t \cdot F^c_{3/2} = \Delta\sigma_1 = \frac{\delta_{3/2}}{\delta_1}\sigma_1 \cdot \Delta t \cdot u$$

Since

$$\Delta t < \left(\frac{1}{|u|}, \frac{1}{|\partial u/\partial x|}\right)$$

the change in σ_1 will be bounded by

$$|\Delta\sigma_1| \le \frac{\delta_{3/2}}{\delta_1}\,\sigma_1$$

which is larger than σ_1 if $\delta_{3/2} > \delta_1$.

But note that the same configuration also leads to negative heights in cell 1, and this is the root of the problem. In configurations such as the one shown, a stronger condition on Δt should be imposed, namely,

$$(3.105) \qquad\qquad \delta t < \frac{\delta_1}{\delta_{3/2}\,u_{3/2}}$$

This condition is stronger than (3.60). In the way in which both heights and salt contents are updated, condition (3.105) would only be necessary in those situations where the surface cell cannot exchange mass with its neighbor below (Fig. 17), which is a configuration that can only very

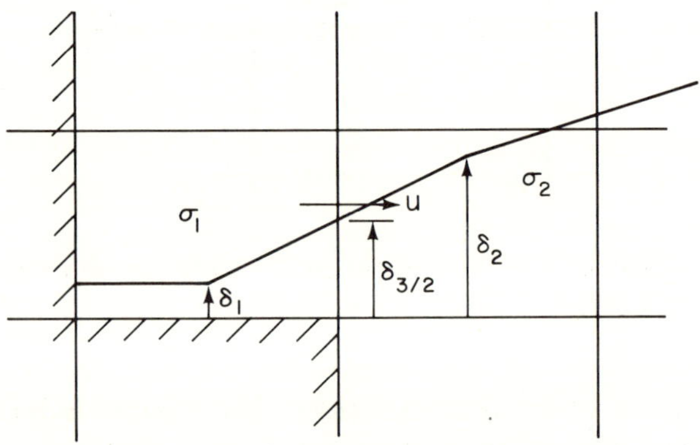

FIG. 17. A dangerous configuration.

coarsely be represented in finite differences. Our approach was to avoid, by decree, the appearance of negative salinities or of the surface going down into the bottom, instead of implementing condition (3.105). Thus, in a configuration of the type shown in Fig. 17, the maximum mass and salt fluxes allowed to flow out of cell 1 are $\delta_1/\delta t$ and $\sigma_1/\delta t$, respectively.

3.1.6. Initialization. The problem of building or adequately smoothing experimental data, or both, to be fed as initial conditions is sometimes a very serious problem. The amount of work to be devoted to this stage is to be compared to the savings resulting from a shortened transient for the particular problem under consideration.

Our initialization technique is based on the solution of the steady-state, vertically integrated horizontal flux equations [Eqs. (3.6) and (3.7)]. They are reduced to a single Poisson equation in terms of a transport stream function ψ, which is solved with the boundary conditions of no flux through the boundaries, specified fluxes through the IN faces, and an average flux (sum fluxes through IN cells divided by the number of OUT cells) through each of the OUT faces. The assignment of boundary values for ψ is done automatically for arbitrary geometries, by a subroutine that sweeps interior cells along the boundary leaving it to the left.

The initial solution of the ψ field is used to determine the surface elevation distribution in the interior, and these values are used to compute a modified source term for the Poisson equation for ψ. The double procedure is iterated until the maximum difference between old and new values of the surface deviation at any place is less than a specified bound.

The final values of ψ are used to determine the u and v fields, assuming a parabolic profile in depth. Values of w are then determined by continuity over all the grid.

The procedure is considered to work properly in all cases tested, and it does reduce the transients that appear at the beginning of the three-dimensional computation.

Modifications and additions to incorporate a similar procedure for the initialization of the salinity field distribution have also been incorporated. These modifications comprise two stages. The first one solves for a vertically integrated salinity distribution consistent with the vertically integrated fluxes and horizontal mixing coefficients. In the second stage, the u and v profiles are modified so that they reflect the existence of the horizontal salinity gradients while at the same time maintaining the vertically integrated fluxes unchanged. In this second stage, final values for the vertical fluxes are determined by continuity, and the salinity field is relaxed vertically.

Some possible refinements in this second stage (iteration of the procedure, salinity induced modification of the surface slope) have also been tried, but although the initial velocity distributions obtained in this manner are consistent with the salinity distribution, this last one still departs significantly from the three-dimensional steady state distribution (cf. Section 4.2).

3.2. The Computer Code

The model presented in Section 3.1 has been codified for its numerical solution by a digital computer. The "Grand Plan" and a few notes that may help to the identification of the various subroutines are given in the first subsection. The second one deals very briefly with some practical considerations.

The program has been written in FORTRAN IV, with very few features of FORTRAN V, and has been run in the UNIVAC 1108 of the Computer Science Center at the University of Maryland. Complete listings are available (Caponi, 1974).

3.2.1. Organization of the Code. Structurally, the code consists of two main regions, the "setup," which reads in the problem and runs through the initialization routines if required, and the "main computational cycle," which updates the values of the variables from t to $t + \Delta t$.

3.2.1.1. Setup. The first card in the data deck establishes the maximum number of time steps desired for the run, and the nondimensional time intervals that are going to separate successive detailed printouts (Fig. 18). The values of two switches are also given in this card: one specifying

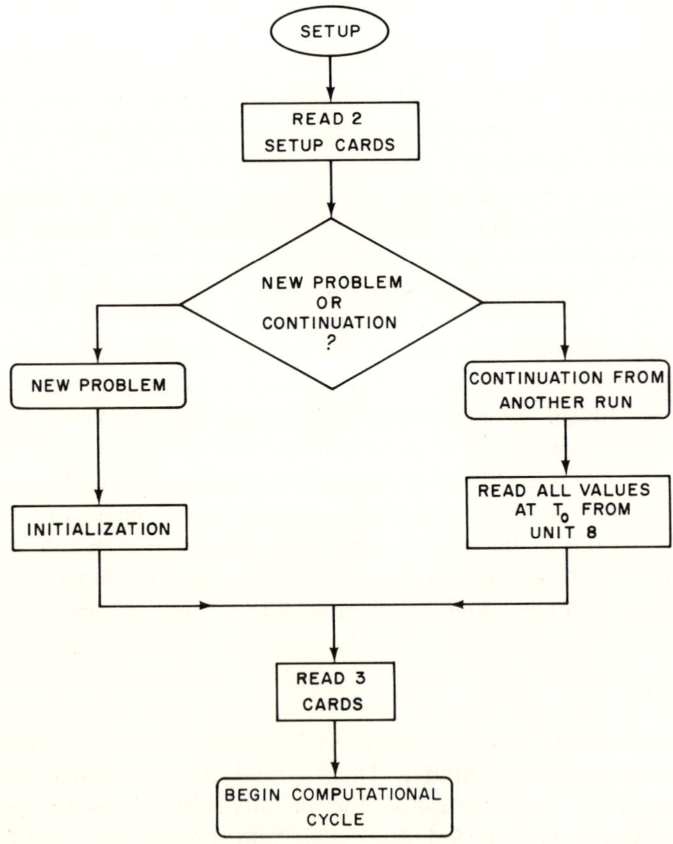

Fig. 18. Flow chart of the setup section.

whether or not data are to be stored in unit 10 at the end of the run, and the other specifying whether tides will be "on" or "off."

The second card specifies whether this is a run to be continued from another or this will be the beginning of a new problem. In the first case, the data at t_0, stored at the end of the previous run in unit 8, are read in. Otherwise, the initialization procedure, schematically shown in the flowchart of Fig. 19, is called.

The last required card for input contains the value of the nondimensional time at which the run should end (the run ends when either this value of t is assumed or when the maximum number of allowed time steps is reached, whichever comes first).

Two additional cards are optional. The first one specifies up to ten locations of (i, j) columns whose surface elevation values are printed at every time step. The second one gives maximum core size required after initialization and is used as parameter of an executive request to free unneeded core space.

3.2.1.2. Main computational cycle. The first part of the cycle computes the divergence deviation, the acceleration terms, and the source terms for the pressure equation. It also updates the surface, extrapolates the u, v values outside, reflags the cells, and solves for the pressure field.

Before incrementing the value of t, a test is made to determine if digital or graphical output is required at this stage. If so, the corresponding routines are executed.

The value of t is then advanced. The new velocity field is computed dynamically everywhere in the interior and according to the specified algorithm at OUT and IN cells. Velocities are extrapolated outside of the surface, salinity values are updated, and markers (if present) are moved.

The flowchart of this section of the code is given in Fig. 20.

3.2.2. A Few Practical Considerations. (a) All routines have been written in FORTRAN, with as few features of FORTRAN V as possible. They have been compiled under the University of Maryland RALPH FOR/MAD 1108 compiler, mainly because of its infinitely better error messages and debugging aids as compared to Univac's FOR/V compiler.

The feature that RALPH treats worse than FOR are multidimensional arrays, for which RALPH computes the index in the equivalent linear array every time a multidimensional one is called, with the consequent cost in computation time. To bypass this, we define all the arrays as being linear and build, once and for all, a table containing the necessary products to find the index in the linear array that corresponds to given indices in the multidimensional arrays. By the use of suitable DEFINE statements, the index for the linear array is obtained as a simple sum, as is done in FOR.

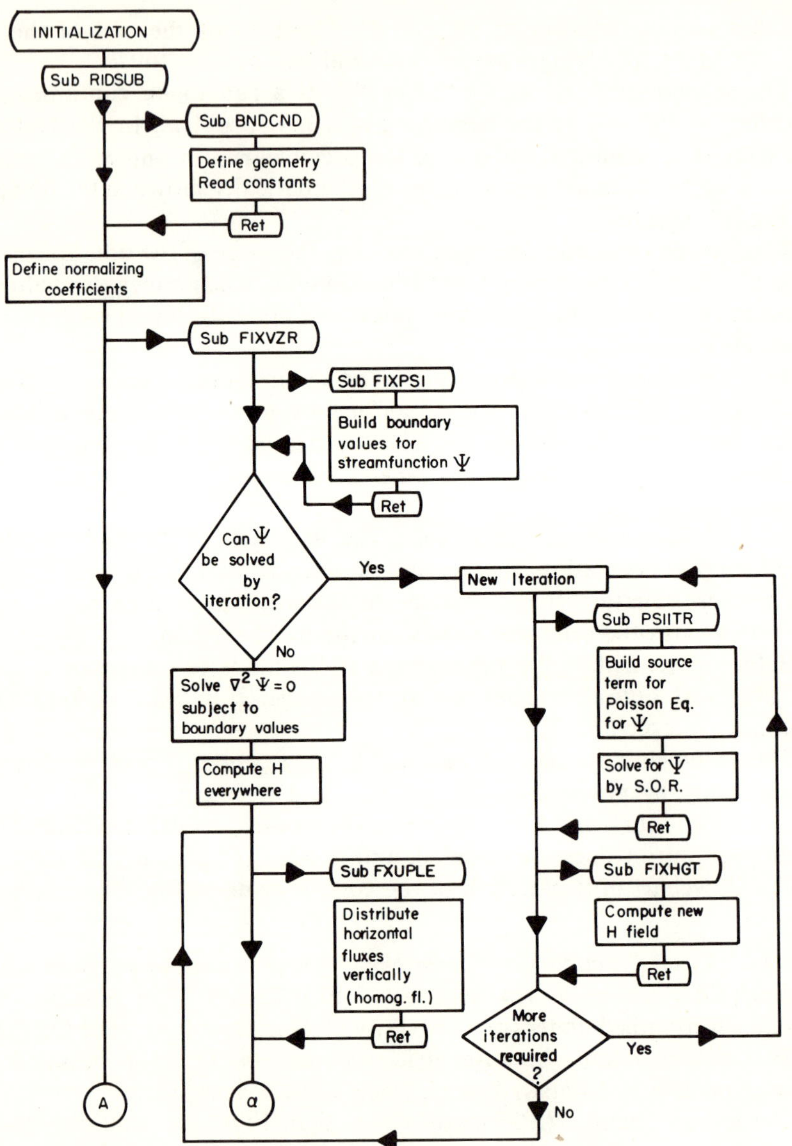

FIG. 19. Flow chart of the initialization procedure.

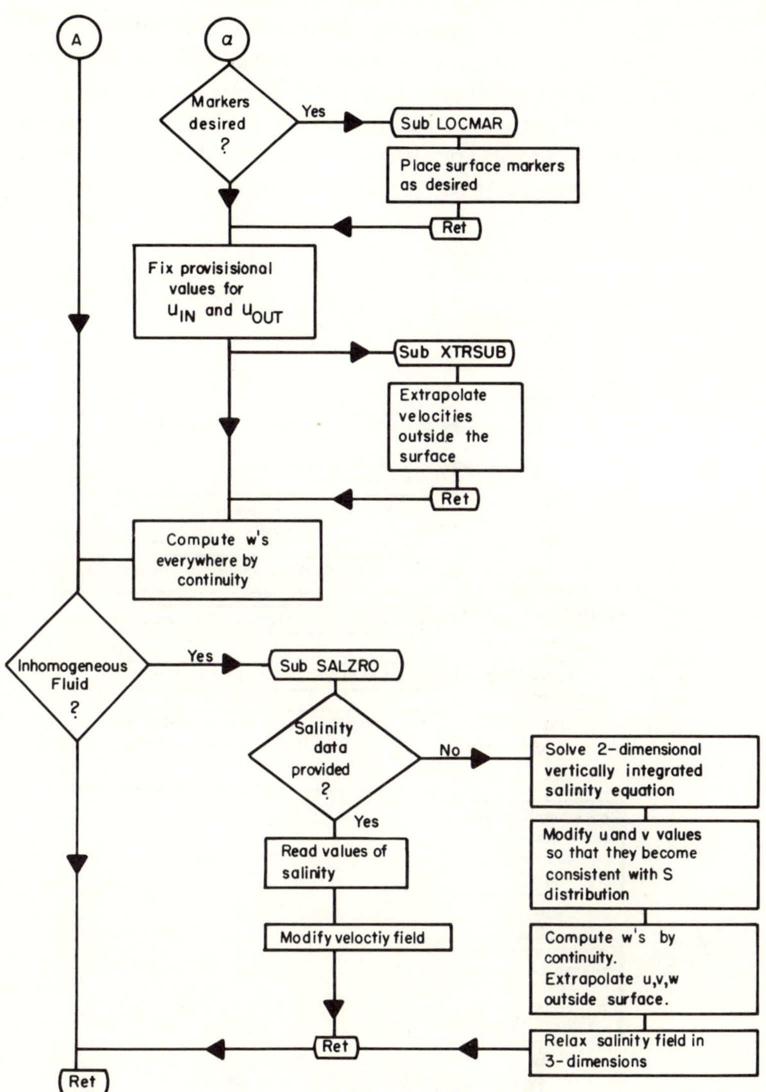

FIG. 19 (*continued*)

(b) Results from a three-dimensional simulation are better understood graphically, but the two-dimensional representation of three-dimensional structures is not a trivial problem. This code includes subroutines for the on-line plotting of surface levels, salinity levels at the various layers, and velocity profiles. Their use was very helpful in the initial stages of developing, debugging, and interpretation of results for the test cases, but are not

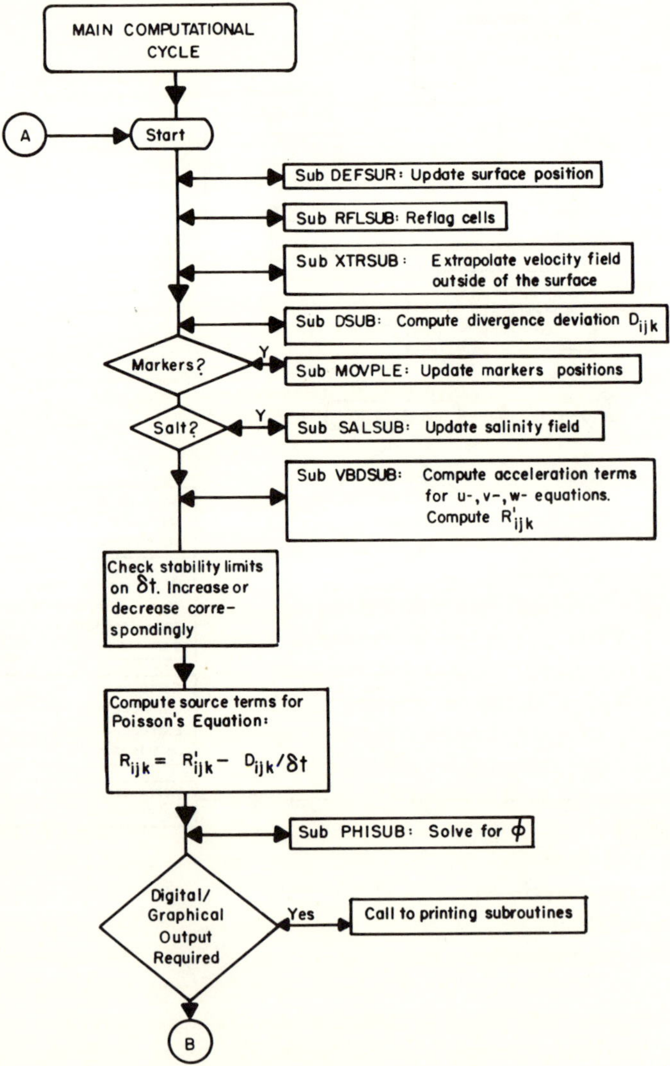

FIG. 20. Flow chart of the main computational cycle.

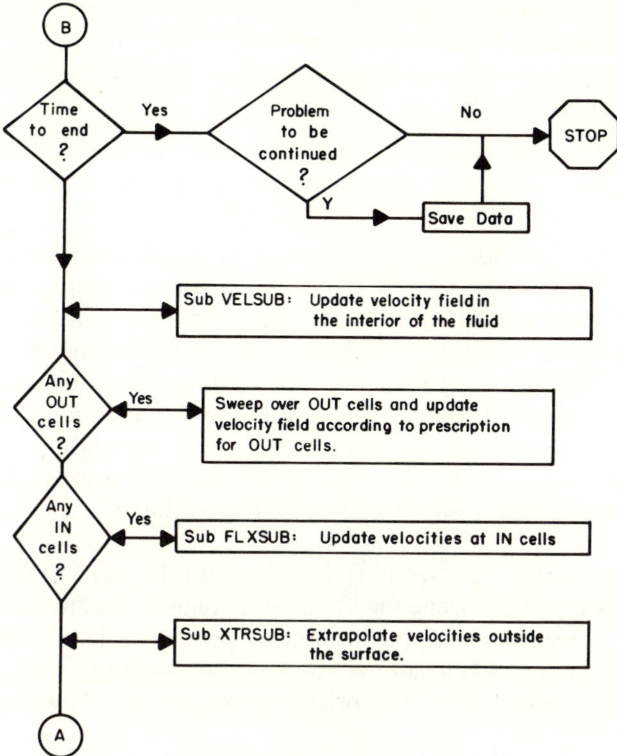

FIG. 20 (*continued*)

adequate for a clear presentation of results, especially for complicated geometries. Plots presented in Sections 4 and 5 have been laboriously built from the digital output fed by hand to a Hewlett Packard 9810A Calculator/9862 Calculator Plotter System. An average of 30 human minutes per plot, even though conservative, constitutes an extremely urgent argument for the automation of this task.

Stereoscopic plots (Hirt and Cook, 1972) are possibly not required for this kind of problems, but the adaptation of existing routines (Wright, 1973; Robinson and Scarton, 1972) to this code will be useful for perspective views in some cases.

(c) Since many of the subroutines used (those in the setup region) are called only once, segmented mapping is desirable. Furthermore, even when the setup does not include initialization of the salinity field, the instruction bank for this segment is a few thousand words larger than the instruction bank for the main computational cycle. Therefore, it is worthwhile to release unneeded core.

(d) The data storage required for irregular geometries as that used in Section 5 is needlessly high in the present version, due to the storage of so many "dead" cells which are never used. An indirect labeling algorithm that will require storage only for cells actually used will shortly be incorporated.

4. Test Cases

In order to show some of the possibilities of this numerical model and its apparent usefulness, several test cases in a very simple geometry are shown here. Each test case introduces a new feature so that a most desirable property of the model, that of reflecting all the physics put into it, can be easily seen. Results are, at least near the outflowing boundary, very much dependent on the OUT boundary conditions for longitudinal and transverse fluxes. The configurations shown here seem not to be unduly influenced by the presence of such boundary. This is stressed here because this problem was perhaps the most difficult one encountered during the development of this model.

In the first part of this section, the effect of tides, river inflow, Coriolis, and winds on a homogeneous fluid in a regular basin, are separately examined. The second part of this section presents a few examples of the resulting circulation when salinity is taken into account.

All examples model a regular prismatic basin (Fig. 21) with dimensions

$$(4.1) \qquad L = 288 \text{ km}$$
$$W = 96 \text{ km}$$
$$H_0 = 8 \text{ m}$$

The depth H_0 is approximately the average depth of the Chesapeake Bay. Our model basin is as long as this estuary, but four times as wide. A small grid was used $(7 \times 6 \times 7)$ allowing five interior cells in the longitudinal direction and four in the transverse. Thus, $X_0 = 5.76 \times 10^6$ cm, $Y_0 = 2.4 \times 10^6$ cm.

Six cells in a vertical column could in principle be occupied by fluid. In general, the undisturbed height was taken as lying at most at the top of the fourth cell, so that there was ample space for waves.

For the homogeneous cases of the section 4.1, maximum storage requirements were: for the Instruction Bank: 14,300 decimal words (only 12K required after initialization); for the Data Bank: 10,300 decimal words (only 9500 required after initialization).

When salinity was included, the storage requirements went up 3K and 1K decimal words for the I- and D-banks, respectively. After initialization, storage requirements for the I-Bank were reduced by 3000 words.

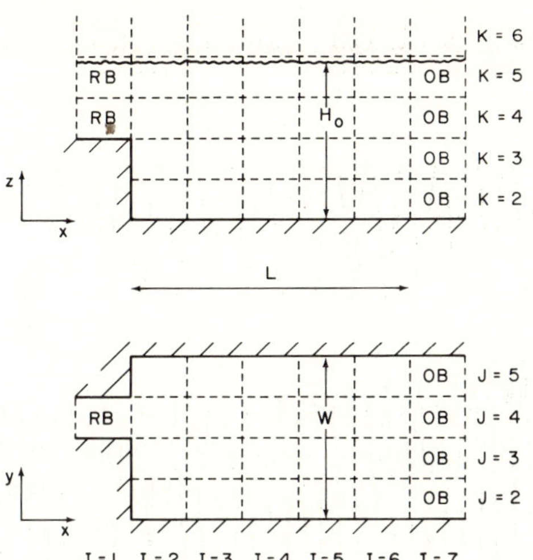

FIG. 21. Idealized basin used for the test cases. RB designates the river boundary cells and OB the ocean boundary cells.

The mixing coefficients have been taken constant throughout the fluid volume. The values of the horizontal eddy coefficients were estimated from results obtained by Okubo (1970; Okubo and Ozmidov, 1970) from a wide collection of surface measurements, and their apparent dependency on the distance from the boundary (Konovalova, 1972) was neglected. Their use here is objectionable, but they serve an illustrative purpose. For the resolution X_0, Y_0 of this grid, the horizontal coefficients for all these runs were set to:

$$(4.2) \qquad A_1 = \frac{v_1 \tau_0}{X_0^2} = 0.138 \times 10^{-2} \qquad A_2 = \frac{v_2 \tau_0}{Y_0^2} = 0.645 \times 10^{-3}$$

where v_i stands for the dimensional eddy mixing coefficient in each direction, $\tau_0 = 1$ day, and X_0, Y_0 are the horizontal dimensions of each computational cell. The values used for the vertical mixing coefficients are given below for each case.

4.1. Homogeneous Fluid

4.1.1. Test Case 1: Symmetric Case. It is important to verify that under symmetric forcing and geometry, a symmetric flow will result. The method of solution must have the property of not introducing spurious asymmetries for times at least as long as those of the proposed use for the model.

A symmetric geometry was obtained by blocking the river inflow in the basin of Fig. 21. A symmetric forcing was insured by switching off the

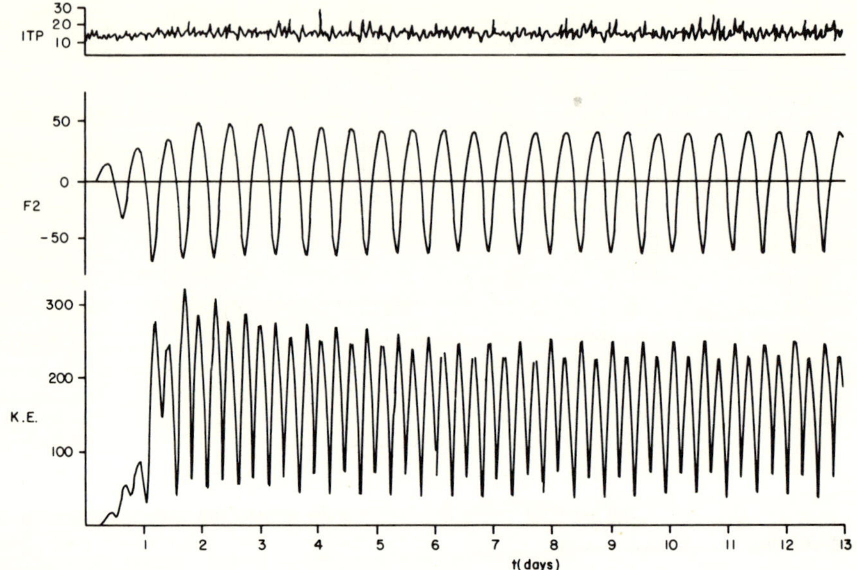

FIG. 22. Number of iterations (ITP) required for solution of the dynamic pressure field, flux from the system into the ocean ($F2$), and kinetic energy ($K.E.$) as a function of time, for a symmetric tidal excitation in a symmetric prismatic basin.

Coriolis effect (i.e., $f^u = f^v = 0$) and assigning the same amplitude and frequency to each of the consecutive OUT cells. The mean heights across these cells were made to vary as

$$(4.3) \qquad \frac{\partial \overline{H}}{\partial t} = a\omega \cos \omega t$$

In order to avoid a discontinuity in $\partial \overline{H}/\partial t$ at $t = 0$, the amplitudes were linearly varied from 0 to a between $t = 0$ and a specified t. After that, the full excitation was applied.

For this case, the mean depth of 8 m was assigned to 3.3 cells, thus defining $Z_0 = 2.42$ m. In units of Z_0, the nondimensional amplitude was $a = 1$, and the vertical-mixing coefficient, taken as 10 cm^2/sec, led to $A_3 = 14.7$. The ω value used corresponded to a nondimensional period $T = .5208$ (or 12.5 hr).

Figure 22 shows the temporal traces of the flux $F2$ through the OUT boundaries and the total kinetic energy ($K.E.$) for some 13 days. The top trace indicates the total number of iterations required for the solution of the dynamic pressure field.

An examination of the kinetic energy trace seems to indicate that a steady state has been attained, for practical purposes, after $t = 6$. Two peaks in the *K.E.* are presented for each tidal cycle, one corresponding to flood and the other to ebb. Minimum values for *K.E.* correspond very closely to zero values of *F2*.

The total mass transport through the OUT walls ($\sum_n F2^n \cdot \Delta t^n$) has also been monitored. The results indicate that a steady mass interchange regime between the basin and the ocean is established after a few cycles [i.e., the integral of *F2* on a semicycle where *F2* > 0 (outflow), equals the integral on the other semicycle, *F2* < 0 (inflow)].

Values of transverse velocities are everywhere very small and cannot be plotted conveniently. If P and S denote two points symmetric one to the other with respect to the longitudinal cross section representing the symmetry plane, the conditions

$$v_P + v_S \simeq O[10^{-m} v_P]$$

$$u_P - u_S \simeq O[10^{-m} u_P]$$

$$w_P - w_S \simeq O[10^{-m} w_P]$$

$$H_P - H_S \simeq O[10^{-m} H_P]$$

(where $m = 7$ for single precision computations), are everywhere and at all times satisfied, as required by symmetry.

Figure 23 is a trace of the surface elevation values at the centers of consecutive cells along a longitudinal cross section. The top trace corresponds

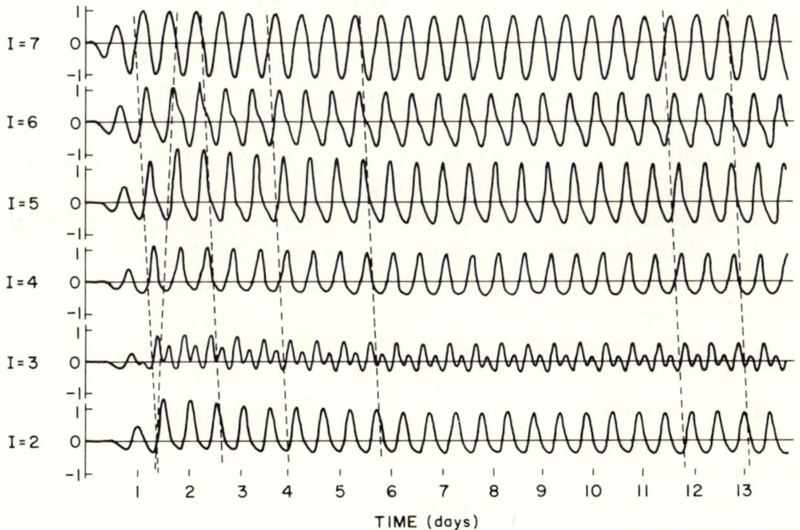

FIG. 23. Surface elevation traces along the symmetric basin as a function of time. Top trace ($I = 7$) is the tidal excitation at the opening to the sea. Characteristic lines from the shallow water approximation are shown by broken lines.

to the $I = 7$ (OUT) cell, and others to the consecutive interior cells at $I = 6, 5, \ldots, 2$. The interaction between the primary wave and that reflected at the wall between $I = 1$ and $I = 2$ is apparent in all the interior cells.

In the shallow water approximation, small amplitude waves should propagate with a speed

$$(4.4) \qquad c = (gH)^{1/2} = 8.85 \times 10^2 \text{ cm/sec} \simeq 13.3 \, U_0$$

and their typical wavelength, for the given forcing frequency is

$$(4.5) \qquad \lambda = \frac{c}{v} = \frac{2\pi c}{\omega} = cT = 3.98 \times 10^7 \text{ cm} \simeq 6.9 \, X_0$$

Thus, the time that a monochromatic wave would take to travel the $5 \, X_0$ distance between the excitation region and the center of the interior cells adjacent to the closed end is

$$\Delta t \simeq 0.376 \, \tau_0$$

The total time for a wave to be reflected and get back at the OUT walls is in this approximation,

$$\Delta t \simeq 0.752 \, \tau_0 \simeq 1.5T$$

Some characteristic lines are superposed to Fig. 23, in order to show the good agreement between the numerical results and those of this approximation.

Figure 24 is an attempt to describe the resulting velocity field and surface elevation fields in a cross section along the basin. The x and z components of the vectors plotted are proportional to the nondimensional horizontal and vertical fluxes interpolated at centers of cells. The surface is delineated by arbitrarily joining with straight lines the H values obtained at the cell's centers. The unperturbed, original reference surface is given by broken lines. Comparison with the top trace of Fig. 24 shows that the times chosen for representation correspond to lowest and highest surface elevations at the ocean (plots at the left-hand side of Fig. 24) and to maximum positive and negative vertical accelerations at that site (right-hand side plots).

Execution time for this 13.9-day run (600 time steps) was 276 sec in the University of Maryland UNIVAC 1108.

4.1.2. Steady State, River-Driven Circulation in a Rotating Basin. The equilibrium height of 8 m is now assigned to 3.7 cells ($Z_0 = 2.16$ m) and a vertical mixing coefficient of 10 cm^2/sec leads to $A_3 = 18.5$.

The nondimensional river inflow was set at $F = 0.289$, or, in dimensional

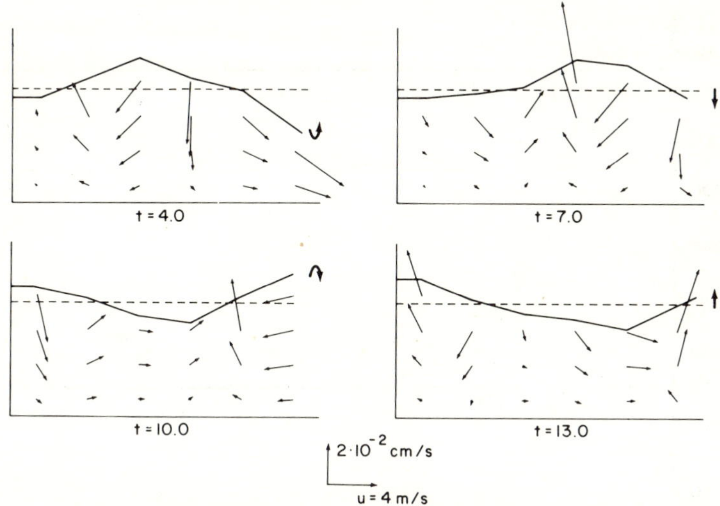

FIG. 24. Longitudinal cross sections for the tidally excited symmetric idealized basin (no Coriolis effect) at instants of the maximum, minimum, and average tidal heights at the opening to the sea. The dashed line represents the unperturbed free surface.

terms, 10^{10} cm^3 sec^{-1}, about ten times stronger than the mean influx from the Susquehanna (Prichard, 1968; Andersen *et al.*, 1973).

Figure 25 shows the traces of the required number of pressure iterations and of the flux through the OUT boundaries and total kinetic energy for the transient. Several things are to be noticed:

(a) For $t > 0.5$, the top trace presents a sharp relative minimum every $\Delta t \simeq 0.126$, which is the nondimensional time that it takes for a wave to travel from one lateral bank to the other.

(b) The initiation procedure slightly underestimates the absolute value of the surface slope. The total flux through OUT walls is during the first day less than the total inflow (maximum variation is less than 14%). This slowing down of the flow is the way in which the extra required surface slope is built up.

(c) During the adjustment process, lateral waves mentioned in (a) seem to account for very little of the kinetic energy. The change in *K.E.* (about 14% of its final value) is mainly due to the slowing down of the flow during the process of adjustment to the full three-dimensional calculation.

Note that the discrete jumps in the curves for *F2* and *K.E.* in Fig. 25 result from using as data diagnostic values with only three significant digits, printed at every time step. From the full data, printed at integer values of t, it was calculated that *F2* differs from the given inflow in less than 0.5% at $T = 4$.

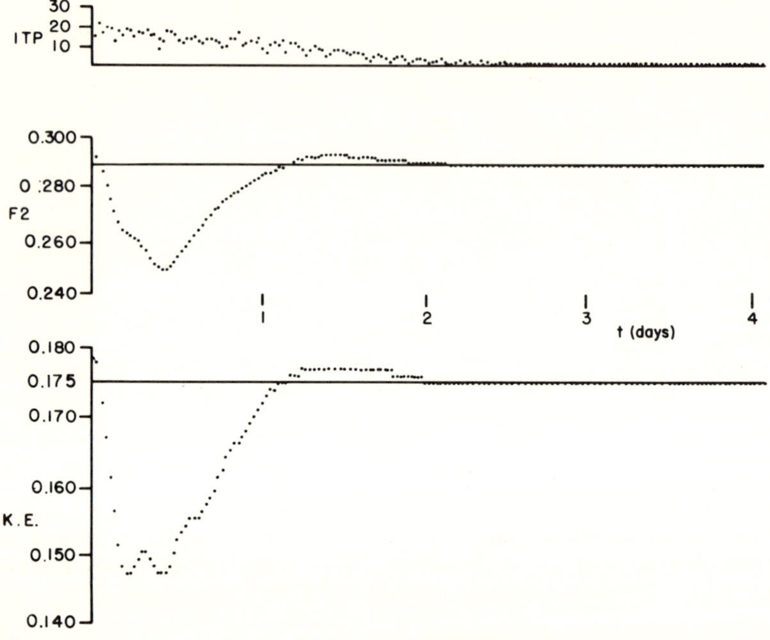

FIG. 25. Number of iterations (ITP) required for solution of the dynamic pressure field, flux from the system into the ocean ($F2$), and kinetic energy ($K.E.$) as a function of time, for the simulation of a steady-state, river-driven circulation in a rotating basin.

Figure 26 shows the interpolated values of the $u - v$ components of the flux field at the cells' centers at different depths. Due to the Coriolis effect, an Ekman spiral is generated as shown, with surface flows toward the right bank and bottom flows toward the left, as expected for the northern hemisphere. The characteristic depth associated to the Ekman spiral as given by (Neumann and Pierson, 1966)

$$D_{\mathrm{E}} = \pi\left(\frac{v_3}{\Omega \sin \lambda}\right)^{1/2}$$

is, in nondimensional units:

$$D_{\mathrm{E}} = \left(\frac{\pi A}{2 \sin \lambda}\right)^{1/2}\left(\frac{T}{\tau_0}\right)^{1/2} \simeq 6.87\left(\frac{T}{\tau_0}\right)^{1/2} \simeq 4.96$$

The helical character of this three-dimensional flow is more easily visualized in Fig. 27, where the v and w components, interpolated at the cells' centers, are plotted in three transsections. The left top one corresponds to cells adjacent to the wall where the river discharges, and consequently the flow is very much influenced by its location.

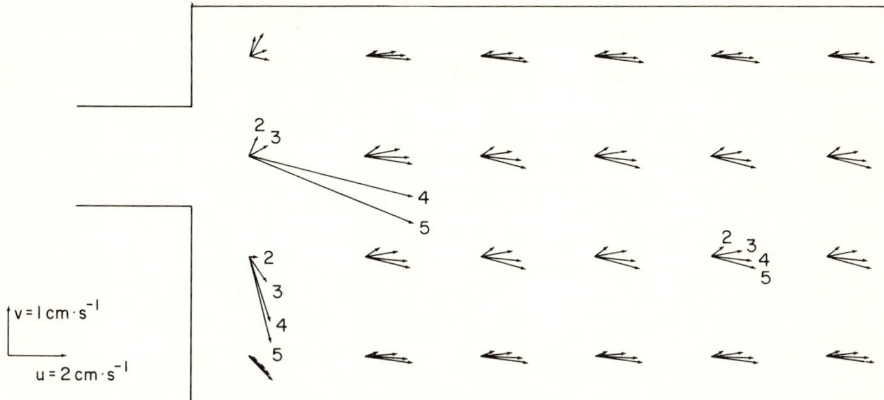

FIG. 26. A steady-state, river-driven circulation in a rotating basin open to the ocean. Examples of the velocity variation with depth are shown by the values of K at selected locations.

For completeness, Fig. 28 shows the four longitudinal cross sections. Data for Figs. 26, 27, and 28 correspond to $T = 7$, when the steady state has been reached within all practical purposes.

One of the possible uses of the markers is shown in Fig. 29, corresponding to the same geometry and approximately the same parameters given above. At $t = 0$, four markers are located at the surface in each (i, j) column. Initial positions for markers assigned to the row of cells adjacent to the IN wall and to the left half of the fourth interior cell have been identified

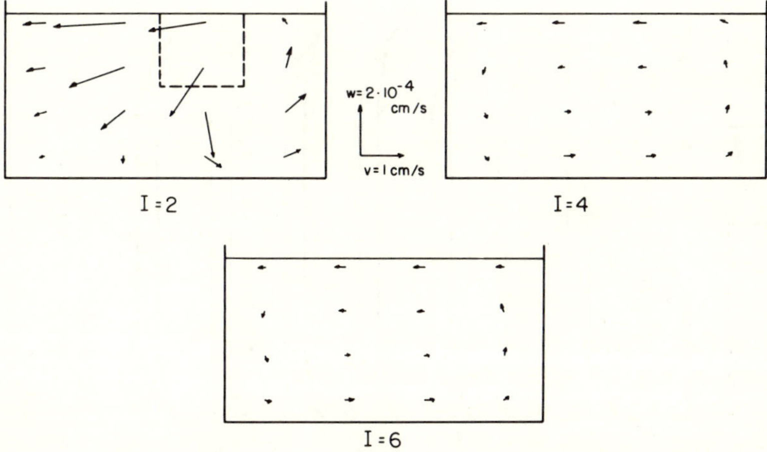

FIG. 27. River-driven circulation: transverse cross sections.

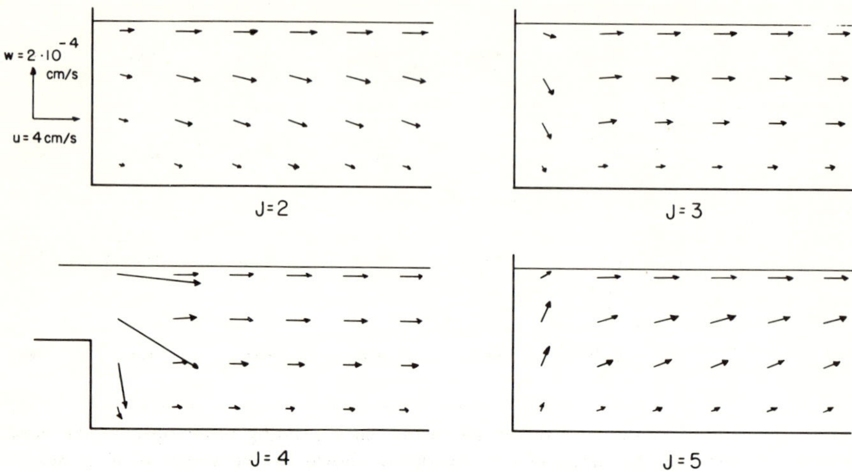

FIG. 28. River-driven circulation: longitudinal cross sections.

by dots. Positions at $T = 5$ and $T = 10$ are shown for all markers, and their locations joined by straight lines to coarsely describe the evolution in the shape of surface lines. Trajectories for some selected positions are indicated by broken lines. Positions at the walls have been kept unmoved, consistently with the no-slip condition on u and v. At the end of the tenth day, six particles are missing: they were destroyed when crossing the boundary into the OUT cells. Markers created near the IN boundaries have not been plotted.

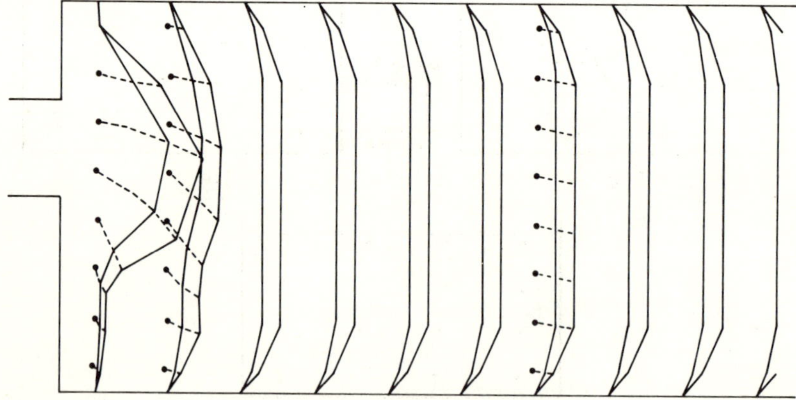

FIG. 29. Displacement of surface tracers subject to the river-driven steady-state circulation of Figs. 26 to 28, after 5 and 10 days.

4.1.3. Wind-Induced Circulations in a Rotating Basin with River Inflow. The configuration was the same as that used in the previous test case, except that the river flow was reduced by a factor of ten, and a boundary condition for the vertical variation of the horizontal fluxes was imposed at the surface. Three examples are shown of steady-state, wind-induced circulations as supporting evidence for the ability of the algorithm used in the updating of the OUT velocities to properly handle these stimuli.

When plotting the results in cases involving a noticeable $\Delta u/\Delta k$ at the surface, it is convenient to use for the values of u and v at the uppermost position below the surface, the total mass flux, rather than the mass flux per unit area which is what u and v represent. Thus, in the plots that follow the interpolated surface value for the u component is $F^{u,\,\text{sur}}_{i,\,j,\,k} = \frac{1}{2}(F^{u,\,\text{sur}}_{i-1/2,\,j,\,k} + F^{u,\,\text{sur}}_{i+1/2,\,j,\,k})$, rather than $u_{i,\,j,\,k} = \frac{1}{2}(u_{i-1/2,\,j,\,k} + u_{i+1/2,\,j,\,k})$. The way in which the $F^{u,\,\text{sur}}_{i\pm1/2,\,j,\,k}$ values are obtained from the u values is shown in Fig. 30.

4.1.3.1. Upstream wind. A value of $\tau_x/(\rho v_3)_{\text{water}} = -0.25$ sec^{-1} was imposed over all the computational grid. From the trace of the flux $F2$ through the outgoing cells as a function of time (Fig. 31), it is apparent that in this case the initialization procedure resulted in an excessive value for the absolute value of the surface slope. The first day of the three-dimensional computation is used in expulsing from the system the extra amount of fluid.

Figure 32 shows a composite plot of the $u - v$ components at the various locations covered by fluid for $T = 7$, almost the steady state. Surface

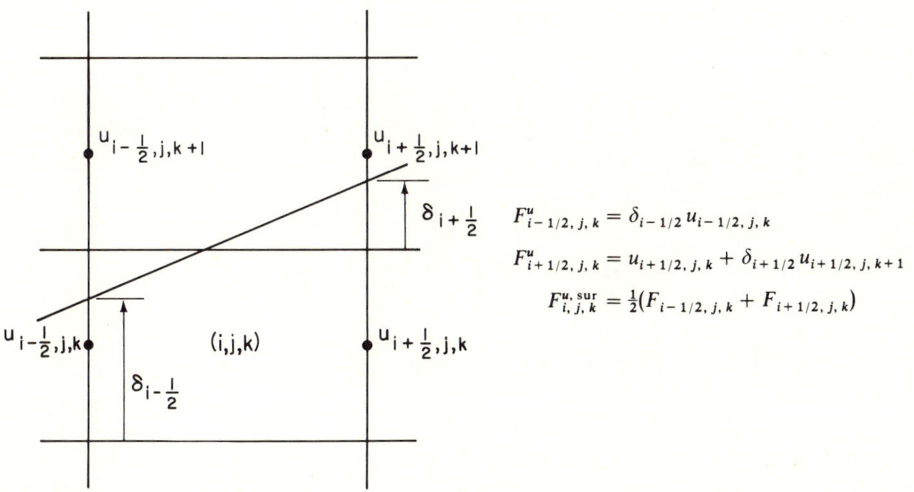

FIG. 30. Algorithm employed to compute the total u flux F^u at surface cells.

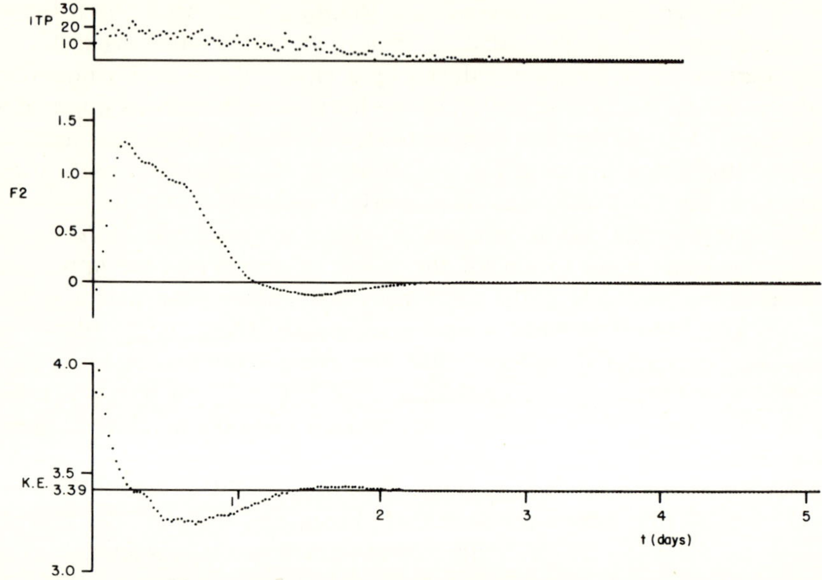

FIG. 31. Number of iterations (ITP) required for solution of the dynamic pressure field, flux from the system into the ocean ($F2$), and kinetic energy ($K.E.$) as a function of time, for a wind induced circulation (upstream wind).

velocities are to the right of the direction of the wind, as expected for the northern hemisphere. Note that at the river boundary, where two K levels are covered by the fluid, the boundary conditions have allowed a reversal of the flow at the surface due to the wind stress, keeping the total river flux constant.

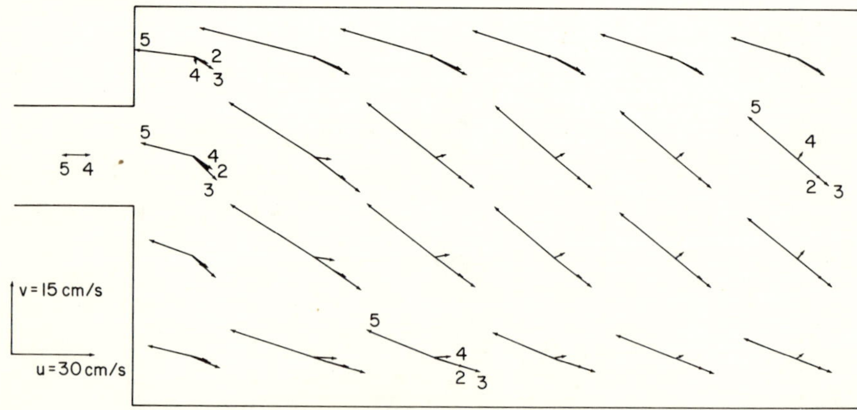

FIG. 32. Wind-driven circulation (upstream wind).

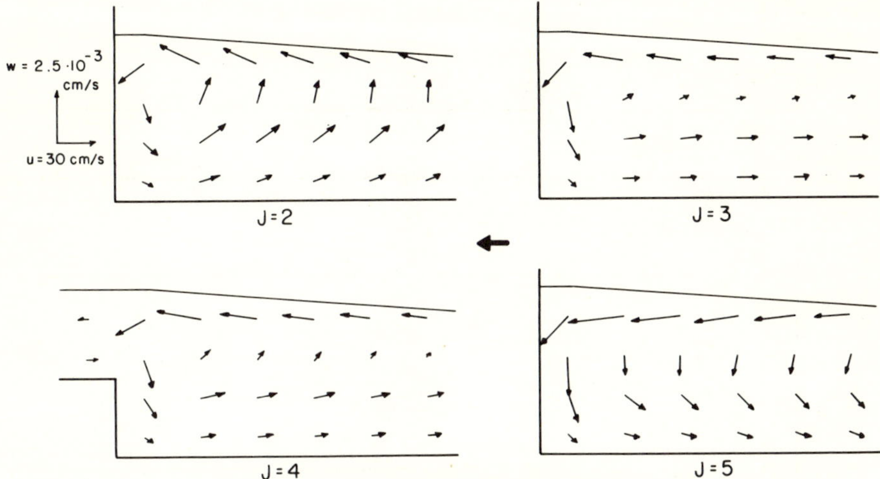

FIG. 33. Wind-driven circulation (upstream wind): longitudinal cross sections.

Figures 33 and 34 show longitudinal and transverse sections. It is apparent that the wind is dominant for the chosen values and the influence of the river flow is negligible (e.g., the right-handed helical character of the river-dominated flow has been changed into a left-handed one).

For the conditions of this example, the surface elevation varies longitudinally by about 15% of the unperturbed height at the OUT walls. The lateral variation is in the average of order 0.3%, heights being larger at the left-hand side of the basin (the right of the wind direction).

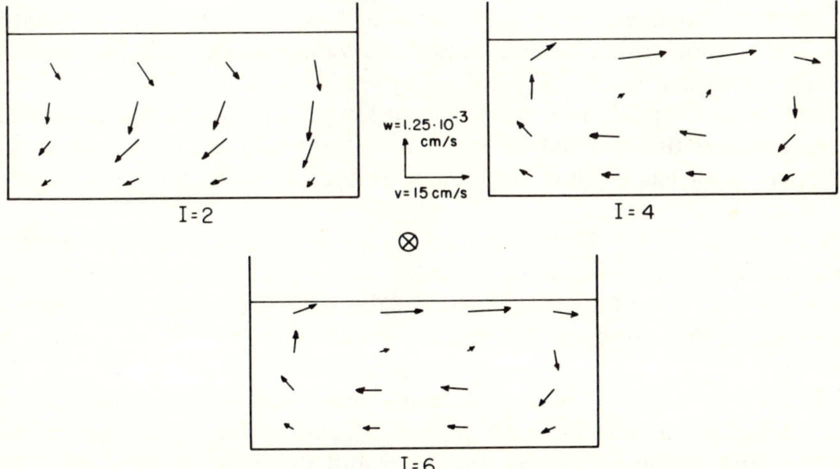

FIG. 34. Wind-driven circulation (upstream wind): transverse cross sections.

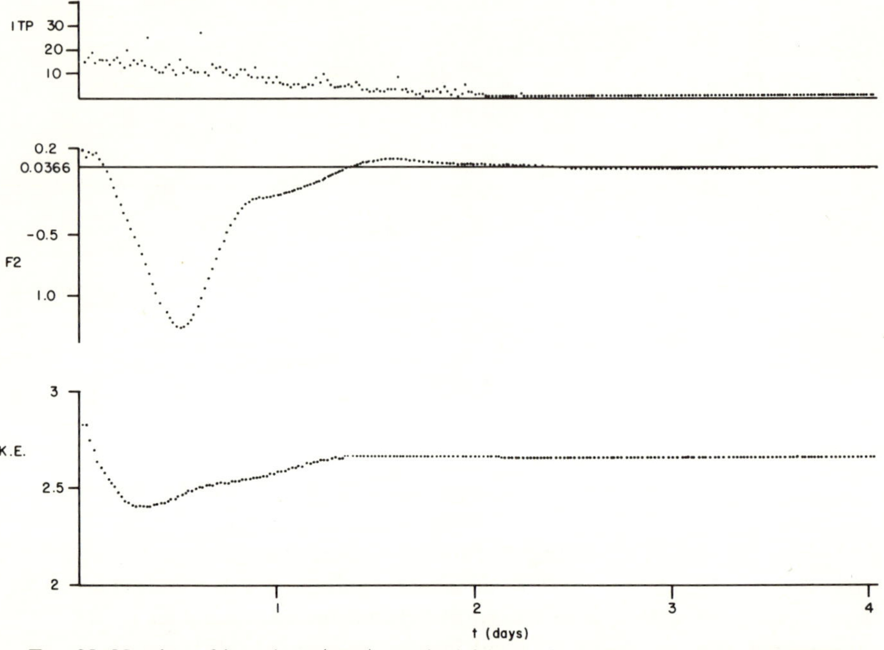

FIG. 35. Number of iterations (ITP) required for solution of the dynamic pressure field, flux from the system into the ocean $(F2)$, and kinetic energy $(K.E.)$ as a function of time, for a wind driven circulation (downstream wind).

4.1.3.2. Downstream wind. In this case, the wind blows toward the ocean: $\tau_x/(\rho v_3)_{\text{water}} = +0.25 \text{ sec}^{-1}$. The mean height of 8 m was now assigned to 4.3 cells so that there remained water in at least 3.5 vertical cells all along the basin. (Since this change modifies the value of $\Delta Z = Z_0$, the vertical mixing coefficient is also changed to $A_3 = 24.96$.)

The initiation procedure establishes a longitudinal slope for the surface that is smaller than needed for the steady state. This can be seen from the fact that water has to enter the system, as shown by the temporal trace of $F2$ (Fig. 35).

Figures 36, 37, and 38 show the composite $u - v$, $u - w$, and $v - w$ cross sections for the quasi-steady state arrived at $T = 7.0$. For this configuration, the surface level is about 17% lower at the river end of the basin than at the ocean end. Left bank heights are lower than right bank heights by less than .5%.

4.1.3.3. Cross wind. The final example deals with a cross wind, blowing from the right to the left bank $[\tau_y/(\rho v_3)_{\text{water}} = .25 \text{ sec}^{-1}]$. In this case, a larger water exchange between the basin and the ocean is required before reaching a steady state (cf. Fig. 39). The final circulation is shown in Figs. 40, 41, and 42.

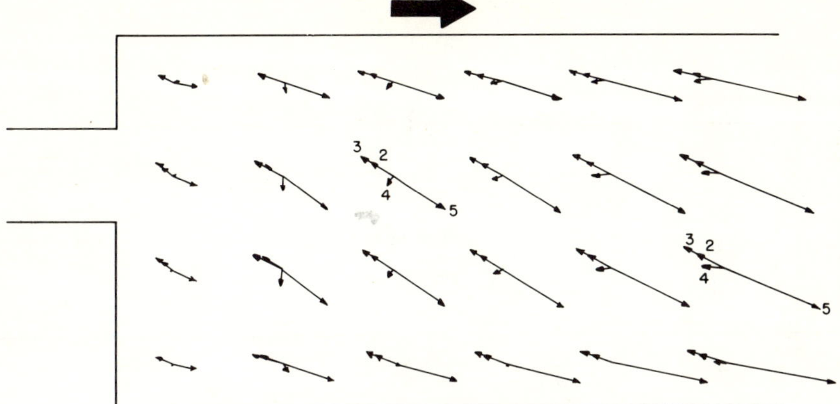

FIG. 36. Wind-driven circulation (downstream wind).

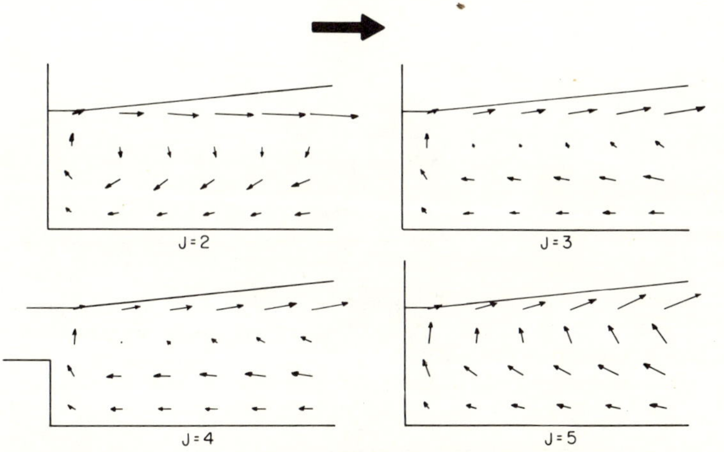

FIG. 37. Wind-driven circulation (downstream wind): longitudinal cross sections.

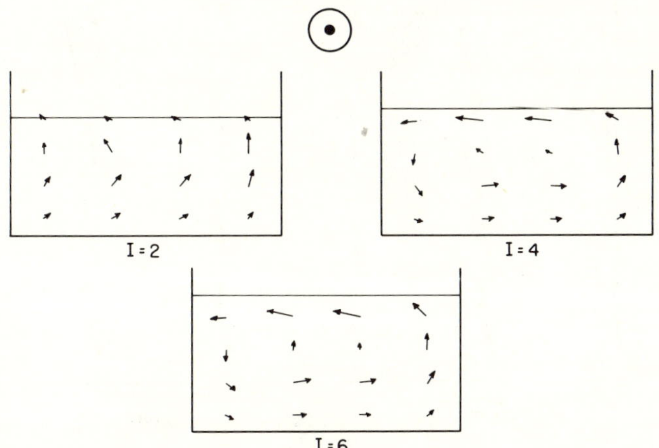

FIG. 38. Wind-driven circulation (downstream wind): transverse cross sections.

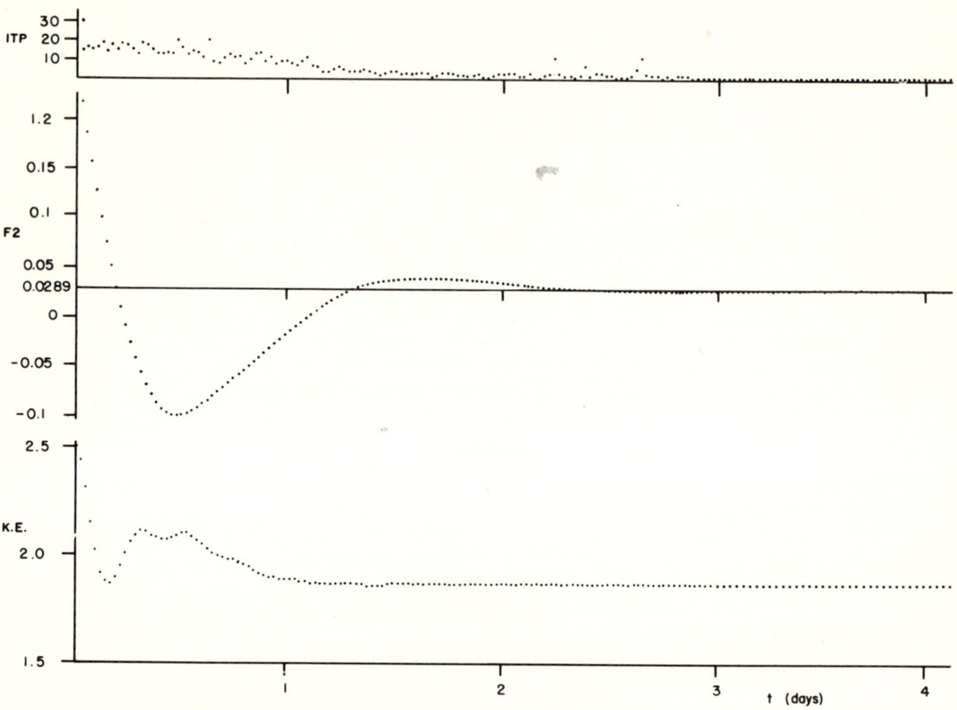

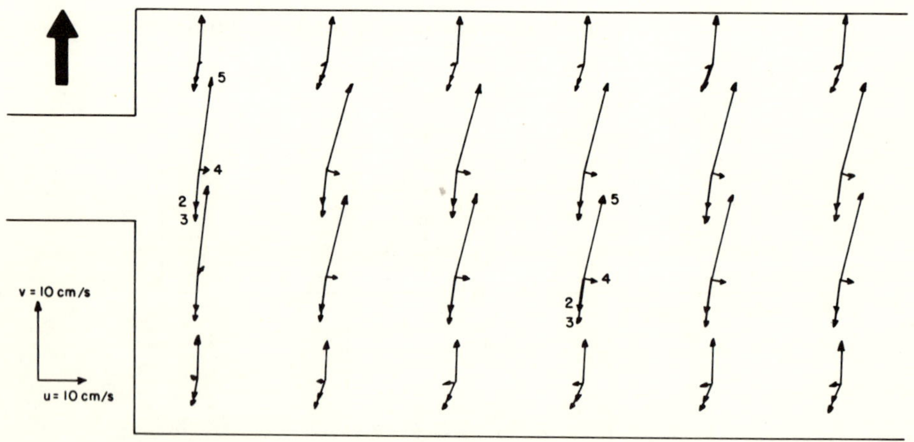

FIG. 39. Number of iterations (*ITP*) required for solution of the dynamic pressure field, flux from the system into the ocean (*F2*), and kinetic energy (*K.E.*) as a function of time, for a wind-driven circulation (cross wind).

FIG. 40. Wind driven circulation (cross wind).

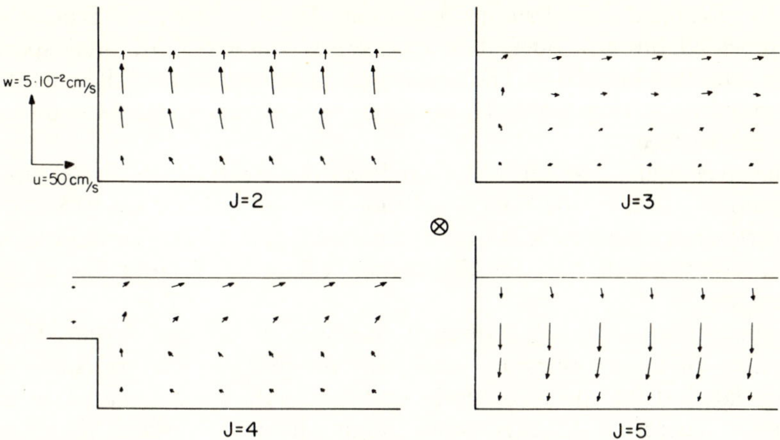

FIG. 41. Longitudinal cross sections of the circulation generated by a cross wind in a prismatic rotating basin.

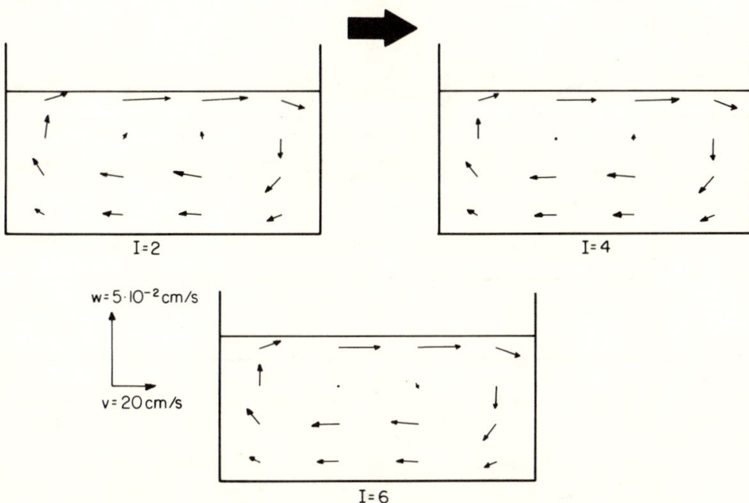

FIG. 42. Transverse cross sections of the circulation generated by a cross wind in a prismatic rotating basin.

4.1.4. Rectified Tide-Driven Circulation. The experiment of Section 4.1.2 was repeated, but now tides of an average amplitude of $0.5Z_0$ were specified at the outgoing boundary. The object was to determine the tidally averaged flow patterns as they would be obtained by instruments at fixed distances from the surface.

The river influx was set at 10^9 cm^3/sec, the mean depth of 800 cm was assigned to 3.3 cells, the tidal amplitude was set at 121.2 cm ($0.5\ z_0$), and the vertical mixing coefficient at 10 cm^2/sec ($A_3 = 17.7$). The fluid velocity at the river mouth was, for the chosen values of X_0 and Y_0, of about 1.1 cm/sec.

Figure 43 shows tidally averaged velocities at selected depths. For the collection of these values, the code was modified so that velocities were interpolated at the specified depths and their values accumulated, in order to mimic recordings of instruments hanging from the surface. Since typical maximum fluxes through the outflowing boundary were 10^3 times stronger

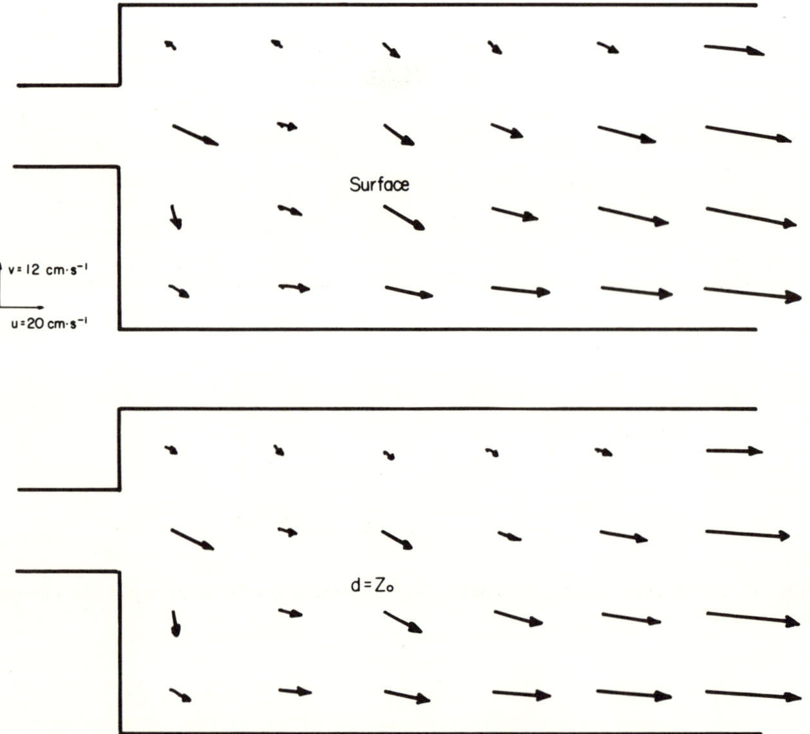

FIG. 43. Tidally averaged fluxes for the tidally excited idealized basin with river inflow in a rotating system. From top to bottom, averages at depths $d = 0$ (surface), Z_0, $2Z_0$, $2.5Z_0$, and $3Z_0$. (Depths are measured from the oscillating free surface.)

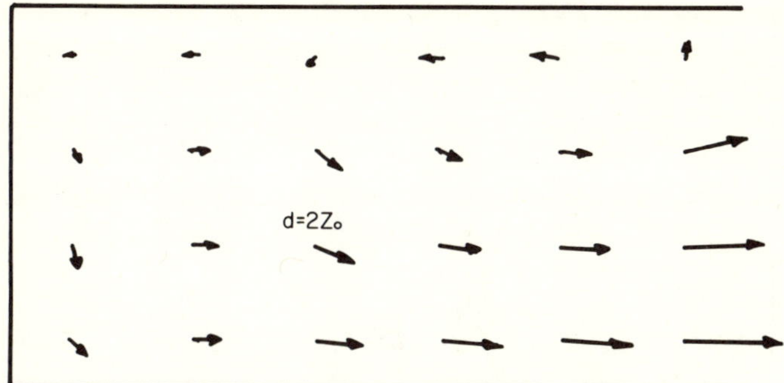

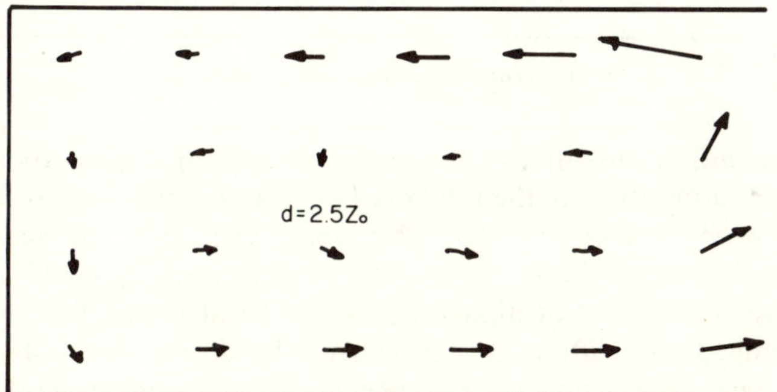

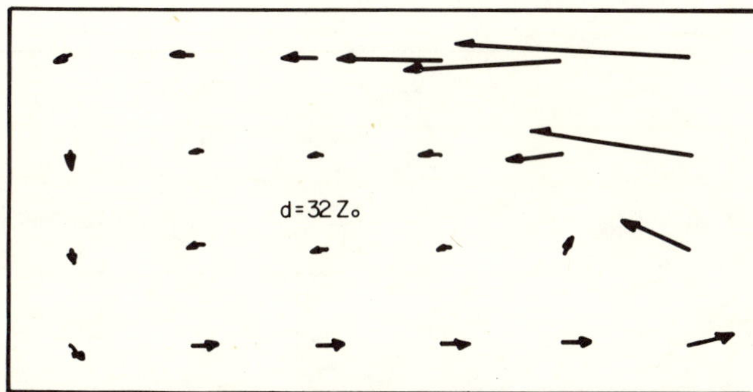

FIG. 43 (*continued*)

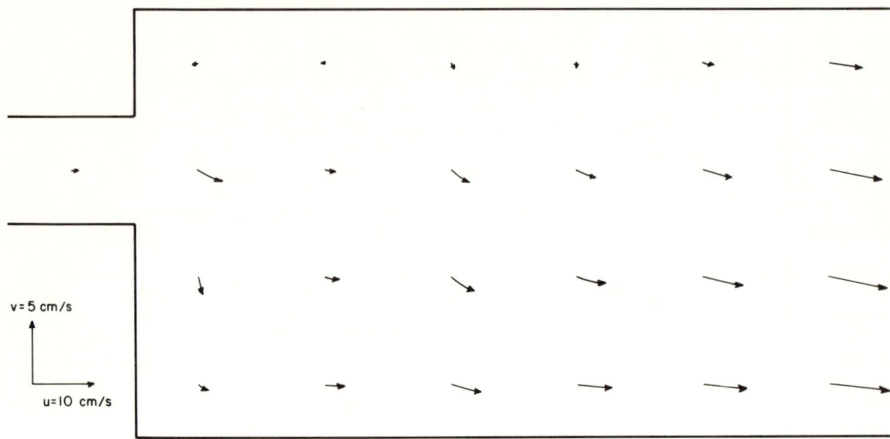

FIG. 44. Surface velocities averaged over a tidal cycle.

than the river inflow, the value of δt used for the temporal integration had to be an integer submultiple of the tidal cycle if averages obtained for different tidal cycles were to be comparable. This would yield sufficient accuracy.

Although the values chosen represent a very extreme case, this example is illustrative of the kind of difficulties encountered in the field when the tidally averaged circulation is experimentally determined through the continuous recording of flow velocities during a short number of tidal cycles.

Due to the Coriolis effect, tides are stronger on the left bank during the flood stage of the cycle, and are stronger on the right bank during the

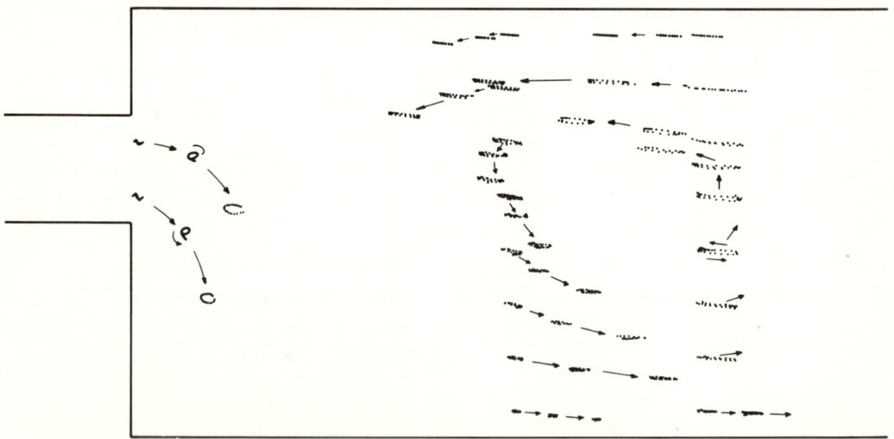

FIG. 45. Motion of surface markers in the rotating prismatic basin with river inflow excited by tides. Positions shown are consecutive markers positions within three different tidal cycles.

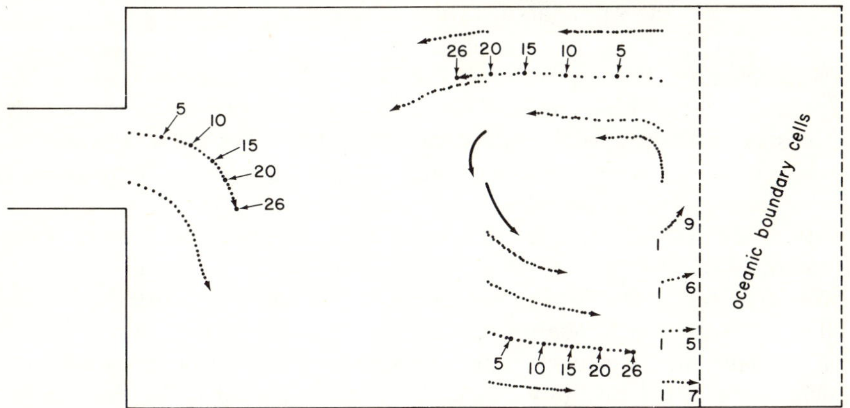

FIG. 46. Markers positions at the end of 26 consecutive tidal cycles from $t = 5$ to $t = 19$ days for the case described in Fig. 43.

ebb stage. Of particular note in Fig. 43 is the comparatively large inward flow at $d = 3Z_0$, a depth of 3 cell thickness below the oscillating surface. This is explained by the fact that at the time of inward flow (maximum tidal height) the current is measured well above the bottom while at the time of outward flow (minimum tidal height) the current is measured very close to the bottom where the flow is restricted by viscous drag.

It is important to realize that the knowledge of the average surface velocities at fixed (x, y) locations is not enough to predict the fate of surface pollutants. The reason for this is that these should be followed in a Lagrangian manner.

Our code has such a possibility and the difference just mentioned is illustrated by comparing Figs. 45 and 46 against Fig. 44. Eighteen markers were followed in a simulation extending for 19.5 days. To minimize transient effects, they were "released" at $t = 5.5$ and followed for the rest of the run. Figure 45 shows their positions within tidal cycles starting at $t = 5.5$, 10.5, and 15.5. Figure 46 shows the positions at the end of the 26 tidal cycles extending from $t = 5.5$ to the end of the simulation. The most striking feature is the strong inward transport at the left bank, which could not be predicted uniquely from the averaged surface velocities (Fig. 44).

4.2. Salt-Driven Circulations

In an estuary, because of the horizontal salinity gradient, positive in the direction to the sea, there is an upstream saline acceleration on the u fluxes. Consequently, for the steady state to include a net seaward flow, the surface slope is required to be larger in absolute value than for a homogeneous flow.

The fact that the saline acceleration is larger at lower depths, tends to produce a circulation consisting of an upper layer of fresh water (due to the river discharge) flowing toward the sea over a bottom salt wedge. Depending on the importance of the vertical mixing processes, the vertical salt distribution in a real estuary can vary from the extreme case just described of a sharp saline discontinuity to an almost homogeneous vertical distribution of salt (the "well mixed" estuary), resulting in characteristic associated circulations. (Prichard, 1955; Hansen and Rattray, 1966; Bowden, 1967).

We have run a few illustrative examples to convince ourselves of the ability of the present model to mimic the characteristic circulation pattern of a partially mixed estuary. Since the salinity distribution responds very slowly, none of the resulting distributions are expected to be final steady state distributions.

The case of a fresh water $(S = 0)$ discharge in the prismatic basin, assuming it in communication with a well-mixed ocean $(S = 30\,\%_{oo})$, was simulated using a value of

$$\alpha \equiv \left(\frac{1}{\rho}\frac{\partial \rho}{\partial S}\right)_{S=0} = 8 \times 10^{-4}$$

and a turbulent Prandtl number $(A_3^S/A_3^{u,\,v,\,w})$ of 1.

For the set of parameters used, small vertical salinity gradients were obtained. Maximum values for the nondimensional ratio $\Delta S/S_b$ of vertical variation ΔS to bottom salinity S_b, ranged from $5\,\%_{oo}$ for $v_3 = 1$ cm^2/sec $(A_3 \simeq 1.8)$ to about $0.08\,\%_{oo}$ for $v_3 = 20$ cm^2/sec $(A_3 \simeq 37)$.

As expected from physical considerations, the effect of the value of the vertical mixing coefficient on the amount of salt admitted in the system is very pronounced (see Figs. 47 and 48 for bottom salinity distributions when that coefficient is increased by a factor of ten).

In Fig. 47 the saline buildup in the left bank as a result of the Coriolis effect is apparent, but the boundary condition of a laterally (and vertically) well-mixed ocean tends to erode such lateral variations.

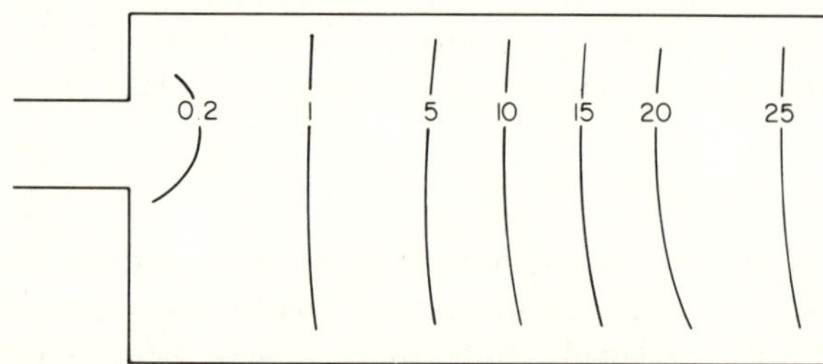

FIG. 47. Bottom salinity distribution at $t = 10$ days for $v_3 = 1$ cm^2/sec $(A_3 = 1.8482)$.

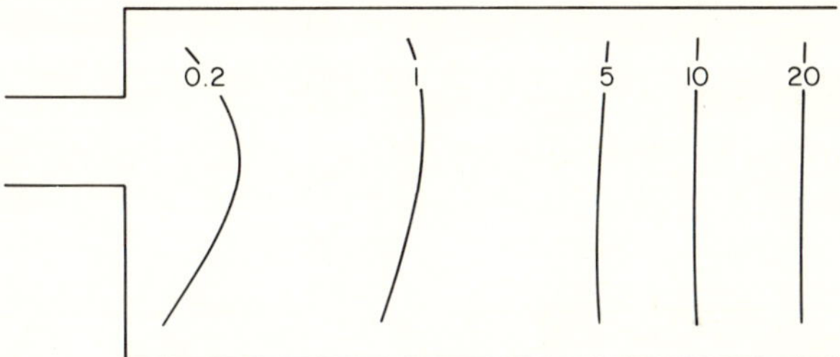

FIG. 48. Bottom salinity distribution at $t = 10$ days for $v_3 = 10$ cm^2/sec ($A_3 = 18.482$).

The circulation resulting for $v_3 = 10$ is shown in Figs. 49, 50, and 51, where the typical configuration of surface fluxes toward the sea and bottom, saline fluxes toward the interior, is shown.

The effect of tides on mixing is shown in Fig. 52. During 5 days, the system is allowed to evolve without tides. The total salt contents inside the basin keeps increasing slowly during this process for the two cases shown ($v_3 = 1$ and $v_3 = 10$). At $t = 5$, tides are turned on, and the full excitation ($a = 0.2$) is reached at $t = 6$. The tidally generated longitudinal advection brings into the system denser water during the flood cycle. Due to the vertical shear, upper layers increase their salinity more than lower layers during this part of the cycle. Not all the salt introduced in this manner is removed during ebb because of the vertical mixing and advection processes. The composite effect is to increase the amount of salt admitted to the system. Although the lower traces of Fig. 52 have already begun to flatten, the system is still far away from a quasi-steady state at $t = 10$.

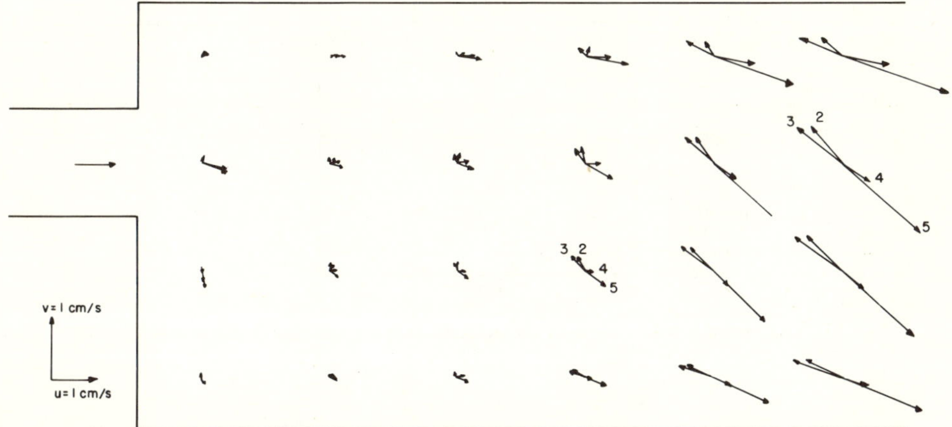

FIG. 49. Salt-driven, steady-state circulation at $t = 10$ days ($v_3 = 10$; $A_3 = 18.482$).

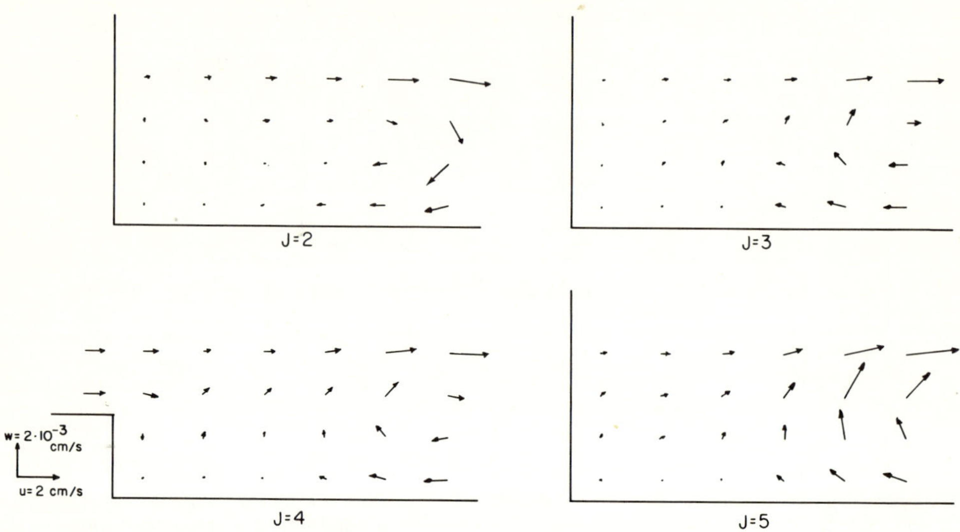

FIG. 50. Longitudinal cross sections of the velocity field for the steady-state, salt-driven circulation in the idealized rotating basin open to the sea at $t = 10$ days. ($v_3 = 10$ cm²/sec; $A_3 = 18.482$).

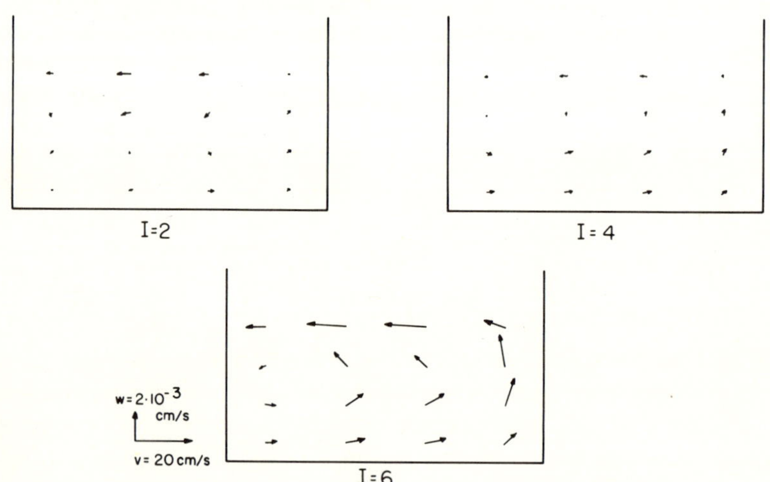

FIG. 51. Transverse cross sections of the velocity field for the case described in Fig. 50.

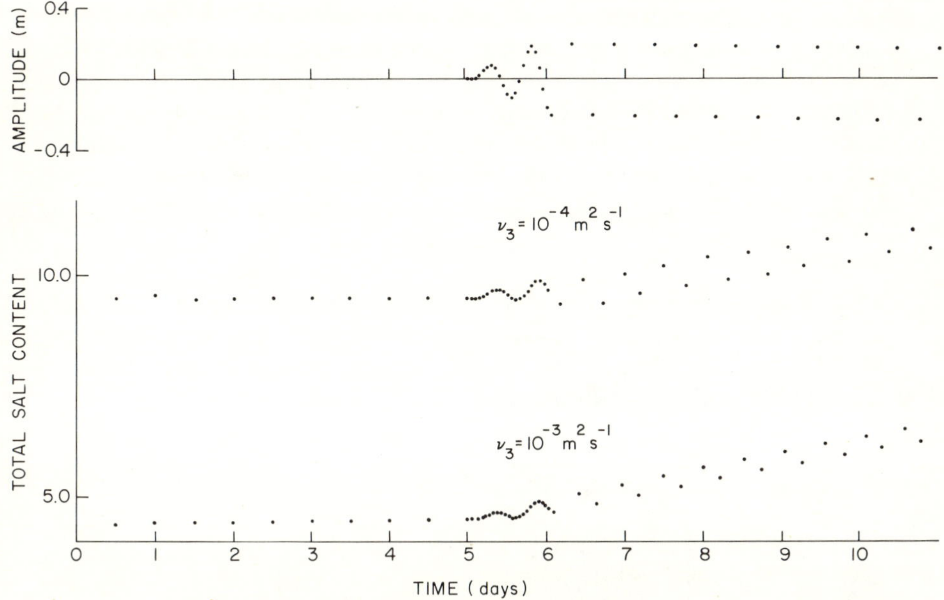

FIG. 52. Influence of tides on the exchange of salt with the ocean in the idealized basin. The top trace is the average excitation amplitude at the opening to the sea. The lower traces show the total salt content Σ in arbitrary units for $v_3 = 1$ and 10 cm^2/sec.

5. SIMULATION OF CHESAPEAKE BAY

As a more complicated geometry to test the possibilities of our model, we chose the Chesapeake Bay. This system is approximately 290 km in length, about 25 km in average width, and its mean depth is of the order of 8 m.

5.1. Geometry and Parameters Used

5.1.1. Geometric Characteristics of the Model. In order to avoid the need for temporary storage, this first application to a real estuary was done on a small grid 20 × 10 × 8. Approximate boundaries were drawn on it resembling the shape of the bay, as shown in Fig. 53. The bottom topography was based upon rough areal estimates on maps published by the Coast and Geodetic Survey. The three most important tributaries to the bay are represented: the Susquehanna, the Potomac, and the James at $(I, J) =$ (1, 3), (11, 2), and (19, 3), respectively.

For the given grid, the horizontal grid size was $X_0 = 13.7$ km, $Y_0 = 8.7$ km. Maximum average depth represented was $H = 14.8$ m. Depth at rest was assigned to 4.3 cells. Thus the vertical resolution was $Z_0 = 3.44$ m. The number of cells in the vertical at each (i, j) column is shown in Fig. 54.

The resulting geometry consists of two interacting basins of variable depth, interconnected by a deep narrow channel at $(I, J) = (8, 4)$ and $(9, 4)$. The smaller one receives only one tributary (the Susquehanna), while the larger receives fresh water at two locations and is also in communication with the sea (cf. Fig. 54). In what follows, depths are indicated by the value of the K-index, as shown schematically in Fig. 55.

5.1.2. Parameters Used. The following values (Andersen *et al.*, 1973) were used for river discharges: Susquehanna, 1.50×10^9 cm^3/sec; Potomac, 0.57×10^9 cm^3/sec; James, 0.48×10^9 cm^3/sec; which corresponded to 0.3159, 0.1201, and 0.1011 in nondimensional units.

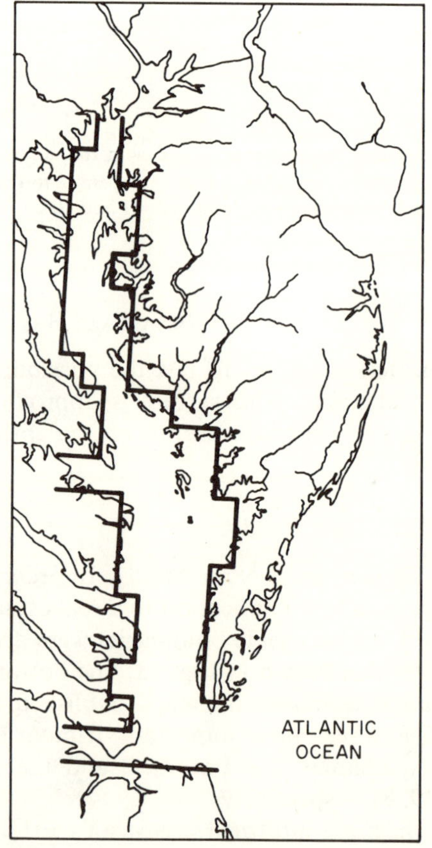

FIG. 53. Chesapeake Bay: outline and approximate boundaries used in this simulation.

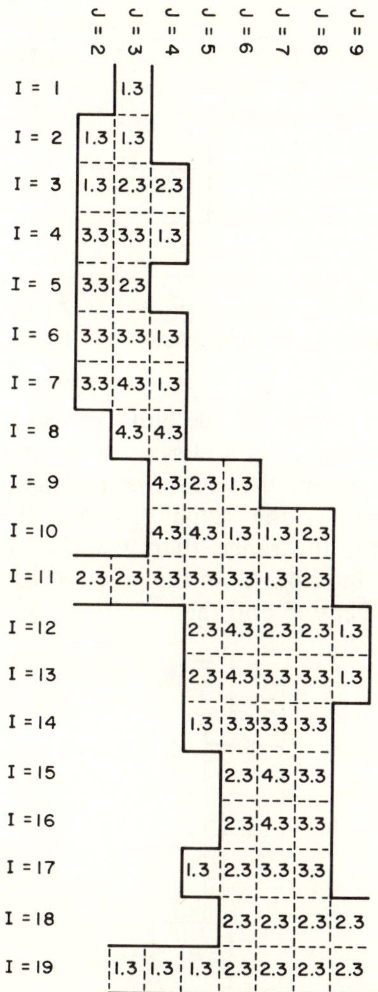

FIG. 54. Bottom topography for the Chesapeake Bay Simulation. Numbers are assigned depths in units of Z_0.

The mean amplitude of the tidal oscillation at the ocean end was chosen to match the mean amplitude given by the Tide Tables (Tide Tables, 1974) for Hampton Roads, Virginia: $a = 62.5$ cm $= 0.182\ Z_0$. The frequency was chosen as that corresponding to a period of 12.5 hr.

Values for the horizontal mixing coefficients, determined as in the cases of Section 4, were $v_1 = 9.3 \times 10^5$ cm^2/sec, $v_2 = 5.7 \times 10^5$ cm^2/sec, and correspondingly, $A_1 = 0.0043$, $A_2 = 0.0065$.

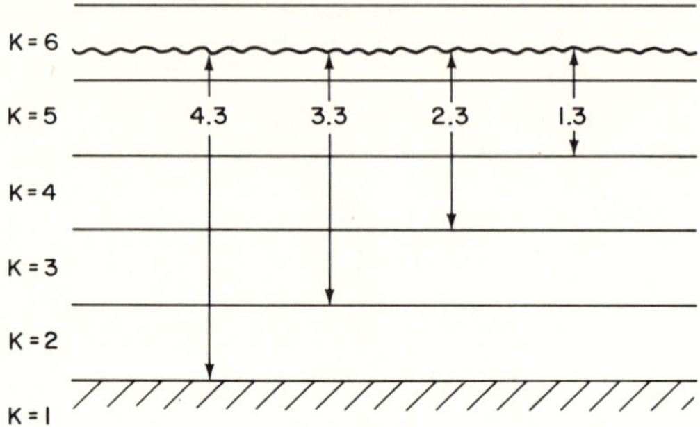

FIG. 55. Relation between depth from reference level and value of K index in the computational mesh.

In order to estimate a value for the bulk vertical mixing coefficient, two short runs were made with $v_3 = 1$ and $v_3 = 10$, monitoring surface elevation values along the bay. From those results, and assuming an exponential decay proportional to v_3, the value $v_3 = 20$ cm^2/sec $(A_3 = 14.6)$ was estimated to be within the range that would give appropriate surface elevation traces in the interior. As in the previous examples, this value was used both for the momentum and the salinity equations.

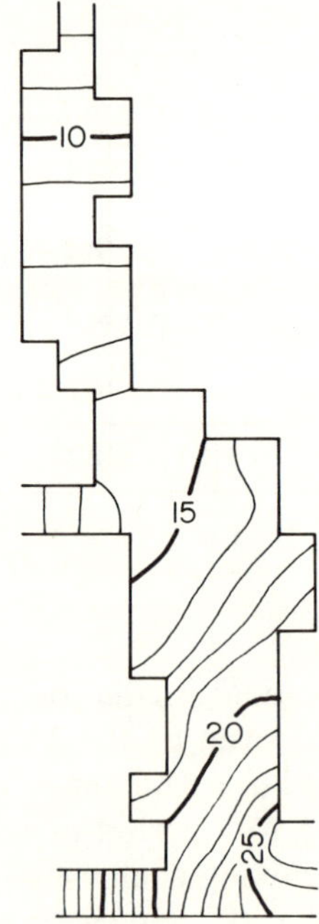

FIG. 56. Surface salinity values for the basin of Fig. 54. Initial conditions.

The initial salinity distribution (Fig. 56) was interpolated from the summer distribution given in Prichard (1968) and using some data from Andersen *et al.* (1973). The salinity boundary condition at the opening to the sea was applied according to the prescription given in Eq. (3.57):

$$S_{i,\,j,\,k} = \frac{1}{2}\,(S_{i,\,j-1,\,k} + S_F)$$

where $S_F = 31.5\,\%_{oo}$ was used for all OUT walls. The boundary conditions at the inflowing cells were given by Eq. (3.36), with λ taken as a constant for each IN cell, and determined from the initial salinity distribution. Values used were $\lambda = 0.9$, 0.9, 0.75 for the Susquehanna, the Potomac, and the James, respectively.

5.2. *Experimental Results*

A simulation extending over a period of 12 days was performed. Total computer time was approximately 90 min on the UNIVAC 1108 of the Computer Science Center of the University of Maryland.

No tides were applied during the first 4 simulated days. The circulation pattern achieved at the end of this period is shown in Fig. 57 for two depths. Most of the features are intuitively consistent with the prescribed bottom topography. Salinity-driven bottom flows were obtained as expected. However, these results do not quite represent a steady-state circulation, because no final salinity distribution was attained and the net outflow through the ocean boundary still exceeded the total river inflow by almost $2\,\%$ at $t = 4$ days.

The expected transverse salinity gradients at the end of this period are found to be in qualitative agreement with observations, as shown in Fig. 58. However, the vertical mixing coefficient was too large to maintain the stratification specified in the initial conditions, which is found to be completely eroded at the end of this period.

From $t = 4$ to $t = 12$ days a tidal excitation of characteristics given in Sec. 5.1.2 was applied. The calculated data show that the predicted tidal amplitudes and velocities at selected locations in the upper bay are about twice as large as expected from observational data. This result suggests that the vertical mixing coefficient chosen for the momentum equations may have been too small to produce the required tidal dissipation.

The total salt contents Σ within the basin as a function of time is shown in Fig. 59. The initial 4-day period is characterized by a transient where the amount of salt within the system increases until $t \cong 0.65$, followed by a downward trend. This observation is consistent with the values obtained for the volume flux through the aperture to the ocean. In effect, from $t = 0$ to $t = 0.668$ there is a net inflow of ocean water into the basin, while

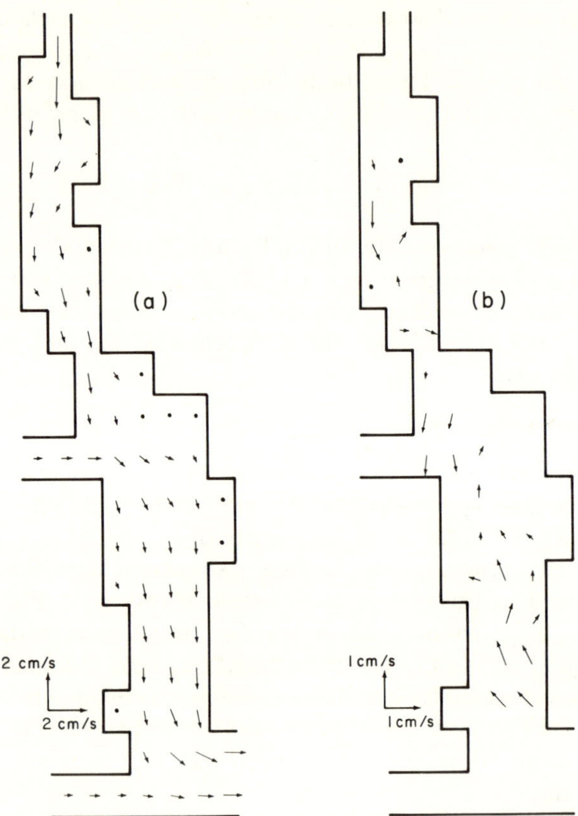

FIG. 57. Steady-state circulation pattern for the geometry of Fig. 54. (a) Surface fluxes ($K = 5$). (b) Fluxes at a lower level ($K = 3$). (Note that the scales are different in each case).

for $t > 0.776$ the outflow through that boundary exceeds the total river flow. The maximum outflow is attained at $t = 1$ day.

The tidal oscillation, imposed at $t = 4$, was modulated from $t = 4$ to $t = 5$ as specified in Sec. 4.1.1 and began with a small depression of the surface followed by a larger height increase. Due to this modulation, the integrated flux through the ocean boundary over the first tidal cycle was a net volume flow into the basin. This net flux was responsible for the immediate 10% jump in Σ. Only after several tidal cycles was a state attained where the tidally integrated flux through the ocean boundary differed very little from the tidally integrated river inflow. The upward trend of Σ maintained through these latter cycles was a result of the transport of salt from the ocean into the system and the enhanced effective lateral mixing within the estuary by tidal action. Figure 60 illustrates the surface salinity

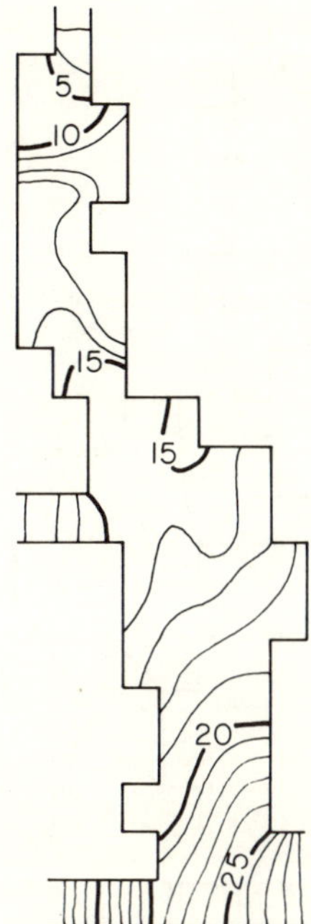

FIG. 58. Surface salinity values for the basin of Fig. 54 at $t = 4$ and with no tides.

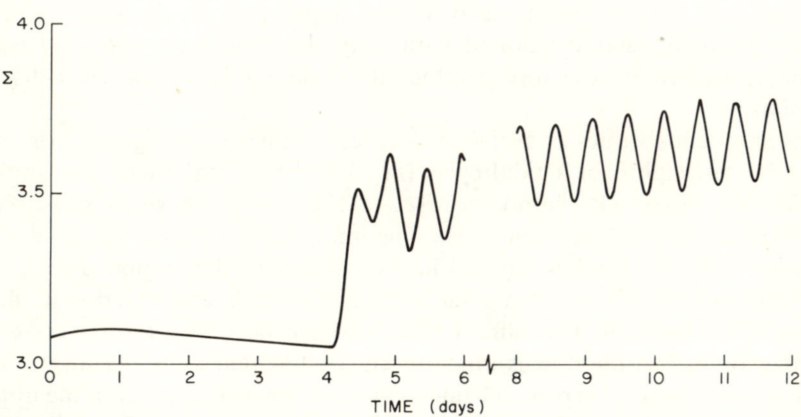

FIG. 59. Total salt content within the basin of Fig. 54 as a function of time. Tides are turned on at $t = 4$ days.

     ENRIQUE A. CAPONI

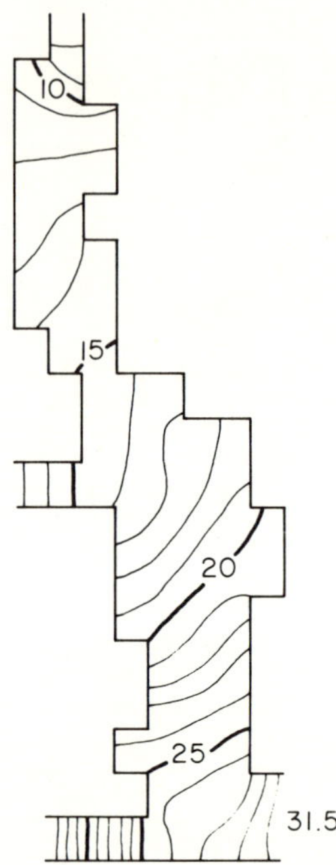

FIG. 60. Surface salinity values for the basin of Fig. 54 at $t = 12$, after 8 simulated days with tides.

distribution after 8 days with tidal action. The overall increase in salinity from Fig. 58 is apparent and the smoothing of the salinity distribution by tidal mixing is clear, particularly in the upper part of the estuary. The strong gradients that developed with only the river flow (Fig. 58) were probably caused by the topographic effects acting upon the river-driven circulation.

Instantaneous values of surface velocities at times differing one from the other by one-eighth of a tidal cycle (i.e., 1.56 hr of real time), are shown in Figs. 61 to 68. The companion set of Figs. 69 to 76 show subsurface velocities ($K = 4$) at the same times. The initial time corresponds to almost high water in the smaller basin. Fluid flows out of this region during the next half of a cycle, comes to a standstill at $t_0 + T/2$, and reverses its flow during the other half. The same general pattern occurs in the lower basin at other times, but the flow is much more complicated there because of the interaction between the primary tidal wave and the discharge from the upper basin. A similar description is valid for the subsurface flows (Figs. 69 to 76).

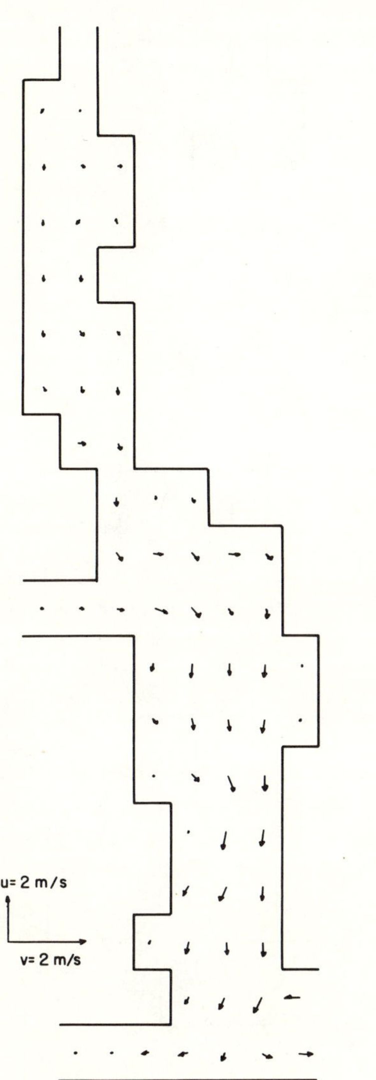

FIG. 61. Surface velocities at $t_0 + T/8$.

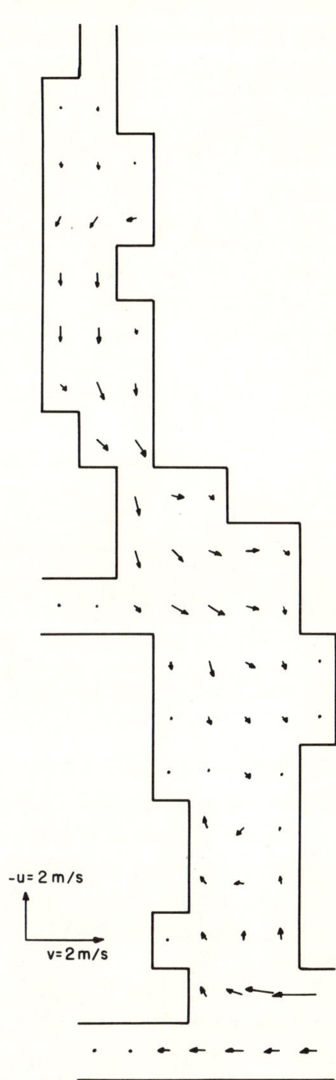

FIG. 62. Surface velocities at $t_0 + 2T/8$.

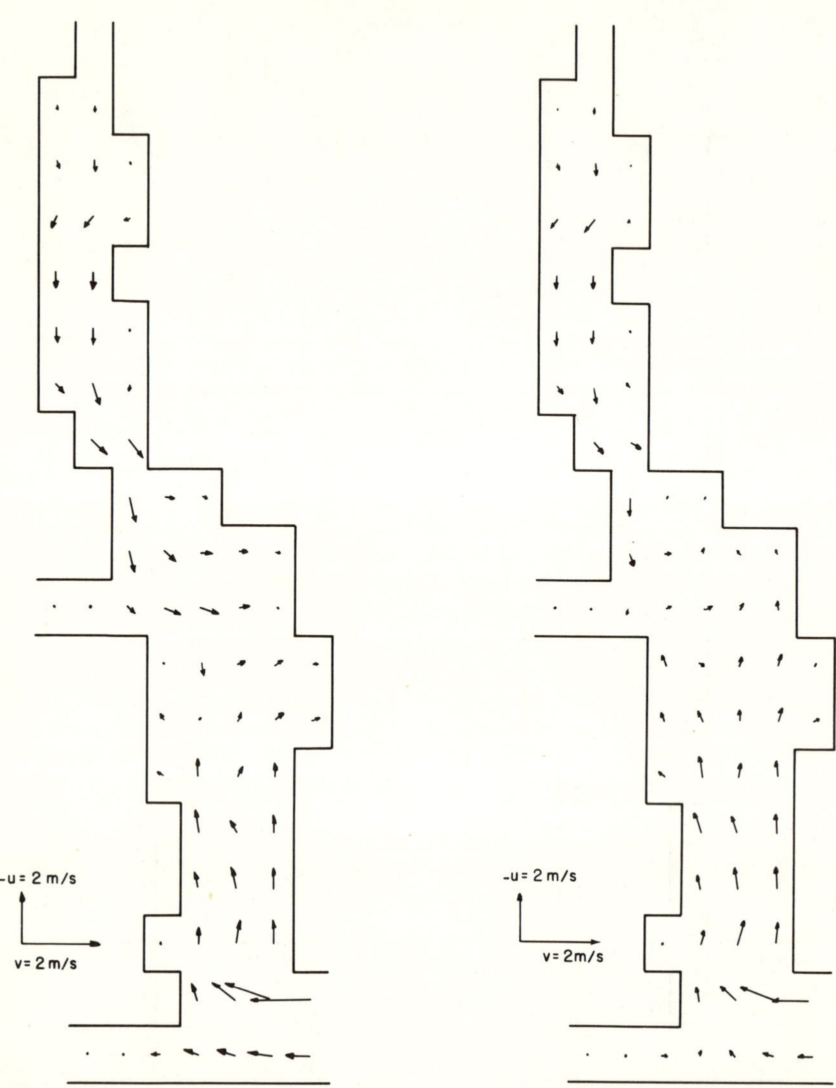

FIG. 63. Surface velocities at $t_0 + 3T/8$.

FIG. 64. Surface velocities at $t_0 + 4T/8$.

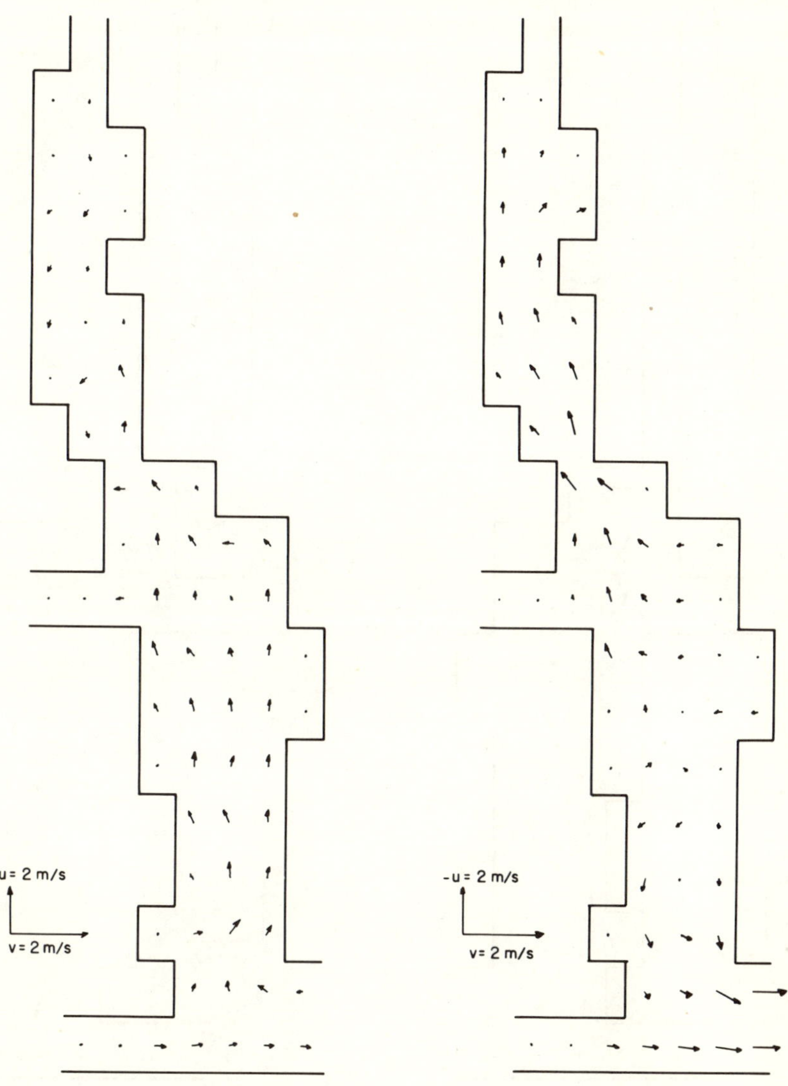

FIG. 65. Surface velocities at $t_0 + 5T/8$.

FIG. 66. Surface velocities at $t_0 + 6T/8$.

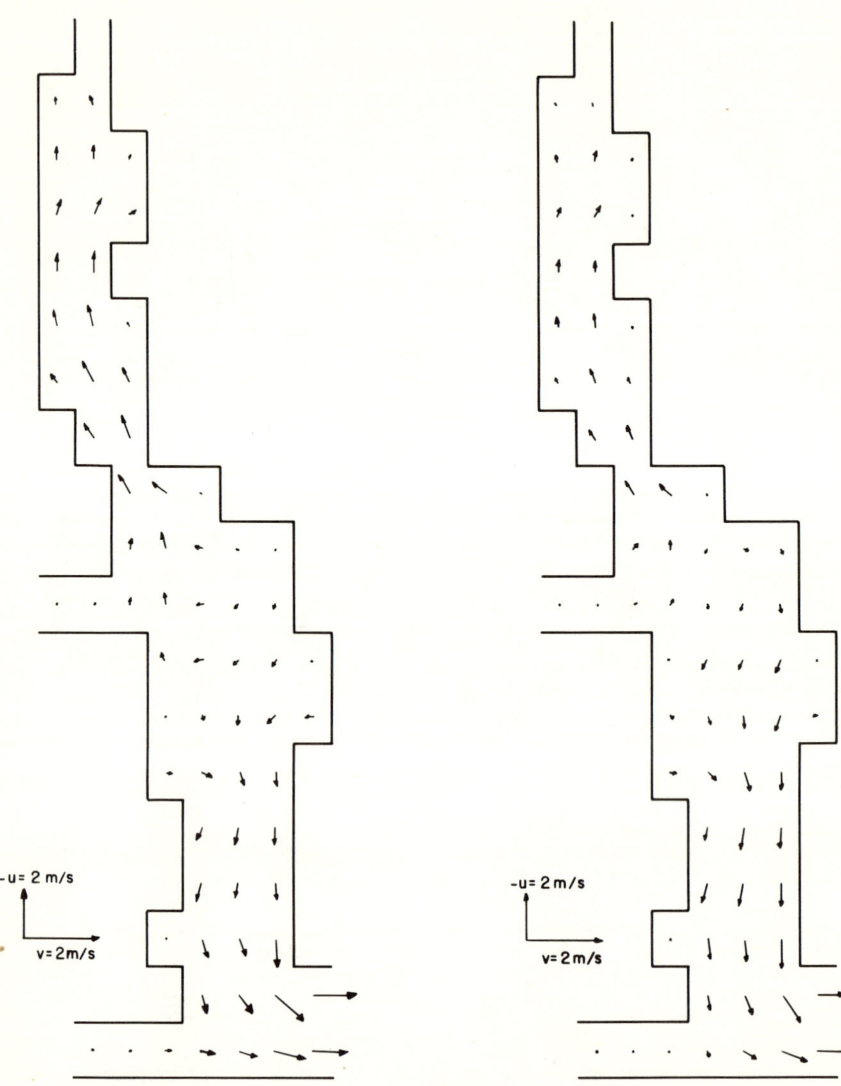

FIG. 67. Surface velocities at $t_0 + 7T/8$.

FIG. 68. Surface velocities at $t_0 + T$.

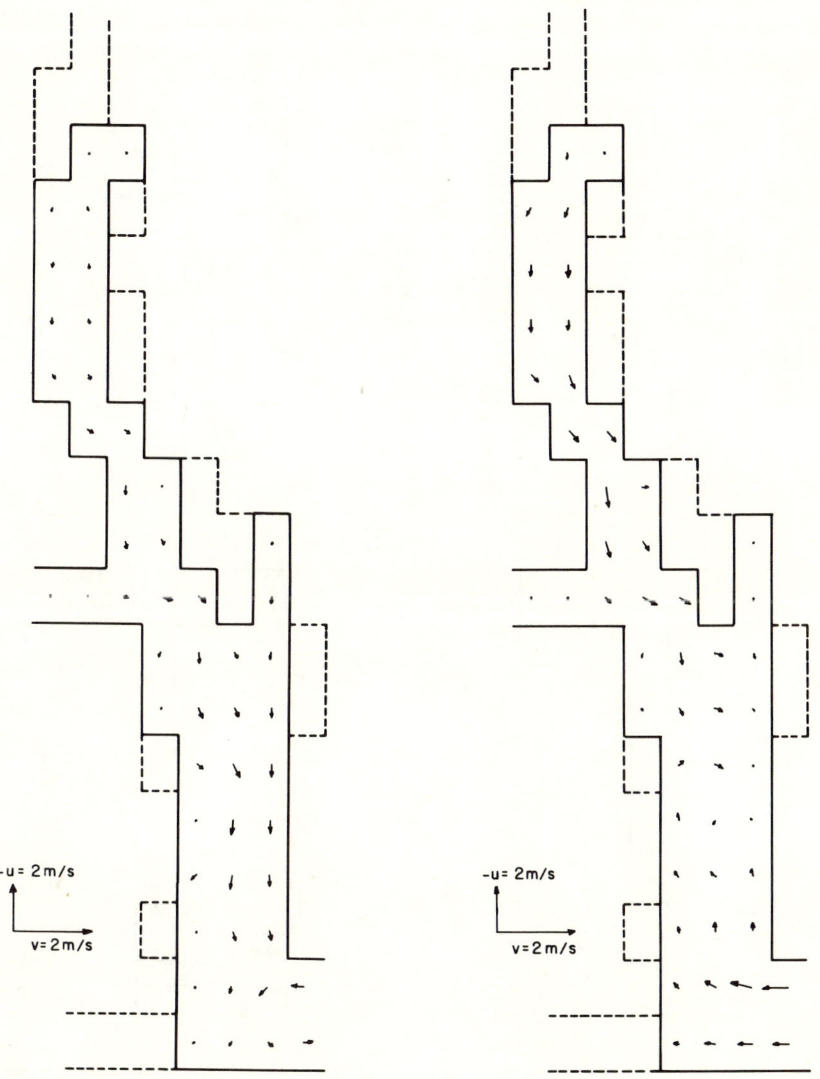

FIG. 69. Subsurface velocities at $t_0 + T/8$.   FIG. 70. Subsurface velocities at $t_0 + 2T/8$.

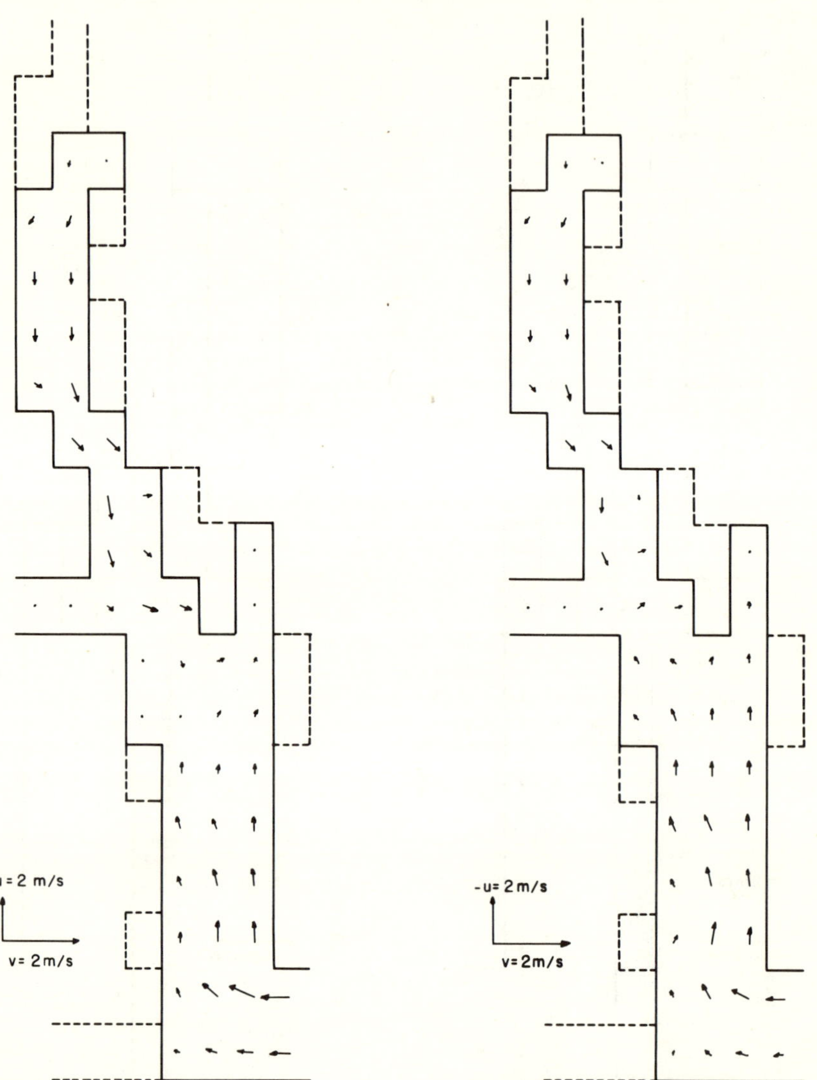

FIG. 71. Subsurface velocities at $t_0 + 3T/8$.

FIG. 72. Subsurface velocities at $t_0 + 4T/8$.

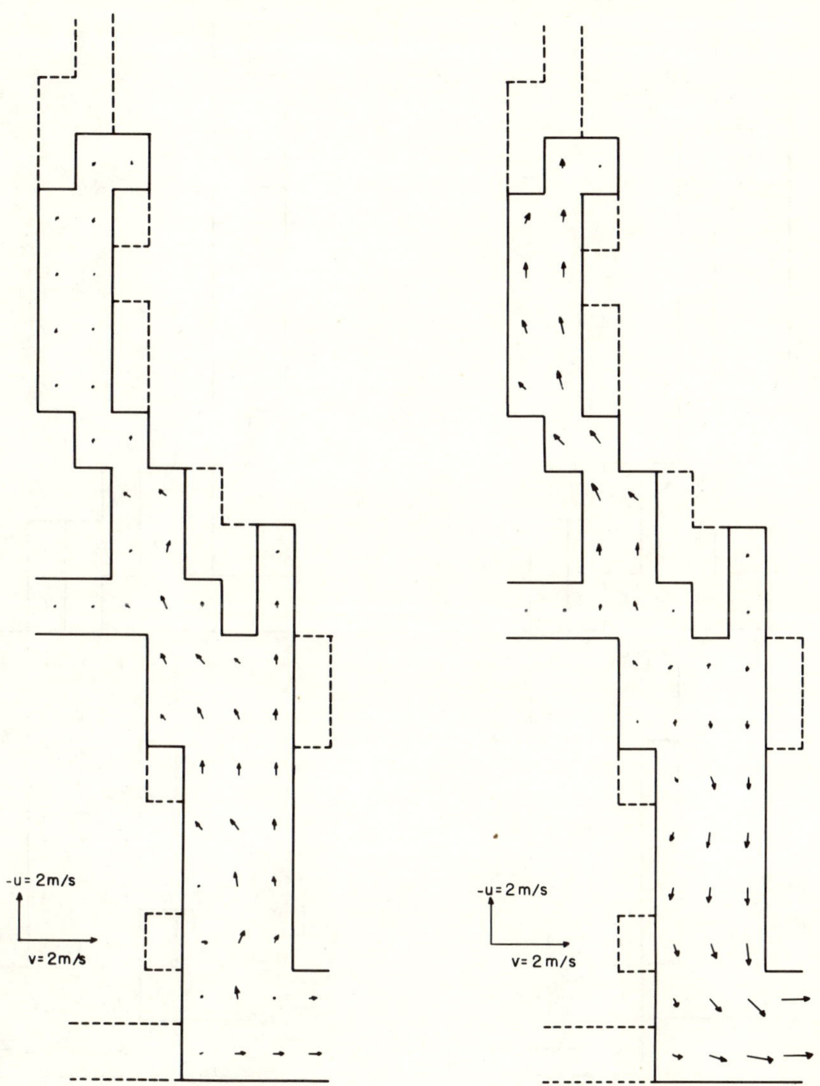

FIG. 73. Subsurface velocities at $t_0 + 5T/8$.   FIG. 74. Subsurface velocities at $t_0 + 6T/8$.

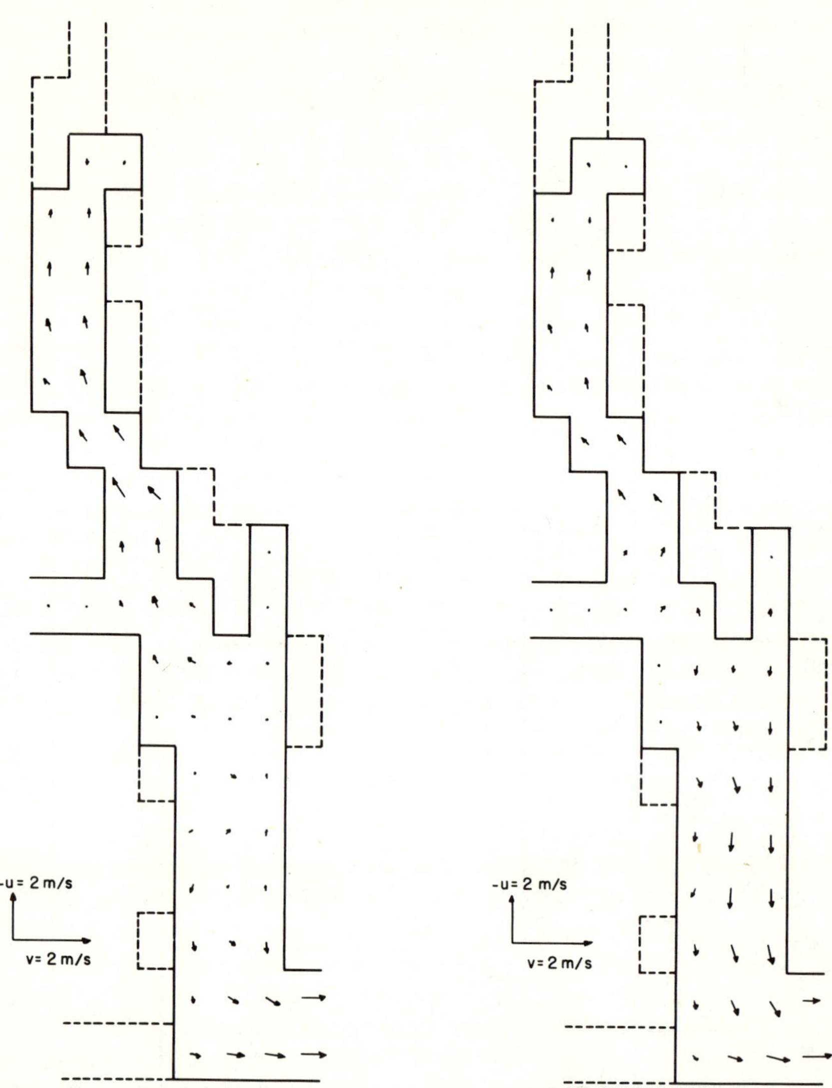

FIG. 75. Subsurface velocities at $t_0 + 7T/8$.   FIG. 76. Subsurface velocities at $t_0 + T$.

Maximum values obtained for surface currents are about twice (and in some cases up to four times) as large as those observed (Tidal Current Charts, 1974). Tidal amplitudes in the interior are also larger than expected. In particular, the predicted amplitude at Pt. Lookout, at the mouth of the Potomac, is about twice the value given in Tide Tables (1974). These results suggest that, as far as the momentum equation is concerned, the chosen value for the vertical mixing coefficient is smaller than needed to model the observed hydrodynamics.

Part of the discrepancy results necessarily from the carelessness with which the bottom topography was determined for this model. In order to be able to obtain meaningful results, this stage in the modeling process should be carefully approached and mean depths assigned in such a way that areas between adjacent (I, J) columns be faithfully reproduced. Perhaps more important is the fact that too many details have been packed in the adopted geometry. Because of this, no computational points are really far enough away from the walls for good results to be expected.

At this stage of development of the model, the most important source of discrepancy is the representation of the subgrid scales. For a formulation of the type (3.1), the effects of v_3 are dominant, and—in the absence of a more meaningful procedure—the introduction of its parametric dependence on the velocity field and the stratification will allow obtainment of more reasonable results. In particular, the dependence of v_3 on the velocity is expected to significantly reduce tidal flows.

5.3. Conclusions

Although the enormous amount of data generated by this experimental application to a real estuary has not been completely analyzed and evaluated, the preliminary results shown here indicate that the fully three-dimensional simulation of estuarine systems is possible by the method presented here. The ability to describe the influence of arbitrary lateral and bottom boundaries, arbitrary number and position of fresh water tributaries, and connections to the sea, make this a very versatile model, easily adaptable to different configurations, including lakes and eventually coastal seas.

Final application of this model for practical purposes is, of course, dependent upon the correct parameterization of subgrid scales both due to unresolved motions and to unresolved features in the geometry. The results described in the last section, where the use of the same value for both $v_3^{u, v, w}$ and v_3^S resulted in larger flow and surface elevation values but weaker stratification than observed, show that different mixing coefficients (at least in the vertical) should be used for the momentum and the salinity equations. Quantitative agreement with observations is expected to be obtained as a result of the incorporation of less naive models for the simulation of the effect of the SGS on the dependent variables.

Although computer costs for the development of this model have been fairly high, operational costs for the examples presented here are not excessive, especially in view of the present availability of much faster machines where the computations could be carried out.

Because of the general applicability of this model, it has the capacity for expansion to modeling of ecosystems. The inclusion of temperature effects and the addition of equations to follow the temporal evolution of biological parameters or water quality quantities, or both, is a simple task that would require a relatively small increase in storage requirements and a very small increase in running time.

This preliminary version included only the influence of salinity in the density, as this is the determining factor for density-driven estuarine circulations. In effect, a change in salinity of only $1\%_{00}$ results in a density change equivalent to that obtained by a $4°C$ temperature variation. This, together with the fact that salinities range from about 0 to about $30\%_{00}$, while temperatures only fluctuate at most by some $10°C$ in a typical estuary, justifies this approach as far as the hydrodynamics are concerned. Although the overall heat budget is determined by solar radiation, local perturbations might be important, mainly near power plants, for example. In such cases, an equation for the temperature and its influence on the water density would be necessary.

That would indeed be the case when data generated by this type of models are used in biological modeling, because biologic and chemical rates are very much sensitive to small temperature fluctuations.

Surface tracers have been shown useful for certain purposes. In the way they are presently used, they are able only to predict the motion of the center of mass of a dump, but not its spread due to the subgrid motions. These effects might in the future be simulated by the appropriate inclusion of a random number generator, the output of which will determine the values of the u and v perturbations to be added to the bilinearly interpolated mean u and v values at the markers' positions. The ability to follow tracers can be extended to the interior of the fluid; this might become useful for sediment deposition studies.

The use of presently available digital computers, faster by a factor of ten than the one used in this work, would allow the horizontal resolution for the estuary simulated here to be refined by a factor of 2, while keeping costs below 5 min of computer time per simulated day.

Finally, many further refinements can be made on the code itself that will hopefully result in an increase in its economy.

ACKNOWLEDGMENTS

The author wishes to express his gratitude to Alan J. Faller, for guidance, encouragement, interest, and helpful discussions.

The research reported on in this paper was supported, in part, by the National Science foundation, under grants number GA-4388 and GA-26028. Computer usage was supported in full by the Computer Science Center of the University of Maryland.

LIST OF SYMBOLS

a — Amplitude of the tidal oscillation

A_1 — Nondimensional horizontal mixing coefficient for the x direction in the nondimensional momentum equations

A_2 — Nondimensional horizontal mixing coefficient for the y direction in the nondimensional momentum equations

A_3 — Nondimensional vertical mixing coefficient for the z direction in the nondimensional momentum equations

C — m-Dimensional concentration vector, the ith species being described by C_i

C_h — Chezy coefficient

D_E — Characteristic depth of an Ekman spiral

D^i — Diffusion or mass transfer coefficient for the ith species

$D_{i,j,k}$ — Divergence-free deviation in cell (i, j, k)

f — Coriolis parameter

f^u — Nondimensional Coriolis parameter for the nondimensionalized u equation

f^v — Nondimensional Coriolis parameter for the nondimensionalized v equation

F — River inflow

$F2$ — Total outflow through the ocean boundary

g — Acceleration of gravity

H — Value of the free elevation surface

$H_{i,j}$ — Average of the surface elevation value over the fluid column (i, j)

i — Computational index for the x direction

$I(S, T)$ — Contribution to the hydrostatic pressure due to vertical variations in fluid density

$(I_k)_{i,j}$ — Salt contents in the fluid column (i, j) from the k level to the surface

j — Computational index for the y direction

k — Computational index for the z direction

$\hat{n}$ — Normal unit vector to a specified surface

p — Pressure field

q — General hydrodynamic variable

r_i — Volume rate of creation/destruction for the ith species

R_e — Reynolds' number

$\mathbb{R}_{i,j,k}$ — Source term for the finite-difference Poisson equation for the dynamic pressure φ for cell (i, j, k)

$R^u_{i+1/2,j,k}$ — Contribution of the spatial subgrid scales to the dynamic equation for the mean flux in the x direction, at $x = (i + \frac{1}{2})\delta x$, $y = j\delta y$, $z = k\delta z$

$\mathscr{R}^u_{i+1/2,j,k}$ — Contribution of the *spatial and temporal* subgrid scales to the dynamic equation for the mean flux in the x direction, at $x = (i + \frac{1}{2})\delta x$, $y = j\delta y$, $z = k\delta z$

S — Salinity

$S_{i,j,k}$ — Volume average of the salinity over the computational cell (i, j, k) (cf. Section 3.1.1 for the correct interpretation of $S_{i,j,k}$ in cells partially filled with fluid)

T — Temperature or period of tidal oscillation

U_0 Nondimensionalizing unit for u

$\mathbf{v}$ Velocity vector with components (u, v, w)

V_0 Nondimensionalizing unit for v

W_0 Nondimensionalizing unit for w

X_0 Nondimensionalizing unit for the x axis

Y_0 Nondimensionalizing unit for the y axis

Z_0 Nondimensionalizing unit for the z axis

α Relaxation parameter for the solution of the Poisson equation for φ or coefficient of saline compressibility

β Coefficient of thermal compressibility

δ Temporal averaging interval

$\delta_{i,j}$ Krönecker delta

$\delta x, \delta y, \delta z$ Dimensions of a computational cell along the x, y, and z axes

Δ Spatial averaging interval

ε_{ijk} Completely antisymmetric Levy-Civita tensor

η First coefficient of dynamic viscosity

φ Ratio of pressure to (reference) density

φ_a Applied dynamic pressure at the surface

φ^d Dynamic part of φ

$\varphi_{i,j,k}$ Volume average of the dynamic φ over the computational cell (i, j, k)

Φ_0 Nondimensionalizing unit for φ

χ Coefficient of thermal diffusivity

λ Geographical latitude

v Coefficient of kinematic viscosity

v_1, v_2, v_3 Eddy coefficients of viscosity (used also for eddy diffusivities)

ψ Geopotential or transport stream-function

Π_{ij} Total momentum flux density tensor

ρ Fluid density

σ_{ij} Viscous stress tensor

Σ Surface bounding fixed control volume or total salt contents within the basin divided by the volume of the unit computational cell

τ_0 Fixed control volume or nondimensionalizing unit for the temporal variable

τ_{ij} Wind-stress tensor

ζ Second coefficient of dynamic viscosity

ω Frequency of tidal oscillation

Ω Earth's angular velocity

Averaging Operators

Let q be any of the hydrodynamic variables. Then, the following symbols are used for different averaging operators:

$\bar{q}$ Running average

q_{klmn} Lorenz notation for spatial-temporal running averages (see Sec. 2.2.2 for definition), where all indices assume values of 0, 1, or 2

$\{q\}_{i,j,k}^{lmn}$ Average of q over a volume $l\delta x$, $m\delta y$, $n\delta z$ centered at $(x_i = i\delta x,\; y_j = j\delta y,\; z_k = k\delta z)$. A deviation from such average is defined by

$$\overset{lmn|i,j,k}{q}(x - x_i, y - y_j, z - z_k)$$

$$\equiv q(x, y, z, t) - \{q\}_{i,j,k}^{lmn}$$

$\langle q \rangle$ Temporal (running) average

References

Amsden, A. A., and Harlow, F. H. (1970). "The SMAC Method: A Numerical Technique for Calculating Incompressible Fluid Flows," Los Almos Sci. Lab. Rep. LA-4370, Los Alamos, New Mexico.

Andersen, A. M., Davis, W. J., Lynch, M. P., and Schubel, J. R. (1973). "Effects of Hurricane Agnes on the Environment and Organisms of Chesapeake Bay," The Chesapeake Bay Research Council, Baltimore, Maryland.

Arakawa, A. (1966). Computational design for long-term numerical integration of the equations of fluid motion: two-dimensional incompressible flow. Part I. *J. Comput. Phys.* **1**, 119.

Bowden, K. F. (1967). Circulation and diffusion. *In* "Estuaries," AAAS, Washington, D.C.

Browning, G., Kreiss, H-O., and Oliger, J. (1973). Mesh refinement. *Math. Comp.* **27**, 29.

Buzbee, B. L., Golub, G. H., and Nielson, C. W. (1970). On direct methods for solving Poisson's equation. *SIAM J. Numer. Anal.* **7**, 627.

Caponi, E. A. (1974). "UMTEC: University of Maryland Three-dimensional Estuarine Circulation Model. Documentation and Listings." Tech. Note BN-802, Inst. Fluid Dynamics Appl. Math., Univ. of Maryland, College Park, Maryland.

Caponi, E. A., and Faller, A. J. (1973). A three-dimensional model for an estuary. *Bull. Am. Phys. Soc.* [2] **18**, 1477.

Carnahan, B., Luther, H. A., and Wilkes, J. O. (1969). "Applied Numerical Methods." Wiley, New York.

Chambers, A. J., Mangarella, P. A., Street, R. L., and Hsu, E. Y. (1970). "An Experimental Investigation of Transfer of Momentum at an Air-Water Interface," Tech. Rep. No. 133, Dept. Civ. Eng., Stanford Univ., California.

Chan, R. K. C., and Street, R. L. (1970a). "SUMMAC—A Numerical Model for Water Waves," Tech. Rep. No. 135, Dept. Civ. Eng., Stanford Univ., California.

Chan, R. K. C., and Street, R. L. (1970b). A computer study of finite-amplitude water waves. *J. Comput. Phys.* **6**, 68.

Christiansen, J. P., and Hockney, R. W. (1971). Delsqphi. A two dimensional Poisson solver program. *Comput. Phys. Commun.* **2**, 139.

Courant, R., Friedrichs, K. O., and Lewy, H: (1928). Über die partiellen Differenzengleichungen der mathematischen Physik. *Math. Ann.* **100**, 32.

Crowder, H. J., and Dalton, C. (1971). Errors in the use of non-uniform mesh systems. *J. Comput. Phys.* **7**, 32.

Daly, B. J., and Pracht, W. E. (1968). Numerical study of density-current surges. *Phys. Fluids* **11**, 15.

Deardorff, J. W. (1969). "A Three-Dimensional Numerical Study of Turbulent Channel Flow at Large Reynolds Numbers," NCAR Ms. No. 69-19, Nat. Cent. Atmos. Res., Boulder, Colorado.

Dorr, F. W. (1970). The direct solution of the discrete Poisson equation on a rectangle. *SIAM Rev.* **12**, 248.

Douglas, J. (1962). Alternating direction methods for three space variables. *Numer. Math.* **4**, 41.

Easton, C. R. (1972). Homogeneous boundary conditions for pressure in the MAC method. *J. Comput. Phys.* **9**, 375.

Faller, A. J. (1971). On the difference equations for meteorological prediction. *In* "Numerical Solution of Partial Differential Equations—II" (B. Hubbard, ed.) SYNSPADE 1970, p. 215. Academic Press, New York.

Ferziger, J. H. (1973). Effect of time differencing on stability in flow computation. *Bull. Am. Phys. Soc.* [2] **18**, 1476.

Forester, C. K. (1973). A method of time-centering the Lagrangian marker particle computation. *J. Comput. Phys.* **12**, 269.

Forristall, G. Z. (1974). Three-dimensional structure of storm-generated currents. *J. Geophys. Res.* **79**, 2721.

Godunov, S. K., and Ryaben'ky, V. S. (1964). "Theory of Difference Schemes, an Introduction." North-Holland Publ., Amsterdam.

Hadjidimos, A. (1970). Optimum extrapolated ADI iterative difference schemes for the solution of Laplace's equation in three space variables. *Comput. J.* **14**, 179.

Hamilton, H. D., Williams, F. R., and Laevastu, T. (1973). Computation of tides and currents with multi-layer hydrodynamical numerical (MHN) models with several open boundaries. *Proc. 1973 Summer Comput. Simulation Conf.*, p. 634. Simulation Councils, La Jolla, California.

Hansen, W. (1956). Theorie zur Errechnung des Wasserstandes und der Strömungen in Randmeeren nebst Anwendungen. *Tellus* **8**, 287.

Hansen, D. V., and Rattray, M., Jr. (1966). New dimensions in estuary classifications. *Limnol. Oceanogr.* **11**, 319.

Harlow, F. H., and Welch, J. E. (1965). Numerical calculation of time-dependent viscous incompressible flow. *Phys. Fluids* **8**, 2182.

Heaps, N. S. (1973). Three-dimensional numerical model of the Irish Sea. *Geophys. J. R. Astron. Soc.* **35**, 99.

Hirt, C. W. (1968). Heuristic stability theory for finite difference equations. *J. Comput. Phys.* **2**, 339.

Hirt, C. W., and Cook, J. L. (1972). Calculating three-dimensional flows around structures and over rough terrain. *J. Comput. Phys.* **10**, 324.

Hirt, C. W., and Harlow, F. H. (1967). A general corrective procedure for the numerical solution of initial value problems. *J. Comput. Phys.* **2**, 114.

Hirt, C. W., and Shannon, J. P. (1968). Free-surface stress conditions for incompressible flow calculations. *J. Comput. Phys.* **2**, 403.

Hockney, R. W. (1972). The solution of Poisson's equation. *In* "Computing as a Language of Physics." IAEA, Vienna, Austria.

Kálnay de Rivas, E. (1972). On the use of nonuniform grids in finite-difference equations. *J. Comput. Phys.* **10**, 202.

Konovalova, I. Z. (1972). Experimental study of horizontal turbulence in the littoral zone using ocean current data from an aerial survey. *Oceanology (Engl. Trans.)* **3**, 443.

Kurihara, Y., and Holloway, L., Jr. (1967). Numerical integration of a nine-level global primitive equations model formulated by the box method. *Mon. Weather Rev.* **95**, 509.

Lamb, H. (1932). "Hydrodynamics." Cambridge Univ. Press, London and New York. Republished by Dover, New York, 1945.

Landau, L. D., and Lifschitz, E. M. (1959). "Fluid Mechanics." Pergammon, Oxford.

Leendertse, J. J., and Liu, S-K. (1975). "A Three-Dimensional Model for Estuaries and Coastal Seas," Vol. II: Aspects of Computation. Rand Corp., R-1764-OWRT, Santa Monica, California.

Leendertse, J. J., Alexander, R. C., and Liu, S-K. (1973). "A Three-Dimensional Model for Estuaries and Coastal Seas," Vol. I: Principles of Computation. Rand Corp., R-1417-OWRR, Santa Monica, California.

Liggett, J. A. (1969). Unsteady circulation in shallow, homogeneous lakes. *J. Hydraulics Div., ASCE*, **95**(HY4), 1273.

Lilly, D. K. (1966). The representation of small-scale turbulence in numerical simulation experiments. *Proc. IBM Sci. Comput. Symp. Environ. Sci.*, IBM Form No. 320-1951, 195, Yorktown Heights, New York.

Marchuk, G. I., Gordeev, R. G., Rivkind, V. Ya, and Kagan, B. A. (1973). A numerical method for the solution of tidal dynamics equations and the results of its application. *J. Comput. Phys.* **13**, 15.

Monin, A. S., and Yaglom, A. M. (1971). "Statistical Fluid Mechanics; Mechanics of Turbulence," Vol. 1. MIT Press, Cambridge, Massachusetts.

Neumann, G., and Pierson, W. J., Jr. (1966). "Principles of Physical Oceanography." Prentice-Hall, Englewood Cliffs, New Jersey.

Nichols, B. D., and Hirt, C. W. (1971). Improved free surface boundary conditions for numerical incompressible flow calculations. *J. Comput. Phys.* **8**, 434.

Nichols, B. D., and Hirt, C. W. (1973). Calculating three-dimensional free surface flows in the vicinity of submerged and exposed structures. *J. Comput. Phys.* **12**, 234.

Nihoul, J. C. J. (1975). "Modeling of Marine Systems." Elsevier, Amsterdam.

Noh, W. F. (1964). CEL: A time-dependent, two-space dimensional, coupled Eulerian-Lagrange code. *Methods Comput. Phys.* **3**, 117–179.

Okubo, A. (1970). "Oceanic Mixing," Tech. Rep. 62, Chesapeake Bay Institute, Baltimore, Maryland.

Okubo, A., and Ozmidov, R. V. (1970). Empirical dependence of the coefficient of horizontal turbulent diffusion in the ocean and the scale of the phenomenon in question. *Izv., Acad. Sci., USSR, Atmos. Oceanic Phys.* (*Engl. Transl.*) **6**, 308.

Orlob, G. T. (1972). Mathematical modeling of estuarial systems. *In* "Proceedings from the International Symposium on Modeling Techniques in Water Resources Systems" (A. K. Biswas, ed.), pp. 78–128. Environment Canada, Ottawa, Canada.

Phillips, N. A. (1956). The general circulation of the atmosphere: a numerical experiment. *Q. J. R. Meteorol. Soc.* **82**, 123.

Piacsek, A. A., and Williams, G. P. (1970). Conservation properties of convective difference schemes. *J. Comput. Phys.* **6**, 393.

Potter, D. (1973). "Computational Physics." Wiley, New York.

Pritchard, D. W. (1955). "Estuarine Circulation Patterns," Contrib. No. 23, Chesapeake Bay Institute, Johns Hopkins Univ., Baltimore, Maryland.

Pritchard, D. W. (1968). Chemical and physical oceanography of the bay. *In* "Proceedings of the Govenor's Conference on Chesapeake Bay," Annapolis, Maryland.

Pritchard, D. W. (1971). Three-dimensional models. *In* "Estuarine Modeling: An Assessment," Water Pollution Control Res. Ser. 16070 DZV 02/71, Washington, D. C.

Richtmeyer, R. D., and Morton, K. W. (1957). "Difference Methods for Initial Value Problems." Wiley (Interscience), New York.

Roache, P. J. (1971). "A New Direct Method for the Discretized Poisson Equation (Lecture Notes in Physics, Vol. 8)" (M. Holt, ed.), p. 48, Springer-Verlag, Berlin and New York.

Robinson, E. L., Jr., and Scarton, H. A. (1972). CONTOR: A Fortran sub-routine to plot smooth contours of a single-valued arbitrary three-dimensional surface. *J. Comput. Phys.* **10**, 242.

Romov, A. I. (1973). A method of discretizing the dynamic equations for the atmosphere and deriving numerical weather-forecasting schemes. *Izv., Acad. Sci., USSR, Atmosph. Oceanic Phys.* **9**, 61.

Simons, T. J. (1974). Verification of numerical models of Lake Ontario: Part I. Circulation in spring and early summer. *J. Phys. Oceanogr.* **4**, 507.

"Tidal Current Charts for Upper Chesapeake Bay" (1973). National Ocean Survey, Rockville, Maryland.

"Tidal Current Tables" (1974). National Ocean Survey, Rockville, Maryland.

"TIDE TABLES, 1974: East Coast of North and South America," (1974). National Ocean Survey, Rockville, Maryland.

Van Mieghem, J. (1973). "Atmospheric Energetics." Oxford Univ. Press, London and New York.

Viecelli, J. A. (1969). A method for including arbitrary external boundaries in the MAC incompressible fluid computing technique. *J. Comput. Phys.* **4**, 543.

Ward, G. H., and Espey, W. H., eds. (1971). "Estuarine Modeling: An Assessment." Water Pollution Control Res. Ser. 16070 DZV 02/71, Washington, D. C.

Warming, R. F., and Hyett, B. J. (1974). The modified equation approach to the stability and accuracy analysis of finite difference methods. *J. Comput. Phys.* **14**, 159.

Welch, J. E., Harlow, F. H., Shannon, J. P., and Daly, B. J. (1965). "The MAC Method: A Computing Technique for Solving Viscous, Incompressible, Transient Fluid-Flow Problems Involving Free Surfaces," Los Alamos Sci. Lab. Rep. LA-3425 Rev., Los Alamos, New Mexico.

Williams, G. P. (1969). Numerical integration of the three-dimensional Navier Stokes equations for incompressible flow. *J. Fluid Mech.* **37**, 727.

Wright, T. J. (1973). A two-space solution to the hidden line problem for plotting functions of two variables. *IEEE Trans. Comput.* **c-22**, 28.

Yanenko, N. N. (1971). "The Method of Fractional Steps." Springer-Verlag, Berlin and New York.

SUBJECT INDEX

A

Aerosols
atmospheric, properties, 73-188
of cumulus clouds, 3-10
Aerosol particles
extinction coefficient related to mass of,
173-176
mean density of, 96-97
refractive index of, as function of relative
humidity, 97-100
scattering albedo of, 176-180
Atmospheric suspensoids
measuring techniques for, 104-113
model computations on, 120-180
particle size of, 121
relative humidity and, 122-146
sampling methods for, 102-103

B

BET-adsorption isotherm, in studies of
atmospheric suspensoids, 92-94

C

Chesapeake Bay, estuarine modeling of,
287-304
Computer(s)
use in estuarine modeling, 257-264
use in weather-modification studies, 47-48
Cumulus clouds
dynamics and models of, 10-22
interactions, groups, and patterns of, 22-31
model simulation of, 53-64
one-dimensional, 59-64
three-dimensional, 54-57
two-dimensional, 57-59
modification of, 3-31
agents for, 31-36
assessment, 31-53
computer studies, 47-48
measurement systems, 37-47
statistical aspects, 48-52
precipitation from, 1-72

E

Electromagnetic radiation, in calculation of
suspensoid particle size, 147-156
Estuarine modeling, 189-310
of Chesapeake Bay, 287-304
experimental results, 291-303
geometry and parameters used, 287-291
equations governing, 192-220
boundary conditions, 199-201
conservation laws, 192-196
turbulence, 201-205
numerical model, 220-264
computer code, 257-264
solution technique, 220-257
test cases for, 264-287
homogeneous fluid, 265-283
salt-driven circulations, 283-286
Extinction coefficient, aerosol particle mass,
relative humidity and, 173-176

H

Humidity, relative, visual range as function
of, 171-173

I

Index of refraction, mean complex,
measurement of, 112-113, 118-119

J

Jet impactor, for aerosol sampling, 102-103

L

Lambert's law, 146